Computing Technologies and Applications

Demystifying Technologies for Computational Excellence: Moving Towards Society 5.0

KEY WORDS: Cloud Computing, Computing Applications, Database Management, Machine Learning

Series Editors: Vikram Bali and Vishal Bhatnagar

This series encompasses research work in the field of data science, edge computing, deep learning, distributed ledger technology, extended reality, quantum computing, artificial intelligence, and various other related areas, such as natural language processing and technologies, high-level computer vision, cognitive robotics, automated reasoning, multivalent systems, symbolic learning theories and practice, knowledge representation and the semantic web, intelligent tutoring systems, and education.
The prime reason for developing and growing out this new book series is to focus on the latest technological advancements – their impact on society, the challenges faced in implementation, and the drawbacks or reverse impact on society due to technological innovations. With these technological advancements, every individual has personalized access to all services, and all devices can connect with each other and communicate among themselves. Thank you to technology for making our lives simpler and easier! These aspects will help us to overcome the drawbacks of the existing systems and help us to build new systems with the latest technologies that will help society in various ways, proving Society 5.0 to be one of the biggest revolutions in this era.

Data Science and Innovations for Intelligent Systems
Computational Excellence and Society 5.0
Edited by Kavita Taneja, Harmunish Taneja, Kuldeep Kumar, Arvind Selwal, and Ouh Lieh

Artificial Intelligence, Machine Learning, and Data Science Technologies
Future Impact and Well-Being for Society 5.0
Edited by Neeraj Mohan, Ruchi Singla, Priyanka Kaushal, and Seifedine Kadry

Transforming Higher Education through Digitalization
Insights, Tools, and Techniques
Edited by S. L. Gupta, Nawal Kishor, Niraj Mishra, Sonali Mathur, and Utkarsh Gupta

A Step towards Society 5.0
Research, Innovations, and Developments in Cloud-Based Computing Technologies
Edited by Shahnawaz Khan, Thirunavukkarasu K., Ayman AlDmour, and Salam Salameh Shreem

Computing Technologies and Applications
Paving Path Towards Society 5.0
Edited by Latesh Malik, Sandhya Arora, Urmila Shrawankar, Maya Ingle, and Indu Bhagat

For more information on this series, please visit: https://www.routledge.com/Demystifying-Technologies-for-Computational-Excellence-Moving-Towards-Society-5.0/book-series/CRCDTCEMTS

Computing Technologies and Applications

Paving Path towards Society 5.0

Edited by Latesh Malik, Sandhya Arora
Urmila Shrawankar, Maya Ingle, and Indu Bhagat

CRC Press is an imprint of the
Taylor & Francis Group, an **informa** business

First edition published 2022
by CRC Press
6000 Broken Sound Parkway NW, Suite 300, Boca Raton, FL 33487-2742

and by CRC Press
2 Park Square, Milton Park, Abingdon, Oxon, OX14 4RN

CRC Press is an imprint of Taylor & Francis Group, LLC

Library of Congress Cataloguing-in-Publication Data
Names: Malik, Latesh, editor.
Title: Computing technologies and applications : paving path towards society 5.0 / edited by Latesh Malik, Sandhya Arora, Urmila Shrawankar, Maya Ingle, Indu Bhagat.
Description: First edition. | Boca Raton : Chapman and Hall/CRC Press, [2022] | Includes bibliographical references and index. | Summary: "The book focuses on suggesting software solutions for supporting societal issues such as health care, learning and monitoring mythology for disables and also technical solutions for better living. It also has the high potential to be used as recommended textbook for research scholars and post-graduate programs"--Provided by publisher.
Identifiers: LCCN 2021027902 (print) | LCCN 2021027903 (ebook) | ISBN 9780367763701 (hbk) | ISBN 9780367763749 (pbk) | ISBN 9781003166702 (ebk)
Subjects: LCSH: Internet of things. | Cloud computing. | Industry 4.0. | Industries--Data processing.
Classification: LCC TK5105.8857 .C66 2022 (print) | LCC TK5105.8857 (ebook) | DDC 004.67/8--dc23
LC record available at https://lccn.loc.gov/2021027902
LC ebook record available at https://lccn.loc.gov/2021027903

ISBN: 978-0-367-76370-1 (hbk)
ISBN: 978-0-367-76374-9 (pbk)
ISBN: 978-1-003-16670-2 (ebk)

DOI: 10.1201/9781003166702

Typeset in Palatino
by MPS Limited, Dehradun

Contents

Preface

Latesh Malik, Sandhya Arora, Urmila Shrawankar, Maya Ingle, and Indu Bhagat

The trend of digital technology advancement is changing many aspects of society, including administration, industry, healthcare, and personal life. New technologies, such as the Internet of Things, the cloud, robotics, artificial intelligence, machine learning, and big data – all of which can affect the course of society – are continuing to progress. The ultimate aim of technological development is to benefit humankind. By doing so, the society of the future will be one in which new values and services are created continuously, making people's lives more conformable and sustainable. Future society will be a people-centric, super-smart society.

This book aims to cover technological developments, concepts, case studies, research, and solutions for societal benefit using the Internet of Things, cloud technology, machine learning, and data science to handle societal challenges.

A lot of research and development is happening in the field of the Internet of Things, high-performance computing, data science, artificial intelligence, machine learning, the cloud, and many more, which is helping us to move toward Society 5.0. This book aims to fulfill the ultimate aim of all these developments, i.e., to serve humankind in a better way and build a smart society. This book brings together the use of these technologies for societal benefit. Undergraduate students, postgraduate students, researchers, academics, and industry persons will benefit, and it will serve as a guide for them. The research work contributed in this book will open new avenues to help build Society 5.0.

This book focuses on suggesting software solutions for supporting societal issues such as health care, support systems for elderly citizens, learning and monitoring for disabilities, and also technical solutions for better living. Technology enables people to access advances so that there can be benefits in health and technology. System models are discussed here in the book as well as prototype designs, and product development is possible using the given solutions.

This book is divided in three sections: 1. Cloud Computing 2. Internet of Things 3. Data Science, Deep Learning, and Machine Learning

The book is organized in 18 chapters.

Chapter 1 provides an overview of cloud technology and management. It includes types of clouds, cloud computing architecture, common applications hosted in the cloud, attributes of cloud computing, cloud vulnerabilities, and cloud management and operations. A case study on AWS Storage is discussed.

Chapter 2 deals with real-time edge computing. This chapter discusses the various applications, services, technologies, protocols, platforms, and challenges and open research problems in the effective implementation of real-time edge computing in industrial automation as a case study.

Chapter 3 is about fog computing architecture and applications. This chapter comprises a detailed comparison of cloud computing and fog computing as well as fog and edge computing. It explains in detail fog federation, fog computing architecture, benefits and advantages of fog computing, categories and subcategories of optimization problems in fog computing, IoT challenges, and solutions with fog computing.

Chapter 4 deals with mobile web service architecture in the cloud environment. This chapter focuses on the study of various mobile web service architectures that enhance the performance of cloud-based mobile web services and provide reliable services that are available to users. Mobile cloud computing (MCC) is emerging as one of the important branches of cloud computing. MCC is an extension of cloud computing with the support of mobility.

Chapter 5 is an essential Internet of Things (IoT) business guide with different case studies. The chapter is an extensive reference on the different uses of IoT with multiple variations and applications. It also explains the evolution of the IoT. The chapter discusses multiple case studies, like a solar tracker, a robot track follower, and soil moisture-level detection and control based on temperature. The most important parameter is the cloud connected to things for data collection and processing of temperature. All these are explained using actual hardware, sensors, devices, etc., to implement the cases.

Chapter 6 discusses research issues in IoT. This chapter includes IoT challenges, tools, and technologies for facing challenges. It also includes probable solutions to address IoT research challenges. A case study on an SSM model for a home security system based on a context-aware application using context modeling is also covered.

Chapter 7 discusses intrusion detection systems in IoT. This chapter introduces a far-reaching audit of IoT framework-related advances, conventions, and threats arising out of undermined IoT gadgets alongside giving a review of intrusion detection models. Additionally, this chapter covers the investigation of different AI-based machine learning strategies appropriate to identify IoT frameworks related to cyber-attacks.

Chapter 8 provides a case study of smart farming using IoT. Smart farming is an idea brought forward to enhance the basic capabilities of traditional farming using modern-day technologies such as the IoT to provide real-time information in accordance with ad-hoc sensor networks spread across the agriculture field. Data are then maneuvered through application software systems. This chapter focuses on actual implementation module details.

Chapter 9 discusses stochastic computing (SC) for deep neural networks (DNNs). SC can implement DNNs with essentially low hardware overhead and offers good scalability. First, the chapter presents a detailed study of processing elements used in neural networks. Then, it discusses combining several building blocks, such as stochastic approximation of activation functions in SC, saturation arithmetic, and scaling, to propose a new scheme for DNN inference and training in SC. At last, it shows that SC-compatible DNNs can be effectively deployed on low-cost hardware.

Chapter 10 is on convolutional neural networks (CNNs) and their advances. CNN shows its excellence in areas like computer vision, natural language processing, speech processing, and radiology. Based on various scenarios, different dimensions of convolutions like one-dimensional, two-dimensional, and multi-dimensional CNN are presented. The chapter presents applications of CNN in various domains as well as some common issues and future research directions in the field. The aim of the chapter is to provide a detailed understanding of CNN with its advances and applications to help researchers further develop applications in the field of CNN.

Chapter 11 describes CNNs and their application using Keras. CNNs have the capability to capture spatial and temporal dependencies with the application of various filters. Abundant availability of data from various sources and improvements in hardware has accelerated research in CNN. The use of different optimization functions, the activation function, along with different architectural designs has generated a lot of scope in innovation and new ideas in CNN. This chapter introduces the workings of CNNs, an application using Python, and a systematic review of CNNs along with the latest research trends.

Chapter 12 defines big data analytics. Data analytics plays an important role in decision making in academics, research, and industries. It is an essential part of businesses nowadays. In this chapter, the fundamental concepts of big data are discussed. The different challenges in big data analytics are highlighted. How these issues and challenges can be handled using the Hadoop framework is also shown. In addition, different applications of big data analytics are discussed. Finally, one case study on big data analytics is presented. This chapter will help beginners to understand the big data domain and will demystify big data computing.

Chapter 13 explains augmented reality (AR) powered computer science education. With programming education expanding into early education, more effective methods of teaching have to be utilized toward explaining core concepts like data structures and algorithms. This chapter compiles research done at various levels to propose a possible implementation that could inaugurate the long-term use of AR in programming and engineering courses. This chapter provides the research, requirements, and implementation of a tool to help teachers create immersive and interactive visualizations utilizing AR.

Chapter 14 discusses information technology (IT) for student decision support in college planning. This chapter deals with the application of IT in providing decision support to students facing the challenging problem of college degree planning (CDP). During CDP, students need to determine a set of courses and a schedule for completing those courses in order to satisfy requirements for a specific college degree at a given institution.

Chapter 15 is about attention-based image captioning and evaluation methods. This chapter presents a survey on image captioning, which is a challenging task involving an understanding of vision and language. The chapter presents a categorization of attention-based techniques. The techniques that utilize only spatial attention may not fully utilize the attention mechanism capabilities. The multi-layer CNNs are naturally spatial. It is evident from the survey that techniques that fuse different attention mechanisms or utilize reinforcement learning are capable of handling the challenges of the image captioning task. Moreover, the evaluation metrics that are used to evaluate the performance of the captions generated are also discussed.

Chapter 16 describes a smart vehicle monitoring and control system using Arduino and speed guns. In this chapter, the proposed smart vehicle monitoring system uses IoT technology for early accident detection using the sensor-acquainted Raspberry Pi. The system also uses an image classification model based on machine learning to determine the severity of the accident. The system is familiar with GPS and GSM modules to find the location of the vehicle and communicate via cellular networks.

Chapter 17 is about deep learning approaches to pedestrian detection. This chapter presents a dogmatic analysis of a state-of-the-art deep learning framework adopted for pedestrian detection, along with real-time solicitudes and challenges. A well-organized comparison of an advanced deep learning framework utilized for pedestrian detection is covered. This chapter compares resembled benchmark pedestrian datasets and proposes a unique pedestrian database that embraces students' behavior in educational institutions. A case study of a scale invariant approach of pedestrian detection is also proposed.

Chapter 18 deals with crop yield forecast using a hybrid framework of deep CNN with recurrent neural network (RNN) technique. This chapter proposes a hybrid deep CNN with RNN for paddy crop yield forecasting based on weather, soil, crop genotype, and yield output activities. The hybrid framework focuses on a certain design for gathering environmental factor dependency and genetic development span without considering genotype data.

Acknowledgments

This book is the outcome of the initiation of an edited book series on "Demystifying Technologies for Computational Excellence Moving Towards Society 5.0" with our series editors Dr. Vishal Bhatnagar and Dr. Vikram Bali.

Technology enables people so that they can have access to advances and can avail themselves to the benefits of technology in day-to-day life. In a future smart society, any product or service should be optimally delivered to people and tailored to their needs. Technological advancements of Internet of Things, data science, communication, high-performance computing, machine learning, cloud technology, and many others are helping to achieve this. Using these technologies, we will reach a step ahead of information society by converting collected data into a new type of intelligence. As we move into smart society, people's lives will be more comfortable and sustainable. This book addresses core theories related to these technologies, concepts, systems, and their real-life practical applications as case studies. It also explores the aspects of Internet of Things, machine learning, data science, and cloud technology from a research perceptive in various fields such as Society 5.0, healthcare, persons with disabilities, and the elderly. The book will encourage readers and serve as a guideline to academic and industry persons working in these areas for societal benefit.

Many professionals have contributed chapters to this book. To the Taylor and Francis team, who initiated, designed, proofread, and indexed the book, we offer them all our heartful thanks for their editorial and production support. We would like to thank all the reviewers of this book, who helped us through their critical comments and creative suggestions, which eventually helped us to improve the text of the book. Many of the ideas in this book come from cloud computing, Internet of Things, and related courses like data science and machine learning. We thank all our professors who helped and taught us basic concepts about these subjects. Thanks to all contributors who completed all revisions and submitted chapters per the guidelines.

We owe an enormous debt of gratitude to our series editors, who have given us the opportunity to contribute. We would like to take this opportunity to thank our mentors, teachers, and friends for motivating us throughout this journey. We also thank all our college colleagues and principals who have given immense support in molding this work.

Finally, we would like to thank our family members for their support, good wishes, encouragement, and understanding while we wrote this book.

Editors

Dr. Latesh Malik is working as Associate Professor & Head, Department of Computer Science & Engg, Government College of Engineering, Nagpur. She completed her Ph.D. (Computer Science & Engineering) from Visvesyaraya National Institute of Technology in 2010, M.Tech. (Computer Science & Engineering) from Banasthali Vidyapith, Rajasthan, India, and B.E. (Computer Engineering) from University of Rajasthan, India. She is a gold medalist in B.E. and M.Tech. She has 20+ years of teaching experience. She is a life member of ISTE, CSI, and ACM and has more than 160 papers published in international journals and conferences. She is the recipient of two RPS and one MODROBs by AICTE. She guided 30+ PG projects, and 8 students completed Ph.D.s under her. She is the author of two books, *Practical Guide to Distributed Systems in MPI* and *Python for Data Analysis,* on Amazon Kindle Direct Publishing.

Dr. Sandhya Arora is working as Professor, Department of Computer Engineering, MKSSS's Cummins College of Engineering for Women, Pune. She completed her Ph.D. (Computer Science & Engineering) from Jadavpur University, Kolkata in 2012, M.Tech. (Computer Science & Engineering) from Banasthali Vidyapith, Rajasthan, India, and B.E. (Computer Engineering) from University of Rajasthan, India. She has rich teaching experience of 22+ years. She is a life member of ISTE, CSI, and ACM and has published papers in thoroughly acclaimed international journals and conferences. She is guiding PG and Ph.D. students. She has received prestigious awards in the field of computer science. She authored two books, *Practical Guide to Distributed Systems in MPI* and *Python for Data Analysis,* on Amazon Kindle Direct Publishing.

Dr. Urmila Shrawankar received her Ph.D. degree in Computer Sci & Engg from SGB Amravati University and M.Tech. degree in Computer Sci & Engg from RTM Nagpur University. She is the author of two books, six book chapters, and more than 160 research papers in international journals and conferences of high repute. She has published 16 patents. Her biography was selected and published in the *Marquis Who's Who in the World.* She is the recipient of an international travel grant and UGC Minor Project Grant. She participated in many international conferences worldwide as a core organizing committee member, technical program committee member, special session chair, and session Chair. Dr Urmila is a member of IEEE (SM), ACM, CSI (LM), ISTE (LM), IE (LM), IAENG, etc. At present, Dr. Urmila is working as a Professor in the Department of Computer Science and Engineering, G H Raisoni College of Engineering, RTM Nagpur University Nagpur (MS), India.

Dr. Maya Ingle is working as Director, DDU-Kaushal Kendra (A Scheme of Skill Development of UGC) at DAVV, Indore. Also, she worked as Nodal Person, Model Career Centre (Under Ministry of Labour & Employment, Govt. of India) at DAVV, Indore. She worked as Dean, Student Welfare, DAVV, Indore, from 2006 to 2009. She has guided 16 research scholars who have been awarded Ph.D.s, and 6 are presently pursuing them. She has more than 200 research papers published in international/national journals and conferences. She has visited as an Expert (and sometimes Chairman) with the National Board of Accreditation (NBA) all over India for the past 10 years. She is associated with organizations that help to promote higher education for students with visual challenges and organizations that help to promote basic education for street children, and she has engaged in activities to encourage children deprived of education due to financial crisis.

Dr. Indu Bhagat is Principal Member of Technical Staff at Oracle Canada. She currently contributes to the compiler and toolchain components of Oracle Linux. She has a Ph.D. degree in Computer Science specializing in Computer Architecture from UPC, Barcelona, Spain. She received her Bachelor of Technology degree (in Computer Science and Engg) from Indian Institute of Technology, Varanasi, India. She has 14 years of industry and academic experience in system software and hardware development and research. Her contributions have ranged from low-level hardware architecture design, compiler support, to higher level application stack (query execution software, database management systems). Her research and industry contributions have been accepted as ACM and IEEE publications and as a patent. She has also served as a reviewer at various international conferences.

Contributors

Sandhya Arora
Cummins College of Engineering
Pune

Mahip M. Bartere
G. H. Raisoni University
Amravati

Shilpa Bhalerao
Acropolis Institute of Technology and Research
Indore

Sunny Bodiwala
Gujarat Technological University

Pradnya Borkar
Jhulelal Institute of Technology
Nagpur

Mrs. Nutan Hemant Deshmukh
Cummins College of Engineering Pune

Vivek Deshpande
University of Wisconsin–Madison, Madison, Wisconsin, 53706
USA

Snehlata Dongre
G H Raisoni College of Engineering
Nagpur

Ritik Drona
Government College of Engineering
Nagpur

Anil Anandrao Dudhe
P.N. Mahavidyalaya, Pusad

Ujwalla Haridas Gawande
Yeshwantrao Chavan College of Engineering
Nagpur

Yogesh Gangadhar Golhar
G H Raisoni Institute of Engineering and Technology (Autonomous)
Nagpur

Athula D. A. Gunawardena
University Of Wisconsin–Madison, Madison, Wisconsin, 53706
USA

Kamal Omprakash Hajari
Yeshwantrao Chavan College of Engineering
Nagpur

Maya Ingle
Devi Ahilya Vishwavidyalya, University of Indore
Indore

Santosh Balaji Kagne
Government College of Engineering
Nagpur

Anant Kaulage
Computer Engineering Department, Army Institute of Technology, Dighi Hills, Alandi Road
Pune 411 015

Khushboo Khurana
University of Wisconsin–Platteville Baraboo Sauk County, WI 53913
USA

Prasad P. Lokulwar
G. H. Raisoni College of Engineering
Nagpur

Priya R. Maidamwar
G. H. Raisoni University
Amravati

Latesh Malik
Government College of Engineering
Nagpur

Robert R Meyer
University Of Wisconsin–Madison, Madison, Wisconsin, 53706
USA

Nirali Nanawati
Sarvajanik College of Engineering & Technology

V.R. Niveditha
Dr.M.G.R Educational and Research Institute, Maduravoyal
Chennai

Shreya Pandit
Jhulelal Institute of Technology
Nagpur

Shrinit Sanjeev Patil
Government College of Engineering
Nagpur

Kirthi Premadasa
University of Wisconsin–Platteville Baraboo Sauk County, WI 53913
USA

Muskan Qureshi
Jhulelal Institute of Technology
Nagpur

Jyoti S. Raghatwan
RMD Sinhgad School of Engineering, Research Scholar at SKN College of Engineering
Pune.

S. Radha Rammohan
Dr.M.G.R Educational and Research Institute, Maduravoyal
Chennai

Dheeraj Rane
Medicaps University
Indore

Sagar Rane
Computer Engineering Department, Army Institute of Technology, Dighi Hills, Alandi Road
Pune 411 015.

Sobitha Samaranayake
University Of Wisconsin–Madison, Madison, Wisconsin, 53706
USA

Sayali Ashok Sapkal
Smt. Kashibai Navale College of Engineering
Pune

Sakshi Sarile
Jhulelal Institute of Technology
Nagpur

Amandeep Singh K.
Dr.M.G.R Educational and Research Institute, Maduravoyal
Chennai

Jagruti Shah
Oberoi Center of Excellence, RTM Nagpur University, Inurture Education Solutions Pvt. Ltd

Ameya Shahu
Government College of Engineering
Nagpur

Swati Sunil Sherekar
SGB Amravati University

Urmila Shrawankar
G H Raisoni College of Engineering
Nagpur

Rohit Kumar Anil Suryawanshi
Government College of Engineering
Nagpur

Laveena Tahilani
Jhulelal Institute of Technology
Nagpur

Girish Talmale
G H Raisoni College of Engineering
Nagpur

Reena Thakur
Jhulelal Institute of Technology
Nagpur

Chetan Ravindra Wagh
Government College of Engineering
Nagpur

Sanjeev Wagh
Department of Information Technology,
Government College of Engineering,
Karad
MH, India

Kapil Wankhede
Ministry of Electronics & Information
Technology, Indian Computer
Emergency Response Team CERT-In
Delhi

T. Yuvarani
Dr.M.G.R Educational and Research
Institute, Maduravoyal
Chennai

Section I

Cloud Computing

1

Cloud Technology and Management

Dr. Jagruti Shah
Senior Faculty, RTMNU's OCE, Inurture Education Solutions Pvt. Ltd.

1.1 Introduction to Cloud Technology

1.1.1 Definition

Cloud is a buzzword nowadays. With the advent of cloud technology, the computing paradigm has changed drastically. It has given us new insight to see the computing world with new characteristics like on-demand, virtualization, elasticity, and many more. There are many definitions of *cloud computing*, but one definition that has been widely accepted by the IT industry is by the National Institute of Science and Technology (NIST). NIST published two definitions, one in 2009 and a revised definition in 2011. NIST included all the characteristics of cloud computing in one concise framework:

> ***NIST (National Institute of Science and Technology) definition of cloud computing:***
>
> ***"Cloud computing is a model for enabling ubiquitous, convenient, on-demand network access to a shared pool of configurable computing resources that can be rapidly provisioned and released with minimal management effort or service provider interaction."***

Another IT standards giant, IEEE (The Institute of Electrical and Electronics Engineers) IEEE Standards Association (IEEE-SA), proposed two standards related to cloud. The first, which they named P2301 standard (Cloud Profiles), highlights different ecosystems of cloud, such as cloud vendors, service providers, and users. The P2302 (Intercloud) standard provides a definition of topology, functions, and governance for cloud-to-cloud interoperability and federation.

Another definition is from Gartner Group, and it emphasizes IT capabilities to be delivered to the end user through the Internet.

Gartner Group, a global research and advisory group, defined cloud computing as follows:

> ***"a style of computing in which scalable and elastic IT-enabled capabilities are delivered as a service using Internet technologies."***

DOI: 10.1201/9781003166702-1

This chapter provides the following definition of *cloud computing:*

> ***"the use of IT services, namely computing servers, storage, platforms, small utilities, and sets of software via high-speed Internet, that can be provisioned or can be released on demand on a pay-per-use basis".***

Technology enabled the IT infrastructure to have a new dimension – an economic dimension in addition to a technical one.

1.1.2 History of Cloud Computing

1.1.2.1 Evolution of Cloud Computing

Cloud computing is not a new concept. We have been using cloud-like properties such as time-sharing, multitasking, and resource sharing one way or another for years. In the early 1950s, the concept of time-sharing was introduced. The real-time computer for SAGE air defense system was used to capture RADAR data and process that data. This real-time computer was bit slow but amazingly better than the batch processing system that was used. John McCarthy wrote a memorandum proposing the time-sharing system. This memo inspired many researchers, and the first time-sharing system was announced by MIT Group in October 1963. This was the first-generation time-sharing system, and then many versions came with more advancements.

The idea behind the time-sharing system was to develop an interactive computer to use an interactive network like the Internet.

The basis for the Internet was the ARPANET. The ARPANET was based on packet switching technology. The ARPANET was founded by Advanced Research Projects Agency (ARPA) of the U.S. Department of Defense. This major discovery led to the foundation for the Internet. The development was made in the late 1960s.

Time-sharing and network development led to optimal usage of expensive computer resources. Only a small percent of computer resources are being used if only one application is running. The rest of the resources are being wasted. For this, researchers came up with a new solution – virtualization – that enables use of computer resources in an optimal manner. Virtualization provides an abstraction of computer resources. The user is able to create almost all resources virtually. Virtualization refers to the creation of resources like virtual servers, storage, and operating systems. It enabled the scaling of resources so that the traditional IT workload could be distributed optimally. The technology has gained popularity and is supporting almost all levels of virtualization, like desktop virtualization, hardware virtualization, network virtualization, and OS virtualization.

Now, the next step was to enable these virtual resources globally. That meant the resources could be accessible from anywhere and at any point. This required the emergence of the Internet, a network spread across the globe. While developing ARPANET in the 1960s and 1970s, researchers discovered many protocols that established connection between different networks. In 1980, the first education- and research-based network was developed to support and share ideas. The project was named the National Science Foundation Network (NSFNET). The project was funded by the National Science Foundation (NSF). Government agencies and universities were allowed access to the network. Then, commercial Internet service providers started capturing the market. The project extended to use of 56 kbits/sec links, then 1.5 Mbits/sec links, and then 45 Mbits/sec links. The ARPANET led to an evolution of the Internet, and now millions of people are able to connect with each other through a network. The next era alongside virtualization was to efficiently use computer resources. The networked computer was underutilized. Only 20%–30% of resources were in use, and the rest of the resources were just idle. In order to mitigate these challenges, researchers designed a network architecture to use those resources optimally. One idea was grid computing. Grid computing is a network of computers where every computer can share the resources of every other computer to perform a task in an optimal manner. The task is executed in a collaborative manner. It is just like performing a particular task in a group where every responsibility is shared. These collaborative networks of computers act like a single supercomputer that can perform or execute a complex task. The complex task could not be solved by a single computing resource but only by using this approach. The aim to design shared network resources was to solve complex design problems. Grid computing can vary from a set of software and hardware requirements or frameworks to solving complex tasks. The idea of grid computing brought together the shared pool of virtual resources. These resources were pooled to solve a single complex task. A large application is divided into smaller tasks that can run in parallel.

The cloud concept is different. Here, also, the resource pool is available virtually. Those resources, however, can be allocated to an end user but in a unified manner. Every user can access the shared pool of resources individually. The name *cloud* refers to the layer of abstraction the cloud service providers offer. The resources are delivered to the end user, but the implementation details of how the services are delivered to the user are hidden. The service offering shows the cloudy nature that is always hidden. Right from the computing resource, network and storage can be rented or can be used via a pay-per-use model. With the ease of virtualization, virtual machines can be created and allocated within the network. Expanding the same idea, the cloud was first used by the sales force community in 1999. Then, in 2006, Amazon Web Services launched their Elastic Compute Cloud (EC2), which is a virtual machine service. Along with this, they also offered an object storage option and named it S3 (Simple Storage Service). Google allowed Google docs, Spreadsheet, and Forms to be used by the end user on cloud [7].

The cloud nowadays is transforming the IT infrastructure. Almost the entire IT industry is migrating applications, storage, etc., to the cloud. 451 Research also predicts the cloud computing market will reach $53.3 billion in 2021 – up from $28.1 billion in 2017, as stated by William Goddard.

The milestones in cloud evolution are listed below:

- **1950s:** Concept of time-sharing was introduced.
- **1960s:** Development of the ARPANET.
- **1970s:** Concept of virtualization began.
- **1980s:** Emergence of the Internet.
- **1990s:** Growth of the Internet and grid computing.
- **2000s:** Growth of cloud computing.

1.1.2.2 What Are the Advantages of the Cloud?

Every IT firm – from small scale to medium and large IT giants – has to maintain an IT infrastructure. With the advent of new technological requirements, companies have to procure new hardware, software, databases, etc., on a periodic basis, which is cost intensive. The company has to bear huge funds in these types of investments, which will need to be replaced in a few years.

With cloud computing, IT firms do not need to procure anything. Every technology – small word-processing software, high-end licensed software, computing platforms, storage, networking components, and many more – is available on a rental basis just like any household commodity.

Advantages of Cloud Computing

i. **Availability**
   The uptime or availability of almost all cloud service providers is 99.999%. End users do not face any problem while accessing any service. The 24×7 availability feature makes the cloud more attractive than any other technological advancement.
ii. **Low Cost**
   The cloud offers a pay-per-use model. This feature allows the user to utilize the resources at a low cost versus having to maintain the resource on premise. On-premise maintenance or procurement is very costly compared to use on a rental basis that the cloud offers.
iii. **Mobilizing the Workforce**
   Remote storage of data and applications on the cloud allow end users to access relevant content from anywhere with only one requirement – Internet connectivity. This makes the data and applications more flexible.
iv. **Increased Cost Control**
   The capital investment needed to buy and maintain hardware and licensed software as well as upgrade these items is cumbersome in any organization. With new advancements in technology like the cloud, this cost-intensive practice is vanishing.
v. **Enhanced Productivity**
   Businesses are more focused on core application logic and data with these technological advancements. Productivity is thus increasing as maintenance of other issues is taken care of by cloud service providers.
vi. **Reduced Impact on the Environment**

When companies need to individually buy servers, they also need to keep them cool. The cloud is a shared infrastructure, so this drain on power is reduced by half or more.

1.1.3 Cloud Computing Architecture

The cloud computing architecture consists of three layers: front end, middleware, and back end. The front end consists of end users using handheld devices like laptops, cell phones, PDAs, etc. Middleware deals with the logic behind delivering services to the end user. This layer consists of business logic. The last layer is the back end layer that consists of data centers, storage, and the infrastructure that are remotely located or owned by an organization [4].

1.1.3.1 Front End

- The front end of the cloud architecture refers to the client side of the system.
- It includes the network, applications, or programs that are used to access the cloud.

- For instance, while accessing a web-based email application, the web browser acts as the front end.

1.1.3.2 Middleware

- For the smooth communication between the front and the back end of the cloud computing architecture, certain protocols must be followed.
- The part of the system that connects the networked computers and facilitates proper functioning between the front and the back end is called the middleware.
- It is the special software used by the central server to administer the system.

1.1.3.3 Back End

- The back end of the cloud architecture refers to the hardware section, which includes the servers, deployment models, security mechanisms, storage, and the computing systems.
- Based on the requirements, a specific hardware configuration is set up initially to get the organization into the cloud.
- As the requirements increase, additional servers and storage modules are deployed from time to time.

1.1.4 Types of Cloud Services

1.1.4.1 Deployment Model

Cloud services are categorized per the access policies. One category is if services are available on the Internet and are hosted by a third party. Another category is if the services are restricted to one particular network [2, 6].

Basically, cloud services are divided into four broad categories:

1. Public cloud
2. Private cloud
3. Hybrid cloud
4. Community cloud

1.1.4.1.1 Public Cloud

In a *public cloud,* the services (compute, storage, databases, etc.) are hosted remotely by third party vendors. The services are available to end users via the Internet. This pay-per-use model is gaining popularity in recent years. AWS, Microsoft Azure, and Google Cloud Platform are the prominent players in the market in this category. All the services are fully managed by the cloud providers. This utility model of cloud computing provides easy access to a shared infrastructure, storage, as well as other computing resources. These shared resources are hosted in a remote data center, and multiple clients can gain access to them using the Internet. Scalability and cost-effectiveness are the major advantages of a public cloud.

1.1.4.1.2 Private Cloud

Provisioning of the infrastructure for controlled access by a single organization or providing controlled access to restricted data and information to a few organizations in a closed network format is referred to as a *private cloud.* All business units within the organization can access the cloud without having to build their own infrastructure. This deployment model offers all services on-premise. Fully owned data centers can be used to deploy this model. Users have full control over the resources they are using. This category uses the private network in contrast to the public cloud. Therefore, security and flexibility are the major advantages of a private cloud infrastructure. The private cloud can be used within any organization, institute, or private firm where outsourcing of data and applications are security critical. In this case, they can own their cloud infrastructure for maintaining the data and applications.

1.1.4.1.3 Hybrid Cloud

A cloud that is a combination of two or more clouds (public cloud, private cloud, or community cloud) is called a *hybrid cloud.* In a hybrid cloud environment, workloads can be moved between private and the public clouds based on changing computing needs and costs. This allows enhanced flexibility and more options when it comes to data deployment. The combination could be used where security of data is more critical, so data could be saved in a private infrastructure and front end can be deployed in a public network.

1.1.4.1.4 Community Cloud

A *community cloud* is a type of cloud primarily used by a closed group of people or organizations. It is a collaborative effort, and the hardware infrastructure is mutually shared by two or more organizations from a specific community. It could either be managed or hosted internally within an organization in the community or by a third-party cloud service provider. For example, universities like IITs can collaborate to form their own cloud to share their research, resources, etc.

1.1.4.2 Service Model

Cloud computing services can be categorized per the service they provide to the end user. The services vary from using infrastructure to high-end softwares. Depending upon the service the end user is using or vice versa, there are three main delivery models in the cloud [3].

Figure 1.1 details the different layers in each delivery model with the amount of services in each layer that are controlled by whom. In the on-premise infrastructure, all the layers are managed by users, but in the other three models, the ownership of management starts reducing from left to right. From right to left, the flexibility increases, such as custom hardware the user can use instead of dependence on the cloud provider for underlying hardware.

1.1.4.2.1 IAAS (Infrastructure-as-a-Service)

IAAS (Infrastructure-as-a-Service) in the cloud is the service model where infrastructure is offered over the Internet. This includes bare metal, storage, networks, and OS. The whole virtual machine is delivered to the client for use. This eliminates the cost of procurement and maintenance of hardware in any organization, making this delivery model cost effective. IAAS has a scalability feature as well. IAAS forms the base of the

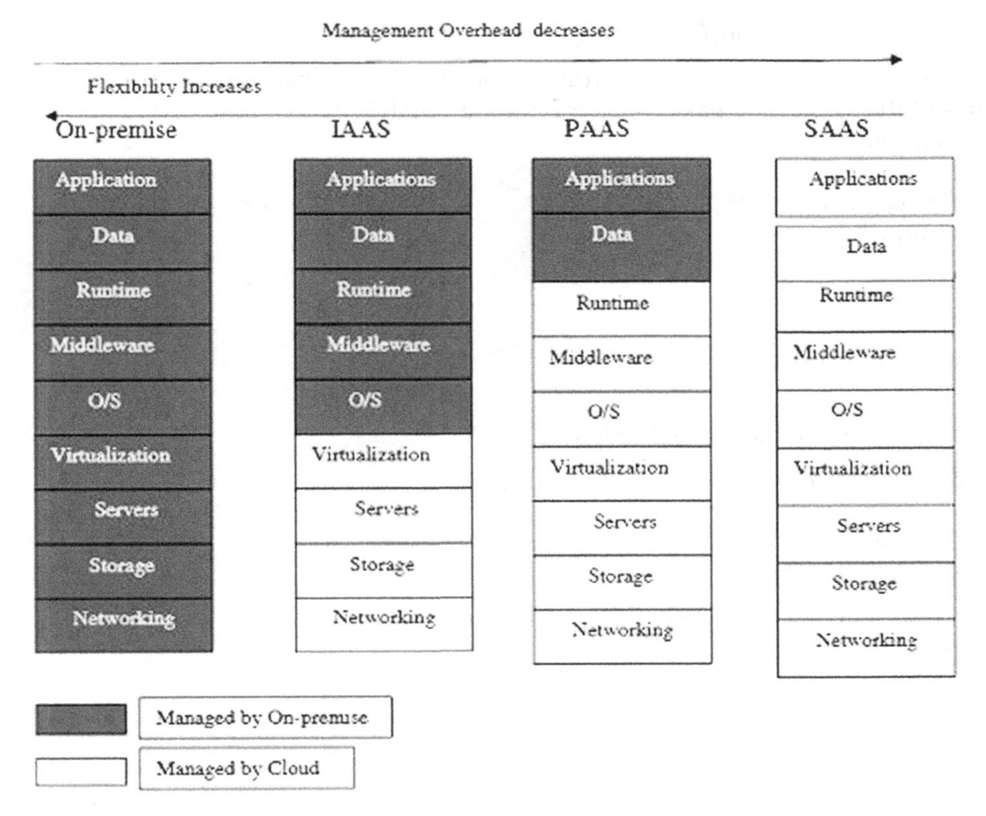

FIGURE 1.1
Delivery Models of Cloud.

cloud computing stacks. It offers the ability to provision a server, storage, networking, and other basic computing infrastructure resources through the Internet as required. Common examples include Amazon EC2, Microsoft Azure Virtual Machine, Google Compute Engine, and many more.

1.1.4.2.2 PAAS (Platform-as-a-Service)

PAAS (Platform-as-a-Service) is a cloud environment where the end user can deploy and develop the applications, and the environment that is needed to run and maintain them is provided by cloud providers. The user needs to focus on business logic and development of applications. The dependencies are taken care of by the cloud providers. After the development of applications, the end user needs to wrap the applications in bundles and upload them to the cloud environment. This offering is known as PAAS.

1.1.4.2.3 SAAS (Software-as-a-Service)

SAAS (Software-as-a-Service) is a cloud offering wherein end users take the use of the application without bothering with the underlying technology details. The end user leverages the application usage only, and all other dependencies are handled by the cloud service providers (CSP). Examples of this model are Gmail, Yahoo, and other applications directly available to the user.

1.1.5 Attributes of Cloud Computing

The following are the important attributes of the cloud. These attributes are the key features of the cloud that make cloud computing possible in the digital world and ensure its popularity.

a. Virtualization
b. Multi-tenancy
c. Network access.
d. Scalability
e. Metering/chargeback

1.1.6 Obstacles for Cloud Technology

The following are obstacles to cloud computing.

a. Data security and privacy issues
b. Network connectivity and bandwidth issues
c. Failed adherence to regulatory and compliance measures
d. Dependency on an outside agency
e. Vendor lock-in
f. Knowledge and integration
g. Lack of performance and uptime
h. Lack of long-term stability of the service provider

1.2 Cloud Management and Operations

Cloud computing is provisioning of resources to end users. In order to control and orchestrate the resources, cloud management plays a vital role. The cloud possesses resources like infrastructure stack consisting of servers, networking components, and storage devices. In addition to infrastructure, the cloud provisioned the services to end users in the form platforms, applications, and data. Management of these resources in a proper and optimized manner is a mandatory requirement in the cloud. With this, customers can leverage the optimum and efficient use of resources in the cloud. Cloud management provides administrative control to resources in a cloud-like infrastructure, platforms, and applications. Cloud management is a combination of software, automation, policies, governance, and people that determine how those cloud computing services are made available [1].

1.2.1 Advantages of Cloud Management

An efficient cloud management strategy improves cloud computing performance. The cloud computing environment follows optimization techniques and practices in the form of performance, cost, and reliability.

Cloud solution providers like AWS, Azurem, and GCP use various optimization tools in various domains. The optimization areas are as follows:

1. Cost optimization
2. Performance optimization
3. Reliability

1. **Cost Optimization**

Cost optimization refers to minimum cost per resource usage. Every organization compares the on-premise cost with resources hosted on the cloud in terms of cost usage. Cloud providers in turn provide attractive features and offers to customers like reserved instances, spot instances, and free tier limit through which users can leverage cloud capabilities. Cloud providers also provide cost calculators for resource usage and the ability to set alerts based on a budget set by an organization.

2. **Performance Optimization**

Performance optimization is running applications and workloads at a faster pace with less response time. The performance parameter is based on the type of application the organization is running, type of service they are using, and the underlying code and technologies tied to the cloud end point.

3. **Reliability**

Reliability is availability of application resources in case of data center failure and cloud-based resource unavailability due to some major disaster. The cloud providers offer preventive measures in the form of redundancy of data and applications in various regions and zones.

All above optimization techniques need cloud management strategy for optimum throughput from cloud providers. Before moving to the role of cloud management, one should know the structure of the shared pool of resources and the abstraction layer located at the data center level that the management layer is going to manage.

1.2.2 Cloud Infrastructure

The multitude of hardware components, like servers; networking components; and software elements, like hypervisor virtualization, that enable cloud computing can be termed *cloud infrastructure*. The cloud infrastructure can be interchangeably be used with *datacenter*. Public clouds are composed of innumerable datacenters located in different geographical regions. AWS maintains multiple geographic *regions*, including regions in North America, South America, Europe, China, Asia Pacific, South Africa, and the Middle East. Each AWS region consists of multiple, isolated, and physically separate *availability zones* within a geographic area [15].

Computing power, storage devices, networking devices, and switches powered with the software stack are termed *datacenter* or *cloud infrastructure*. Cloud infrastructure leverages virtualization to pool resources from the computer, network, and storage services that can be managed per need and can be elastically scaled [8]. As per Red Hat, cloud infrastructure is an IT environment that abstracts, pools, and shares scalable resources across a network. The underlying resources are provided to the end users by an abstraction layer – a virtualization layer that logically segregates the physical hardware and delivers it to the end user through a cloud user interface. The user interface comes in the form of an API (application programming interface), command line interface, or graphical user interface. The cloud infrastructure consists of high-density hardware to handle unpredictable user requirements and provide an optimal, dynamic balance.

The cloud infrastructure stack consists of a layered architecture. Figure 1.2 shows the layered architecture of the cloud infrastructure.

The lowest layer is the datacenter layer, which consists of stacks of servers, storage in the form of images, blocks, and object storage. Storage possesses solid-state drives (SSDs) and hard disk drives (HDDs). The disks in each system are aggregated using a distributed file system designed for a particular storage scenario, such as object, big data, or block. Computer nodes are nothing but high configuration multi-core, multi-socket servers that are accessed by end users as multi-tenant architecture. The high-end networking components are switches and routers. These networking components are responsible for delivering the shared pool of resources to the end user. These network resources can include virtual routers, firewalls, and bandwidth and network management software. The popular public cloud providers like AWS have networking components in the form of VPC (Virtual Private Cloud), Internet Gateway, NACL (Network Access Control List), NAT gateway, and Security Groups. A network device, like Internet Gateway, is the end point through which users connect with the public cloud. To secure the network and the firewalls, security groups play a vital role. The inbound and outbound traffic in the cloud is a fully managed service offered by public clouds.

Above the data center layer, the cloud management layer plays a pivotal role by abstracting the on-demand pool of resources from the infrastructure layer to end customers. There are numerous components of cloud management, ranging from governance to

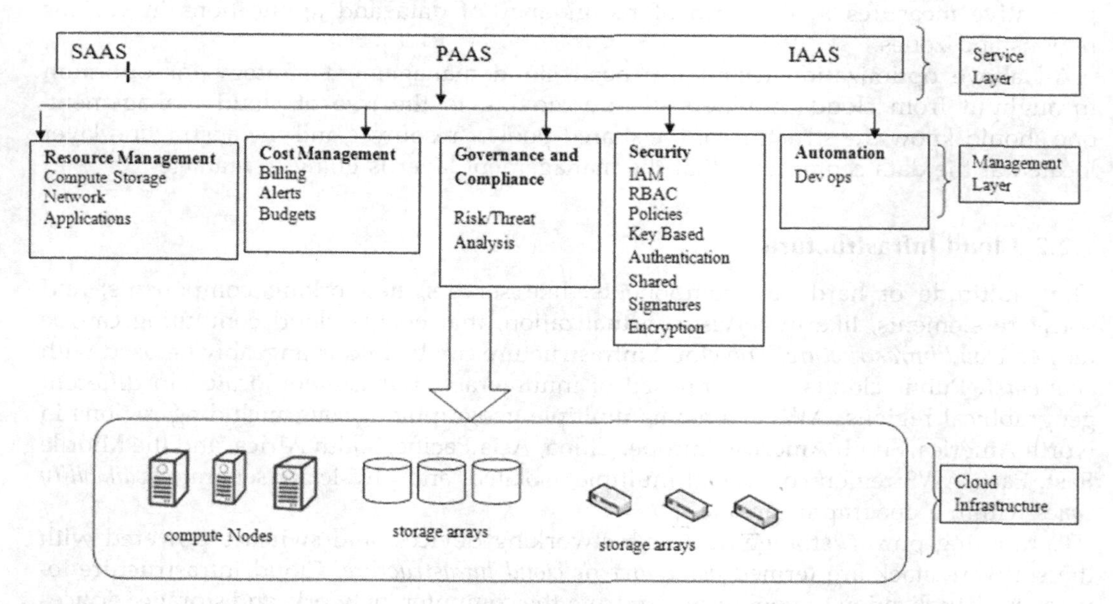

FIGURE 1.2
Layers of the Cloud Infrastructure.

security, that make the cloud a full-fledge service-oriented architecture that focuses on economies of scale. The next section details the components of cloud management.

In the case of the public cloud, the entry point for all end users is cloud dashboards or portals through which cloud users authenticate themselves using their personal credentials. In addition to graphical interfaces like portals, public clouds have numerous options for users. Users can get access to cloud resources through Command Line Interface or through program access. Various tools are available in this context. (AWS CLI is an example.)

With reference to the private cloud, Open Stack offers Horizon as a dashboard for user interface. The user has credentials provided by Open Stack Identity Management that allow the user to log into the private cloud infrastructure.

In the hybrid cloud scenario, the on-premise and remote cloud entry is managed by varying tools. (Azure AD Connect is an example.)

1.2.3 Components of Cloud Management

The key areas of cloud management are [12] as follows:

1. Resource Management
   i. Compute Management
   ii. Storage Management
   iii. Network and Access
   iv. Applications Management
2. Cost Management
   i. Billing
   ii. Alerts
   iii. Budgets
3. Governance and Compliance
   i. Risk and Threat Analysis
   ii. Identity and Access Management
   iii. Role-Based Control
4. Security
   i. Infrastructure Security
   ii. Key-Based Authentication, Shared Signatures, and MFA (Multi-factor Authentication)
   iii. Encryption

1.2.3.1 Resource Management

Cloud resource management and operations require complex policies, decisions, and administrative control for multi-objective optimization [5–7]. The focus of cloud customers is on parameters like high availability, on-demand services, scaling up and down of resources as per need, and elasticity minimizing the cost of this objective function, whereas the focus of cloud providers is to efficiently deliver the resources as per workload and SLAs (service level agreements). In order to meet this criteria, cloud management uses different tools and technologies that provide Quality of Service (QoS) to end users. The resource management can be categorized as per resources available in cloud infrastructure as compute, storage, network, and application management. If we consider the operations of allocation of resources to end users, the resource management can be classified as follows:

1. Resource identification and mapping
2. Resource allocation
3. Resource monitoring

We will explore this classification of resource management in detail.

1. **Resource Identification and Mapping**

Resource identification is the process of choosing the appropriate resource as demanded by the cloud user, and *mapping* is assigning that resource to a user. Theoretically, authors focused on cloud management in terms of scheduling, allocation, provisioning, and monitoring. In this chapter, we will try to explore the capabilities of cloud management area in terms of different categories the public cloud offers.

Different users log into the cloud portal using their credentials or try to access the resources through different tools available in the cloud environment. The customers then enter the requirements as per their workload functioning. As soon as the resource requirement is entered or requested from the client side, the role of cloud management starts. The back end scheduler starts poling the pool of resources demanded by the users. The resource identification can be categorized as discussed above as compute, storage, network, and application. So when the compute resources are demanded, there are a variety of resources available. There are numerous services ranging from EC2, virtual machines (VMs), EC2 scaling, and container services like Kubernetes and AWS Light Sail. The backbone of all services is EC2 that can be termed as a virtual server on the cloud that can be accessed remotely. As per the needs of users, there are a variety of VMs in AWS. Both AWS and Microsoft Azure offer a free tier limit of VMs. Also, both offer different VM sizes as per user requirement. The VM sizes and configurations are available in the form of general purpose VM, compute optimized, storage optimized, and memory optimized [9]. Users can choose among the categories which one they desire. Each category has features like VM size in terms of VCPUs supported, memory, and storage in the form of EBS (elastic block storage). Similarly, AWS Cloud has some offerings in the form of reserved instances, spot instances, and on-demand instance types to deliver users with a varying range of instance types and cost efficiency. This offering falls under cost management, but addressing this feature is a highlight of resource management.

The next category of resource management that works hand to hand with compute is storage. Both AWS and Azure offer storage ranging from object storage, block storage, different disk types (SSDs, HDDs), and structured and unstructured storage types [10]. Cloud storage fulfills all requirements of cloud customers. The requirement of storage mainly is security in terms of encryption and data confidentiality. Another feature needed by users is 24×7 availability of data to applications. The durability is data recovery in case of disaster [10]. All these storage features are managed and provisioned by storage management modules.

Compute with storage cannot be delivered to users without network management module. The networking in the cloud is composed of virtual private cloud or virtual network, collection of subnets within virtual network, IP address ranges, security groups, firewalls, and virtual gateways. The network management module maintains the integrity of data transfer between inbound and outbound traffic to and from the cloud.

Application management is management of applications like serverless computing, functions, website hosting, and many more that use the functionality of the cloud environment to enhance the on-premise workloads. The application management module furnishes and supports all features to run application-oriented workloads in the cloud environment.

2. **Resource Allocation**

With reference to the first stage, the resources that are assigned to users are now allocated. The resource allocation phase is actually the in-use phase of IT resources or infrastructure. After allocation, the resources can be stopped, running, or terminated. These terms (stopped, running, terminated) can be different with respect to other resource categories. The cloud environment provides a resource allocation strategy with user-friendly interfaces. The list of allocated, terminated, stopped, and unused resources can be easily identified through the user interface [11,12].

3. **Resource Monitoring**

Resource monitoring is an important phase in resource management. Cloud monitoring is a method of reviewing, observing, and managing the operational workflow in cloud-based IT resources. There are various ways to monitor on cloud-based resources. The function of monitoring is to confirm the availability and performance of websites, servers, applications, and other cloud-based resources. Continuous monitoring enables cloud providers to predict and estimate possible vulnerabilities and future issues before they arise.

1.2.3.2 Cost Management

Cost management is the process of managing and optimizing the cost of cloud resources. The organization planning to optimize the use of cloud resources in terms of cost is the primary concern for business growth. In order to have effective and efficient cost control in cloud computing services, it is important to analyze and decide the cloud cost and leverage cloud cost management tools to help discover the cause(s) of these inefficiencies in terms of output and cost associated with it. Cost management is essential because the cloud environment offers service benefits in terms of scaling and elasticity. This allows the IT developers to leverage these features without thinking of costs associated with these resources. The parameters to build effective strategies in cost management are as follows:

1. **Knowledge about Cloud Resources**

Knowledge and know-how of cloud resources is essential for effective cost management. The complete information and features of VM instances, memory, and storage will play a crucial role in cost cutting in any organization. Choosing the VM sizes optimal for the proposed workload of an organization, scaling up and down of VMs for cost reduction, and shutting or terminating unused resources are the backbone of cost management.

2. Role-based and Policy-based Access Control

Restricting the access of resources to the users through a role-based mechanism can significantly control the use and ultimately the cost to the organization. Providing access to the administrator only for creating or deleting the resources will definitely control the resource utility.

Knowledge about cloud resources and role-based and policy-based access control is the first step of resource allocation and delivery. After this, cost management includes parameters like billing, alerts, and budgeting [13].

1. Billing

Billing involves payment for cloud services. It also provides insight via a usage meter of resources. It provides analysis of running resources in cloud portals. In this way, it provides options to control costs. Billing in the cloud environment is done through credit card, which is provided by users during the registration process. The public cloud providers offer various cost benefits to users in terms of estimates and resource optimized schemes. AWS and Azure both provide resource estimate calculators for calculating the estimates of the resources to be utilized in proposed workloads.

2. Alerts

Cloud service user can specify the threshold level for each resource category. When this threshold level exceeds, the alarm triggers. It triggers the alarm only when actual billing exceeds the threshold. The estimates of all resources are sent to monitoring tools (Cloud Watch in AWS) to project metric data to cloud users.

3. Budgets

Budgeting is the process of estimating and planning of resource costs. Cloud service providers have the capability to set custom budgets for resources and to track your costs and usage of cloud infrastructure. Each of the cloud provider budgeting tools can be integrated with alert-triggering mechanisms so that when actual usage surpasses the forecasted budget an alert can be sent to organizational authorities in the form of SMS or email.

1.2.3.3 Governance and Compliance

The set of rules for smooth operational process in an organization is governance. It is essential for any IT firm or organization to monitor the operations for a further plan of action in case of any grievances. The compliance comes into picture for this reason. The rules specified by an organization or cloud environment can vary from granting user privilege level, to providing privacy, ownership, and security of data. The decentralized operations in an organization may create communication gaps between IT developers. The cloud environment leverages the IT assets with a click of mouse. A lack of control may affect the business growth of any IT firm. The set of rules and policies can restrict the user from usage of cloud services. The guidelines and policies consist of accessing the resources for only the set of users, deployment of instances in a specific region, and guidelines about what software applications and programs departments can use.

1. **Risk and Threat Analysis**

After applying the specified rules on the set of resources and users, the compliance management console can analyze the compliant and non-compliant resources. With this, the cloud management portal suggests preventive measures in terms of risk of failure, security breaches, data protection, and privacy threats, and it even details costs associated with cloud services.

2. **Identity and Access Management**

As discussed, the set of rules for access management prevents unauthorized use of cloud resources. The identity and access management module provides identification and authorization services in the cloud. Authentication means proving who you are, and authorization is whether the user is allowed to perform a certain task or not. The mentioned module creates a set of users and groups for specific tasks or functions. Access can be granted at the admin level, the tenant level, the management level, or at the subscription or more granularity level, called the resource level.

3. **Role-based Control**

Role-based access control is another level of control that can be imposed on users. These rules allow and disallow users from changing the current state of resources and services. For example, users can be granted an ownership role to add, delete, or terminate resources. A viewer-only role can also be assigned by the administrator so a user can view the resources but cannot edit their current state. In this way, user actions can be restricted.

1.2.3.4 Security

An important and crucial cloud management area is cloud security. Cloud security services are the tools that enable protection, privacy, encryption-based services, and key management of your accounts and applications from unintended and unauthorized access. Security covers a wide area, including infrastructure protection, threat detection and compliance, data protection, and privacy. Cloud services are remote in nature. The sensitive data and infrastructure are hosted in a cloud. Cloud security is the protection of data and workloads or applications remotely, hosted in the cloud environment. For data security and privacy, data leakage, accidental or intentional deletion, data theft, data breaches, and data loss are a few concerns. Cloud security covers and offers security in almost all essential areas.

1. **Infrastructure Security**

With workloads like websites, applications, or VMs running some critical functions, networking security is paramount. Therefore, cloud security includes network firewalls, security groups for allowing and blocking the ports, web application firewalls, and many more tools to use to provide security in the cloud.

2. Key-based Authentication, Shared Signatures, and MFA (Multi-factor Authentication)

In addition to VMs, applications and storage authentication and authorization are crucial in the cloud. Therefore, in addition to username and password-based authentication, users can also use key-based authentication methods to get access to the cloud environment. This is also another way of authentication in the cloud. The identity and access management module discussed above is another user and account protection technique. The root user can grant access to users as per their workloads. A symmetric authentication method to access resources is a shared signature mechanism where communicating parties exchange their signature for gaining access. Another method is multi-factor authentication in the cloud. This is based on the concept that the passwords that we enter can be compromised. Adding another level of security such as scanning a code with a mobile phone and getting an authentication code via SMS or call can add more security.

3. Encryption

The method of converting plain text to cipher text for transmitting it on the network is known as encryption. The cloud offers encryption at rest and at transit. The key management system (KMS) is the service offered by AWS.

1.2.3.5 Case Study on Cloud Storage Service

This case study covers the AWS storage service. AWS offers a wide range of storage, from object storage, to file storage and block storage services, to backup and data migration.

1.2.3.6 Amazon S3

Amazon S3 is easy to use and the simplest form of object storage in AWS. AWS S3 can be accessed easily through a simple web interface. The data storage is done through buckets, which are containers, like folders on local computers. The buckets can be created across the regions and can be available across the different availability zones (AZs). The data storage allows the different versions of data in buckets. With this, data with every version can be restored and recovered in case of accidental deletion. As the data is stored in the form of objects, any type of data ranging from simple text files, image files, videos, and audio files can be stored and can be easily accessed. This means S3 can easily be integrated with any application. Amazon S3 offers different types of S3 storage depending on access patterns. S3 standard is the storage where frequently accessed data is stored. S3 infrequent access stores the data that is not accessed for 30 consecutive days. Archives and deep archives are the S3 types where data are not accessed for 90 and 180 days, respectively. In S3 Intelligent-Tiering, if the access patterns of data change, the objects are moved across S3 access tiers that are from S3 infrequent access tier to S3 standard or vice versa. By default, owners have access to objects they have created but permission of the data at the bucket level or at the fine-grained level can be granted [13].

1.2.3.7 Amazon EBS

Amazon Elastic Block Storage (EBS) is the block-level storage to EC2 offered by AWS. EBS volumes can be attached to EC2 (Elastic Compute Cloud) Instances and offers persistent

TABLE 1.1

Amazon storage types [14]

Sr.no	Storage Service Name	Type of storage	Features
1	Amazon Simple Storage Service (Amazon S3)	Object storage	1. 5 GB free for 1 year 2. Versioning 3. Different access tiers 4. Secure access of data through access keys, Identity and Access Management (IAM) 5. Data encryption
2	Amazon Elastic Block Store (EBS)	Block storage	1. 30 GB Free for 1 year 2. Backup using Snapshots 3. Data encryption at rest and transit 4. SSDs and HDDs 5. Scalable
3	Amazon Elastic File System	File storage	1. 5 GB free for 1 year 2. Can be connected to on-premise file system 3. Elastic

and durable storage. Multiple EBS volumes can be attached to EC2 Instances. Data are retained even if EC2 is stopped or terminated. It is just like disk storage at an on-premise infrastructure. EBS enables backup of data in the form of EBS Snapshots. EBS Snapshots are incremental, which means they add to previous Snapshots rather than overwriting them. There are two types of EBS volumes: SSDs (solid state drives) and HDDs (hard disk drives).

1.2.3.8 Amazon Elastic File System

Amazon Elastic File System (Amazon EFS) provides a fully managed service. It is a simple, scalable, and elastic NFS file system for use with AWS Cloud services and on-premise resources. It enables the connection of an on-premise server to an AWS file system through VPN or AWS direct connect. AWS EFS is designed for high availability and durability.

In addition to this, AWS offers storage services like Amazon Backup for backup, Amazon Storage Gateway Amazon Data sync for online data transfer, and Amazon Snowball for offline data transfer (Table 1.1).

Self Assessment Questions

1. List the advantages of cloud computing.
2. Describe the history and evolution of cloud computing.
3. Explain types and deployment models of the cloud.
4. Explain components of cloud management.

References

1. Abbadi, I. M., 11 August (2014) *Cloud Management and Security Hardcover*. Print ISBN:9781118817094 |Online ISBN:9781118817087 |DOI:10.1002/9781118817087. Wiley.
2. Sosinsky, B. B. (2011) *Cloud Computing*. Wiley.

3. Sher DeCusatis, C. J. and Carranza, A. (2013) "Handbook of Fiber Optic Data Communication (Fourth Edition), A Practical Guide to Optical Networking." *Chapter 15 – Cloud Computing Data Center Networking*, 365–386.
4. Dong, D., Xiong, H., Castañe, G. G., and Morrison, J. P. (2018) Cloud Architectures and Management Approaches. In: Lynn, T. Morrison, J. P. and Kenny, D. (eds.) *Heterogeneity, High Performance Computing, Self-Organization and the Cloud* (pp. 31–61). Palgrave Macmillan.
5. Gonzalez, N. M. , Carvalho, T. C. M. de Brito, and Miers, C. C. (2017) "Cloud Resource Management: Towards Efficient Execution of Large-Scale Scientific Applications and Workflows on Complex Infrastructures. *Journal of Cloud Computing* 6: Article No. 13.
6. Ghutke, B. and Shrawankar, U. (2014, March) Pros and cons of load balancing algorithms for cloud computing. In: Ashish Sharma (ed.) *2014 International Conference on Information Systems and Computer Networks (ISCON)* (pp. 123–127). IEEE.
7. Mishra, P., and Shrawankar, U. 2016.Comparative Study of Cloud and Non-Cloud Gaming Platform: Aperçu. In: International Conference on Research in Intelligent Computing in Engineering (RICE) 2016At: Nagpur. MH, India.
8. https://docs.aws.amazon.com/whitepapers/latest/aws-overview/compute-services.html.
9. https://aws.amazon.com/ec2/instance-types/.
10. https://aws.amazon.com/what-is-cloud-storage/.
11. https://searchcloudcomputing.techtarget.com/definition/cloud-management.
12. https://www.redhat.com/en/topics/cloud-computing/what-is-cloud-management.
13. https://www.vmware.com/topics/glossary/content/cloud-cost.
14. https://aws.amazon.com/products/storage/.
15. https://aws.amazon.com/about-aws/global-infrastructure/regions_az/.

2

Real Time Edge Computing: An Edge of Automation

Girish Talmale[1] and Urmila Shrawankar[2]
[1]*Research Scholar, Computer Science and Engineering Department, G H Raisoni College of Engineering, Nagpur (MS), India*
[2]*Professor, Computer Science and Engineering Department, G H Raisoni College of Engineering, Nagpur (MS), India*

2.1 Introduction

Cloud computing has played an important role nowadays every application switches from the traditional server-based architecture to the cloud for computation, storage, and networking [1]. The emerging applications such as the internet of things, cyber-physical systems are time-sensitive and quick responses [2]. The real-time response is not possible for such applications due to latency in communication as data directly stored on cloud data centers which are located very far from the data generating devices [3]. Edge computing is a computing platform that pushes processing, analytics, intelligence, and communication close to devices that generate the data and not direct data directly to the cloud [4].

The features of edge computing are as follows:

1. **High Computation Hardware**: In complex applications such as machine learning high computational hardware is required. Computing devices with such high computational hardware is a big challenge so edge computing is the solution to solve that problem [5].
2. **Low Latency:** When the applications are time-sensitive then sending the data to the cloud, perform computation and send the response back will have more latency. So to use edge computing platforms the processing is done at the edge computing server which is more near to devices will reduce the latency a lot.
3. **High Data Throughput and Filtering:** The devices produce a huge amount of data if all data stored in the cloud it increases the cost of storage. For example, an autonomous car generates a huge amount of data per second so data filtering take place on edge computing and only require data is stored in the cloud

DOI: 10.1201/9781003166702-2

TABLE 2.1

Difference Between Edge Computing and Cloud Computing

Sr.No	Comparative Parameters	Edge Computing	Cloud Computing
01	Location of Computation	Edge Devices	Remote Server
02	Response Time	Very Less	High
03	Bandwidth Requirement	Less	High
04	Support Real Time Application	Yes	No
05	Computation Capacity	Less	High
06	Storage Capacity	Less	High
07	Power Consumption	Less	High
08	Number of Hops	One or Few	Multi Hops
09	Security	High	Less
10	Communication Latency	Less	High
11	Delay Jitter	Less	High

4. **Reliability, Security, and Privacy:** Edge computing provides reliability, security, and privacy by replacing the failed edge node with another edge node. User privacy is ensured in edge computing as the processing nodes are close to the user.
5. **Scalability:** The edge computing nodes can be easily added or removed from the edge computing framework.
6. **Adaptability:** The edge computing framework easily adapts to the changing dynamic environment [6].

2.1.1 Comparison of Cloud and Edge Computing

The difference between cloud and edge computing is as presented in Table 2.1. The edge computing is performed close to device while in cloud computing the processing is performed on remote server which results in low communication latency. Edge computing is most suitable for applications with real time requirements. The storage capacity, power consumption and computing capacity of edge computing is less than the cloud computing. The edge computing provides higher security as compare to cloud computing.

2.2 Literature Review

The concept of edge computing was introduced in the past with the invention of a content delivery network that is used to boost the performance of the web [7]. The content delivery network performs computation near to end-users and ends to boost the performance of the web using caching techniques. Edge computers offer high bandwidth, real-time response, and very low communication latency [8]. Edge computing provides different services to consumers and enterprise applications [9]. The edge computing is nothing but the independent models of processing consisting of devices with different platforms [10]. As per Cisco, edge computing is the advances in cloud computing that enable the processing and computation at the edge networks [11]. The smart city applications used edge computing platform for real time response system. The security plays an important role in implementation of different edge computing based smart city applications [12]. The lightweight edge computing gateway for Inter of Things(IoT) the IoT is developed using micro-service and flexible virtualization technology. The edge computing is extensively used in healthcare domain using Internet of Things applications[13]. The edge computing with big data analytics and IoT technology now days used in industrial automation applications [14]. The edge computing application for real time remote healthcare monitoring is presented [15]. The virtual machine migration using pareto optimize solution is provided in recent research [16].

2.3 Architecture of Real Time Edge Computing

The architecture hierarchy showed that the computer power increases from bottom to top and communication latency decreasing from top to bottom.

The edge computing architecture consists of the following layers

1. **Cloud:** This is the top layer of edge computing architecture. It has high computation power and storage and the lowest communication latency. The cloud computing layer is used for data storage and large-scale data processing.
2. **Edge Nodes:** The edge processes nodes, processes sensor data of devices and performs routing. Edge computing nodes vary from routers, switches to low computing servers [17].
3. **Edge Gateway:** Edge gateways are low computing devices capable of processing small amounts of data.
4. **Edge Device:** This layer consists of an end device that is used to generate data. Edge devices consist of a sensor and embedded devices with very low computing power [18].

2.4 Case Study on Edge Computing: Industrial Automation

The edge computing emerging application is industrial automation. In the industrial 4.0 revolution industrial automation using a real-time edge computing framework plays an important role. The manufacturing industries perform on-premises processing using PLC and SCADA [19]. Edge computing provides flexible and real-time control of different industrial automation processes. The execution speed, reliability, and secure data transmission is an vital parameter for the implementation of automation in industry which can be achieved using an edge computing framework [20].

The different problems in implementing industrial automation without edge computing are as follows:

1. **The problem of Connectivity**: The industrial internet of things used a large amount of data generated in the physical work and then performs analytics on this data to improve the performance of the physical world. In the industrial world, the different machines are not designed to satisfy this requirement. Industrial automation exists before the invention of the internet. These machines are attached with the sensor, actuators but lack of interface and ethernet jack to communicate to the internet. Industrial machinery also not supporting internet programming languages such as java scripts, rest API, and JSON and CSV, etc. These devices also do not run the operating system and also TCP/IP protocol stack is not installed on these devices.
2. **The problem of Big Data**: The industrial plat consists of a large number of sensors that produce huge data per second. The example of Oil and Gas Industry plant installed 50,000 and more sensors which generate a huge amount of data and the management of this huge volume of data is a Big Data problem. Moving this huge data on the cloud and perform analytics is costly and produces high communication latency which is not acceptable for industry automation where real-time control and monitoring is essential. So there is the requirement of edge computing in the middle to filter that huge data to reduce cost and communication latency.
3. **The problem of Architecture**: The sensors which are used are analog and require software to convert it into digital form. The sensors connected to PLC (Programmable Logic Controller) and these PLC are not upgraded to be used to connect to the cloud. The PLC software, hardware, and programming languages are not designed to operate on the cloud directly. The PLC protocol is not upgraded to connect directly to the cloud. So there is the requirement of edge-enabled open platform communication to solve this problem.

2.4.1 Industrial Automation Edge Computing Use Cases

The different use cases of edge computing in industrial automation.

1. **Remote Industry Asset Monitoring**: Edge computing helps to monitor different assets of the industry remotely. The edge computing framework is used to take real-time action in case of emergencies.
2. **Predictive Machine Maintenance**: Predicative machinery maintenance plays a vital role in the implementation of industrial automation. Edge computing framework is used to process the machine data in real-time to predict the condition of machinery in advance to avoid industry breakdown. Edge computing processes the data near to physical machines which provide real-time data processing and reliability of data transmission
3. **Virtual Reality and Augmented Reality**: Industrial applications nowadays extensively use virtual reality and augmented reality. Creating a virtual environment requires high computational power and low communication latency which is provided with an edge computing framework.
4. **Precision Control & Monitoring**: The main requirement of industry 4.0 is to collect data from various industrial processes and adapt the system control in real-time. Edge computing is used not only to filter processes and send data to

the cloud but also to perform some machine learning tasks to provide real-time analytics.

5. **Testing Quality and Inspections**: The quality of product testing and defect inspection can be done using machine learning techniques which used edge computing for real-time response. Machine-leaning-based defect detection used edge computing to improve the defect detection efficiency over manual testing processes.

2.4.2 Industrial Automation Using Edge Computing Architecture

The physical layer consists of different physical machines and devices using in industry [21]. The physical layer responsible for sensing different machine parameters and get schedules and using real time scheduling and send the data to the edge computing layer. The edge computing responsible for real time data analytics and perform real time action using to actuate the corrective tasks. The edge computing layer is also responsible for data filtering and sends the filtered data to the cloud. The cloud computing layer responsible for large scale big data processing and data warehousing [22].

2.4.3 Industrial Automation Real Time Tasks Allocations on Edge Cloud

The tasks scheduling in edge computing is divided into two main modules.

1. **Tasks Allocation:** This module responsible for the allocation of tasks to edge computing server
2. **Task Scheduling:** Tasks scheduling decides the order of execution of tasks on edge computing servers.

The tasks allocation techniques for edge computing used cluster based approaches where the set of tasks assigned to a group of edge computing machines classes as cluster. The edge computing cluster solves the problem of tasks assignment which is an NP-hard problem in partitioned-based tasks assignment. It also solves the problem of global scheduling by reducing the task migration overheads [23].

2.4.4 Edge Computing Solution for Industrial Automation

Edge computing provides the solution to challenges faced by current industry automation. The different solution to industrial automation challenges are as follows:

1. **The Solution to the scalability of Data**: Industrial automation generates a huge amount of data and to manage that huge scale of data is done using edge computing by filtering sensor data at the edge level and send only filtered data to the cloud.
2. **The Solution to High Communication Latency:** Edge computing pushes the computing on edge devices instead of the cloud data center to reduce communication latency. Edge computing extends the cloud application for real-time application which requires low communication latency and a faster-distributed control system. The applications are installed on edge devices which reduce communication latency and improve the quality of service.

3. **Management of Mixed-Criticality**: Mixed-criticality of application is managed by using virtualization technology, which is applied to real-time embedded applications. In virtualization technology, hypervisors manage different resources on the virtual machine. Each virtual machine has its operating system, so different applications with mixed-criticality are managed on a multi-core platform. Management of the separate operating system and different partition provides the solution to software running on a real-time system.

2.4.5 Real Time Case Study of Industries Using Edge Computing

The case study of different industries using edge computing is as follows:

1. **Dell Use Real Time Edge Computing for Product Defect Detection:** The manual product defect detection is error-prone. The smart product detection system using a machine learning approach to increase production by approximately 50% and improve product defect detection by 90 % as compared to manual product defect detection. The machine learning algorithm used for the detection of defects needs high computation power and less communication latency for real-time response. The edge computing framework provides high computation power and low network latency for such applications. Dell organization gets the IoT-based edge computing analytics solution for smart defect detection from the Daihen Japanese industry. Before the use of edge analytics service, their production manufacturing processes require 200 plus manual inspection which wastes approximately 30% of total manufacturing time. After the installation of an edge-based real-time data analytics platform, they have saved 5000 plus hours that are required for manual data entry in the year. The manual data entry process is also error-prone which leads to huge financial loss. The manufacturing process is now more efficient and intelligent using an edge computing data analytics platform that automatically detects defect detection.
2. **IMS (Integrated Marine System) Uses Real-Time Edge Computing Platform for Optimization of their Refrigeration System:** The IMS refrigeration manufacturing giant company used IoT based real-time edge computing framework for real-time collection and management of refrigeration data for the auto-tuning of temperature as per the standard of food quality which reduces waste and optimizes the cost of refrigeration costs.
3. **Tulip Edge Devices for Machine Performance Monitoring:** Tulip customer uses the IoT gateway to to collect information of their manufacturing machines. The problem in with their machines is that theses machine are not connected to internet. The measurement of machine performance was done manually. The manual collection and measurement of different machine performance parameter leads to error in the measurement which affect the analysis process and further produce difficulty in identifying the main causes. The problem is solved by using Tulip edge devices to convert these analog machine to digital. The Tulip edge devices are used by their customer to measure the different machine performance monitoring parameter online. The machine performance data further analyze on the edge to provide real time analytics to improve the manufacturing process and reduce downtime in manufacturing. The data from different manufacturing machinery is collected analyzed in simultaneously in real time using

Tulip edge analytics. The Tulip customer increased their product manufacturing by approximately by 15% as compare to earlier analog machines.

4. **Intel Used Edge Computing Platform to Boost Performance of Fan Filter Units (FFUs):** The Equipment downtime is a serious manufacturing problem. Uninterrupted manufacturing cannot be possible unless accurate and real-time information about the performance of equipment is monitored. The IoT and edge computing platform ensure real-time predictive maintenance of equipment to reduce equipment downtime. Intel wants to increase the performance of its fan filter units, which are the critical parts of equipment used in preventing infection in clean rooms. To provide the solution to this problem, Intel worked on an edge computing platform based on predictive maintenance. The sensors give information about any variation in the fans' function and then create the baseline for differentiation between the fleets. Intel used edge computing architecture with IoT gateways to measure and analyze the functioning of all the FFUs. The machine learning algorithms used that sensory data to detect faults and send real-time alerts. The cloud-based solution is used to further analyze the large-scale data for all equipment and find the best maintenance schedule to increase the FFUs' performance.

2.5 Challenges in Edge Computing

The cloud computing framework such as Google, AWS, Microsoft support data-intensive applications. The application with real-time requirement such as industrial automation is still a challenging problem [24].

The different challenges in implementing edge computing are as shown in Figure 2.1.

1. **General Purpose Edge Computing:** The edge computing nodes are generally router, switching devices that are specific purpose embedded systems. The challenge in implementing edge computing is how to use it for general-purpose computing environments.
2. **Edge Node Discovering:** Edge computing networks need a discovery mechanism to find the nodes. The discovery mechanism must be automated as the volume of devices in this layer is huge. The fault detection and recovery of nodes are needed to be developed.
3. **Partitioning and Allocation of Tasks:** Edge computing is a distributed network that requires automatic partitioning and allocation of tasks on edge nodes. The development of an efficient real-time scheduler that partitioned and allocates the tasks to edge computing nodes is a challenging task [25].
4. **Maintain Quality of Service:** Quality of edge computing nodes monitored using Quality of Service. The edge computing node must not be overloaded with high overload. The challenge is how to achieve high throughput and reliability for the intended workload. The node management network requires scheduling, monitoring, and rescheduling of edge computing nodes.
5. **Use of Edge Node Publically and Securely:** The use of edge nodes requires challenges to securely and reliably use these nodes. The routers which are used

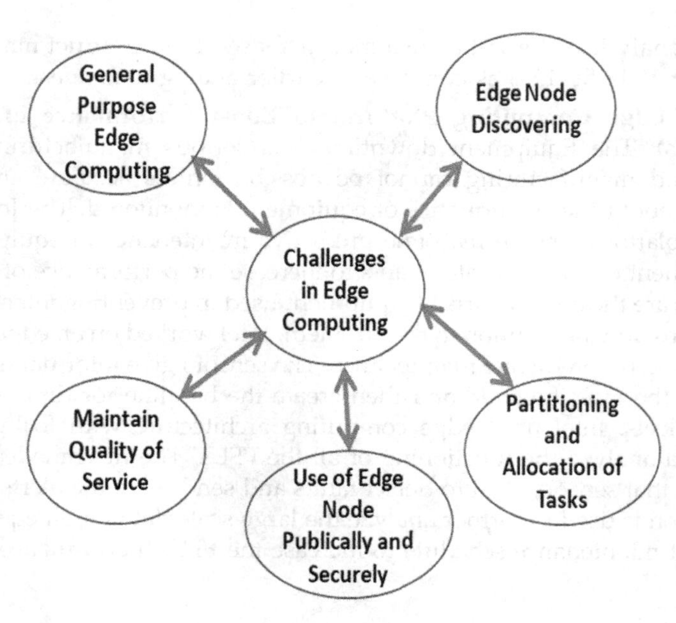

FIGURE 2.1
Challenges in Edge Computing.

to manage internal network traffic must not be compromised for edge computing node tasks [26].

2.6 Scope of Research in Edge Computing

Despite various challenges in implementing edge computing, there are several opportunities and research problems to be solved. The research problems and opportunities in edge computing are as follows in Figure 2.2.

1. **Edge Computing Standard and Benchmark:** The edge computing framework requires standards like cloud computing such as NIST, ISO, and IEEE. There is no such standard is define for edge computing so there is an open research problem to find out a new standard for edge computing platforms. The new SLA (Service Level Agreement) policy development is a new research problem for edge computing. Benchmarking algorithm development to schedule tasks to edge nodes and to monitor the performance of edge nodes is also a new research problem [27].
2. **Edge Computing Framework and Development Languages:** Cloud computing has various computing frameworks and development languages but there is the requirement of such toolkit and development framework and languages for edge computing is an open research problem.
3. **Development Lightweight Libraries and Algorithm:** Unlike cloud computing edge computing nodes require resource constraints lightweight libraries. Edge

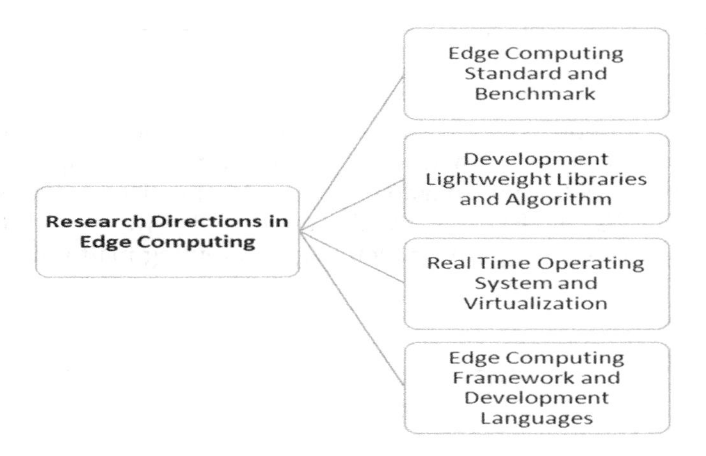

FIGURE 2.2
Research Directions in Edge Computing.

computing nodes are computing and memory constraints so it requires specializing algorithm edge data analytics. Edge computing requires lightweight machine learning libraries like Tensor flow for deep learning computation [28].

4. **Real-Time Operating System and Virtualization:** The development of the new real-time operating system is requiring which manage resources of edge computing nodes efficiently. The new virtualization and container technologies should be developed for creating virtual machines for deploying the application on edge computing [29].

2.7 Conclusions

Edge computing is a distributed computing platform for time intensive application. Edge computing is still an emerging computing model and requires research for efficient distributed computing. This work discusses the various advantages of using edge computing. The real time edge computing architecture is presented with a physical layer responsible [28] for sensing, scheduling and actuation, edge layer responsible for real time data analytics, data filtering and cloud layer responsible for the large scale big data processing. The real time edge computing implementation explained with case study for industrial automation. The different use cases of industry automation using edge computing framework explained. The edge computing architecture for industry automation proposed with a physical layer consist of machines which perform sensing, actuation and scheduling, edge computing layer provide basic data analytics and machine learning tasks on machine data and cloud layer perform large scale big data analytics and data warehousing. The real time tasks allocation for edge computing is proposed by grouping edge computing nodes into clusters to improve resource utilization and reduce migration overheads. The efficient implementation of edge computing must address the various challenges, open research problems in this field.

Acknowledgement

The authors want to acknowledge the support of Rajiv Gandhi Science and Technology Commission, Government of Maharashtra scheme for S&T Applications through University System at Rashtrasant Tukadoji Maharaj Nagpur University Nagpur (MH), INDIA RTMNU/IIL/RGSTC/P/2021/795. The authors also wants to thanks Intel and Tulip for provide information regarding their edge computing platform.

References

1. Wang, Pengfei, C. Yao, Zijie Zheng, G. Sun and L. Song. "Joint Task Assignment, Transmission, and Computing Resource Allocation in Multilayer Mobile Edge Computing Systems." *IEEE Internet of Things Journal* 6 (2019): 2872–2884.
2. Ren, Jinke, Yinghui He, G. Huang, Guanding Yu, Yunlong Cai and Z. Zhang. "An Edge-Computing Based Architecture for Mobile Augmented Reality."*IEEE Network* 33 (2019): 162–169.
3. Ning, Zhaolong, Xiangjie Kong, Feng Xia, W. Hou and Xiaojie Wang. "Green and Sustainable Cloud of Things: Enabling Collaborative Edge Computing." *IEEE Communications Magazine* 57 (2019): 72–78.
4. Stankovski, S., Gordana Ostojić, X. Zhang, Igor Baranovski, Srđan Tegeltija and Sabolč Horvat. "Mechatronics, Identification Tehnology, Industry 4.0 and Education." *2019 18th International Symposium INFOTEH-JAHORINA (INFOTEH)* (2019): 1–4.
5. Saqlain, M., Minghao Piao, Youngbok Shim and J. Lee. "Framework of an IoT-based Industrial Data Management for Smart Manufacturing." *J. Sens. Actuator Networks* 8 (2019): 25.
6. Khan, W.Z., E. Ahmed, S. Hakak, I. Yaqoob, and A. Ahmed. Edge Computing: A Survey. Future Generation Computer Systems 97 (Aug. 2019): 219–235. DOI: 10.1016/j.future.2019.02.050.
7. Satyanarayanan, M. "The Emergence of Edge Computing." *Computer* 50 (2017): 30–39.
8. Satyanarayanan, M., P. Simoens, Y. Xiao, P. Pillai, Z. Chen, K. Ha, Wenlu Hu and Brandon Amos. "Edge Analytics in the Internet of Things." *IEEE Pervasive Computing* 14 (2015): 24–31.
9. Yang, Ruizhe, F. Yu, Pengbo Si, Z. Yang and Y. Zhang. "Integrated Blockchain and Edge Computing Systems: A Survey, Some Research Issues and Challenges." *IEEE Communications Surveys & Tutorials* 21 (2019): 1508–1532.
10. Li, X., S. Liu, F. Wu, S. Kumari and J. Rodrigues. "Privacy Preserving Data Aggregation Scheme for Mobile Edge Computing Assisted IoT Applications." *IEEE Internet of Things Journal* 6 (2019): 4755–4763.
11. Ahmed, E., A. Naveed, A. Gani, S. Hamid, Muhammad Imran and M. Guizani. "Process state synchronization-based application execution management for mobile edge/cloud computing." *Future Gener. Comput. Syst.* 91 (2019): 579–589.
12. Ridhawi, Ismaeel Al, Yehia Kotb and Y. Al Ridhawi. "Workflow-Net Based Service Composition Using Mobile Edge Nodes." *IEEE Access* 5 (2017): 23719–23735.
13. Mutlag, A.A., M.K.A. Ghani, N. Arunkumar, M. Mohammed and Othman Mohd. "Enabling technologies for fog computing in healthcare IoT systems." *Future Gener. Comput. Syst.* 90 (2019): 62–78.
14. Díaz-de-Arcaya, Josu, Raúl Miñón and Ana I. Torre-Bastida. "Towards an architecture for big data analytics leveraging edge/fog paradigms." Proceedings of the *13th European Conference on Software Architecture - Volume 2* (2019): n. pag.

15. Girish Talmale , Urmila Shrawankar. "Real Time on Bed Medical Services: A Technological Gift to the Society." Bioscience Biotechnology Research Communications, 13 (2020):133–137.
16. Dhule, Chetan and U. Shrawankar. "POF-SVLM: pareto optimized framework for seamless VM live migration." *Computing* 102 (2020): 2159–2183.
17. Wang, T., G. Zhang, Anfeng Liu, Md Zakirul Alam Bhuiyan and Q. Jin. "A Secure IoT Service Architecture With an Efficient Balance Dynamics Based on Cloud and Edge Computing." *IEEE Internet of Things Journal* 6 (2019): 4831–4843.
18. Lin, L., P. Li, Xiaofei Liao, H. Jin and Y. Zhang. "Echo: An Edge-Centric Code Offloading System With Quality of Service Guarantee." *IEEE Access* 7 (2019): 5905–5917.
19. Cui, Jie, L. Wei, J. Zhang, Y. Xu and H. Zhong. "An Efficient Message-Authentication Scheme Based on Edge Computing for Vehicular Ad Hoc Networks." *IEEE Transactions on Intelligent Transportation Systems* 20 (2019): 1621–1632.
20. Rapuzzi, R. and M. Repetto. "Building situational awareness for network threats in fog/edge computing: Emerging paradigms beyond the security perimeter model." *Future Gener. Comput. Syst.* 85 (2018): 235–249.
21. Yuan, J. and Xiaoyong Li. "A Reliable and Lightweight Trust Computing Mechanism for IoT Edge Devices Based on Multi-Source Feedback Information Fusion." *IEEE Access* 6 (2018): 23626–23638.
22. Sahni, Yuvraj, J. Cao and L. Yang. "Data-Aware Task Allocation for Achieving Low Latency in Collaborative Edge Computing." *IEEE Internet of Things Journal* 6 (2019): 3512–3524.
23. Du, M., K. Wang, Zhuoqun Xia and Y. Zhang. "Differential Privacy Preserving of Training Model in Wireless Big Data with Edge Computing." *IEEE Transactions on Big Data* 6 (2020): 283–295.
24. Kozik, R., M. Choras, M. Ficco and F. Palmieri. "A scalable distributed machine learning approach for attack detection in edge computing environments."*J. Parallel Distributed Comput.* 119 (2018): 18–26.
25. Ma, L., X. Liu, Qingqi Pei and Y. Xiang. "Privacy-Preserving Reputation Management for Edge Computing Enhanced Mobile Crowdsensing." *IEEE Transactions on Services Computing* 12 (2019): 786–799.
26. Zhao, Zhiwei, G. Min, W. Gao, Yulei Wu, H. Duan and Qiang Ni. "Deploying Edge Computing Nodes for Large-Scale IoT: A Diversity Aware Approach." *IEEE Internet of Things Journal* 5 (2018): 3606–3614.
27. Jia, G., G. Han, Hongtianchen Xie and Jiaxin Du. "Hybrid-LRU Caching for Optimizing Data Storage and Retrieval in Edge Computing-Based Wearable Sensors." *IEEE Internet of Things Journal* 6 (2019): 1342–1351.
28. Chen, Weiwei, Dong Wang and K. Li. "Multi-User Multi-Task Computation Offloading in Green Mobile Edge Cloud Computing." *IEEE Transactions on Services Computing* 12 (2019): 726–738.
29. Talmale, Girish, & Shrawankar, Urmila (2021). Cluster formation techniques for hierarchical real time tasks allocation on multiprocessor system. Concurrency and Computation: Practice and Experience. 10.1002/cpe.6438.

3

Fog Computing: Architecture, Issues, Applications, and Case Study

Reena Thakur[1], Dr. Pradnya S. Borkar[1], and Dr. Dheeraj Rane[2]
[1]*Jhulelal Institute of Technology, Nagpur*
[2]*Medicaps University, Indore*

3.1 Introduction

Fog computing [1] was a strategy employed in 2012 by Cisco to mitigate the weaknesses of cloud computing. It is a highly distributed framework, located at the edge of a network with nodes. These nodes provide the applications operating under this infrastructure with resources like networking, storage, and computing. Almost all efforts, such as Shekhar and Gokhale [2], have intensively reviewed the advantages of fog in the context of Internet of Things (IoT) applications. Works similar to Deng et al. [3] have also addressed issues like allocation of resources, privacy, and safety in fog computing. As the number of users demanding latency-sensitive services increases, the energy usage of fog computing is receiving significant attention from researchers. A few attempts have been devoted to construct energy models, manage workload fluctuations and attempt to achieve effective trade-off solutions between fog computing and Quality of Service (QoS) and energy usage. In research using tools of stochastic optimization [4], the energy consumption of all devices in the home-based fog computing environment and the energy delay trade-off at different levels of fog–cloud interaction have been widely highlighted.

In a paradigm that expands the platform for cloud computing is fog computing [5]. Fog works between the cloud server and also the end devices as a middle layer. It is not really a complete cloud substitution, but instead an enhancement to cloud functionality. Fog works nearer to the devices on the edge and offers these devices computer resources.

There is a conventional need for such a computing paradigm, which really occurs nearer to smart devices to solve the problems of private information, ultra-low latency, high bandwidth, and geographically dispersed application areas. Equally both the academia and industry have suggested fog computing [6,7] to address the above issues and to extinguish the need for a model of computing nearer to connected devices. By allowing computation, networking on network nodes, storage, and data management within the close vicinity of IoT devices, fog computing bridges the gap between the cloud and IoT devices. Furthermore, as data moves from IoT devices to the cloud, decision making, networking, storage, and computation take place along the path between the

DOI: 10.1201/9781003166702-3

cloud and edge devices. The research community has suggested other comparable computing paradigms to fog computing, such as cloudlets, edge computing, mist computing, and cloud of things to solve the issues described. In this chapter, we present a comprehensive fog computing survey and discuss how fog computing with strict latency, privacy, and bandwidth requirements can meet the growing demand of applications. We compared fog computing with other comparable computing paradigms in the survey and believe that fog computing is a more generalized form of computing, primarily because of its comprehensive scope and flexibility of terms.

This chapter comprises a detailed comparison of cloud computing and fog computing as well as fog and edge computing. Followed by federation, fog computing architecture, benefits and advantages of fog computing, categories and sub-categories of optimization problems in fog computing, IoT challenges and solutions with fog computing are explained in detail. Some of the major issues with fog computing are described again. Various challenges and future research directions toward fog computing are elaborated. IoT applications, various case studies, and a framework of fog computing are also described (Figure 3.1).

By allowing computing, storage, networking, and data management on network nodes in the close vicinity of IoT devices, fog computing controls the gap among the cloud as well as end devices. As a consequence, as data moves to the cloud, data management, decision making, networking, storage, and computing occur both in the cloud and along

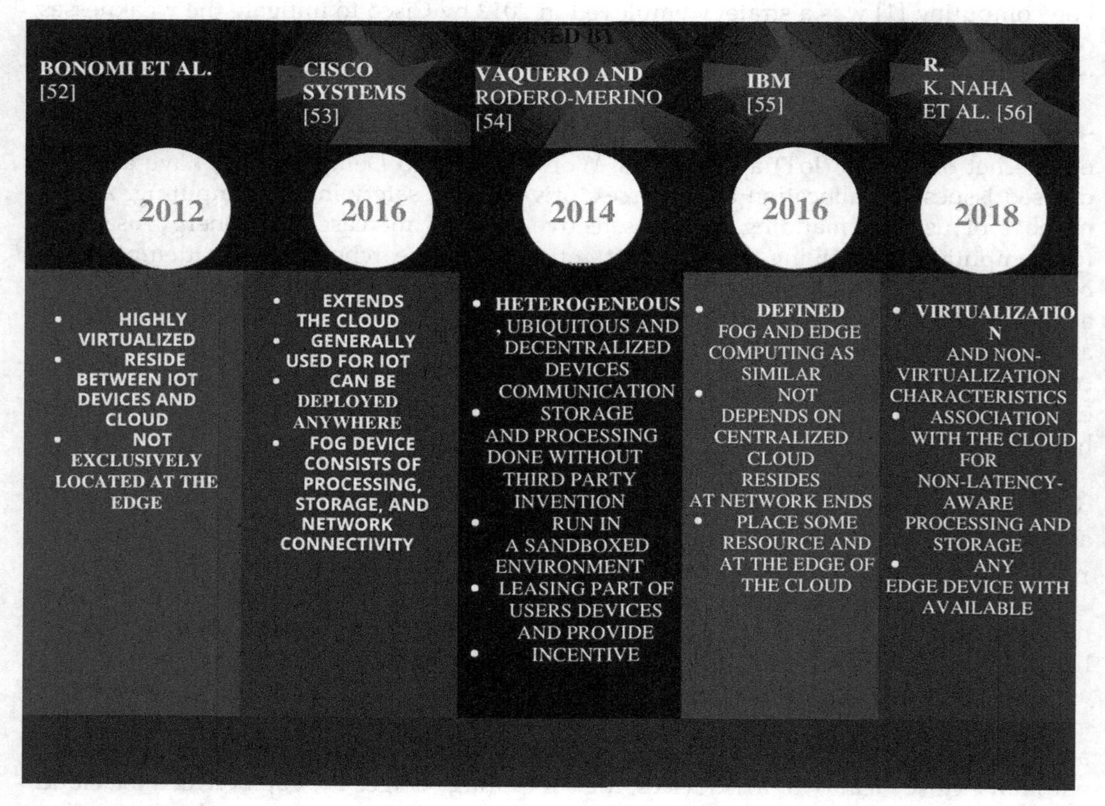

FIGURE 3.1
Summary of Fog Computing Definitions.

the IoT-to-cloud route. In Intelligent Transportation Systems [8], for example, compression of GPS data may exist only at the edge before transmission to the cloud. "A horizontal system level architecture that assigns functions of control, networking, computing, and storage along a cloud-to-thing continuum nearer to the users" explained the OpenFog Consortium [6]. The "horizontal" fog computing platform allows the distribution of computing functions between various platforms and industries, while a vertical platform promotes a platform for siloed fog computing [9]. For a single application category, a vertical framework may also provide tremendous support, but that does not compensate for system interaction in other vertically oriented platforms. Fog computing offers a flexible platform for satisfying the data-driven requirements of operators and customers, in addition to promoting a horizontal architecture. Fog computing is designed to give the IoT strong support.

1. ***Fog versus Cloud***

A projecting instance that is used to distinguish fog and cloud computing is whether it is possible to support latency-sensitive applications while maintaining satisfactory QoS. Fog nodes can be located close to IoT source nodes compared to conventional cloud computing, allowing latency to be greatly decreased. Although this example demonstrates understandable motivation for fog, latency-sensitive applications are only one of many applications that explain the need for fog computing. Fog computing nodes are typically installed in less concentrated locations compared to centralized cloud data centers. Fog nodes are broad and are present in large numbers geographically. In fog computing, protection must be provided at the edge or even in specialized fog node positions, compared to centrally designed security features in dedicated buildings for cloud data centers. The decentralized nature of fog computing allows devices to represent individually as fog computing nodes and to use fog tools as fog users. The size of hardware devices relative to these computing paradigms is due to the number of discrepancies in cloud computing and fog computing. At relatively high energy usage, cloud computing offers high computing resource allocation, whereas fog computing offers modest computing resource availability at lower energy consumption [10]. Big data centers are nearer to consumers. It is typically possible to access fog computing from cloud computing, whereas fog computing uses small servers, set-top boxes, gateways, switches, routers, or access points.

Since fog computing hardware consumes much less space than cloud computing, it is possible to locate hardware closer to users. Fog computing can be accessible again from the network edge to the core of the network by connected devices, whereas cloud computing should be accessible via the network core. Moreover, for fog-based services to operate, continuous Internet connectivity is not necessary. Specifically, with minimal to no Internet connectivity, the services will operate independently and send needed notifications to the cloud until the network is established. Cloud computing, on the other hand, requires that devices be linked while the cloud service is operating. Fog enables devices to measure, monitor, process, analyze, and react closer to IoT devices and distributes processing, decision making, control, storage, and communication [6]. Many industries may use fog for their benefit: energy, manufacturing, smart buildings, healthcare, and transportation, to name a few.

Figures 3.2 and 3.3 show that the characteristics of cloud computing have severe limitations with regard to the QoS needed by real-time applications that require almost immediate server action.

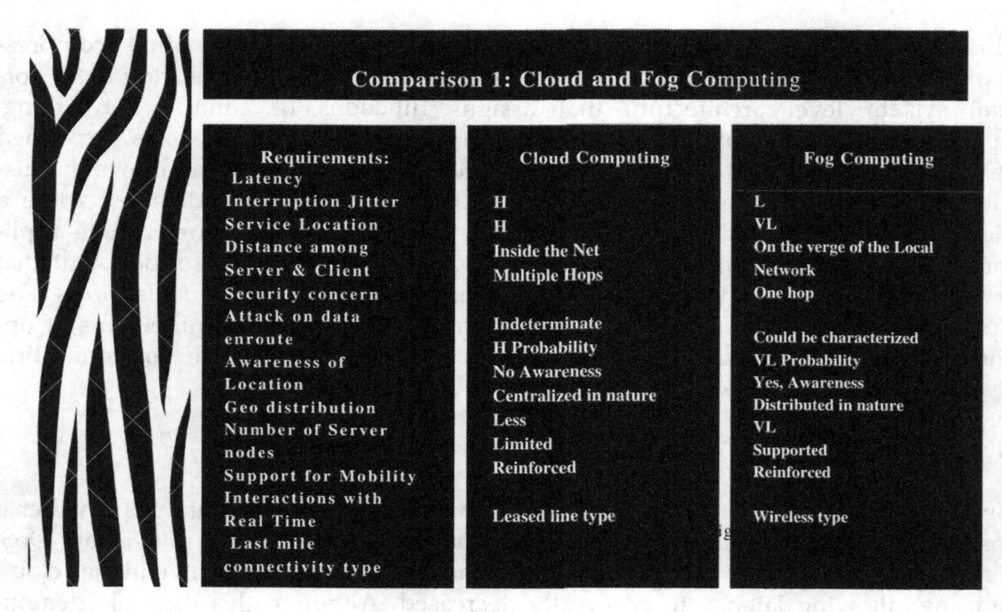

Comparison 1: Cloud and Fog Computing

Requirements:	Cloud Computing	Fog Computing
Latency	H	L
Interruption Jitter	H	VL
Service Location	Inside the Net	On the verge of the Local Network
Distance among Server & Client	Multiple Hops	One hop
Security concern	Indeterminate	Could be characterized
Attack on data enroute	H Probability	VL Probability
Awareness of Location	No Awareness	Yes, Awareness
Geo distribution	Centralized in nature	Distributed in nature
Number of Server nodes	Less	VL
Support for Mobility	Limited	Supported
Interactions with Real Time	Reinforced	Reinforced
Last mile connectivity type	Leased line type	Wireless type

FIGURE 3.2
Comparison 1: Cloud versus Fog Computing.

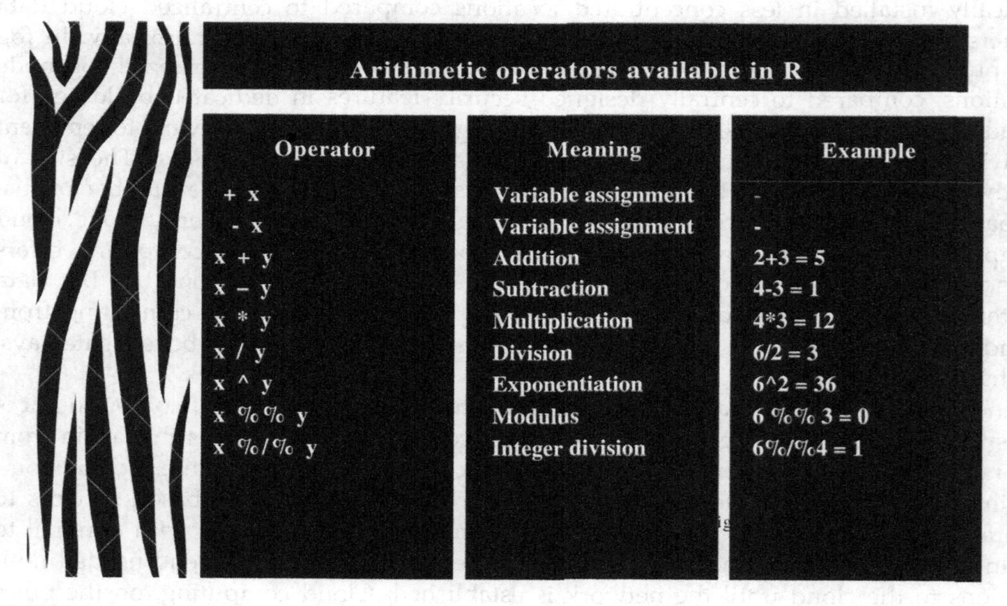

Arithmetic operators available in R

Operator	Meaning	Example
+ x	Variable assignment	-
- x	Variable assignment	-
x + y	Addition	2+3 = 5
x – y	Subtraction	4-3 = 1
x * y	Multiplication	4*3 = 12
x / y	Division	6/2 = 3
x ^ y	Exponentiation	6^2 = 36
x %% y	Modulus	6 %% 3 = 0
x %/% y	Integer division	6%/%4 = 1

FIGURE 3.3
Comparison 2: Cloud versus Fog Computing.

2. *Fog Computing versus Edge Computing*

Computing with the edge and fog are also used synonymously. Edge computing is a more common concept, and as one unified methodology, the concepts behind fog computing are also included. However, when broken down, fog computing was created to accompany edge strategies and act as an additional architectural layer to provide enhanced computing resources that cannot often be accomplished alone by the edge.

Without edge computing, fog cannot exist, whereas without fog, the edge will exist. Fog and edge computing have many similarities, all of which take the computation closer to the data source. The biggest difference between the two, though, is where the processing takes place.

Intelligence is moved down to the local area network architecture stage with fog computing. Data are processed within an IoT gateway or fog node.

Intelligence is guided directly into devices such as edge computing programmable automation controllers (PACs). In fog versus the edge, thus, more sophisticated levels of processing, analytics, and machine learning are possible (Table 3.1).

3. *Fog RAN*

   In the form of radio access networks (RAN), fog computing can be incorporated into mobile technologies to establish what is defined as fog RAN (F-RAN). Computing ability on F-RANs can be used for network edge caching, which allows for faster content retrieval and a lower front-haul burden. F-RAN can be introduced via mobile technologies relevant to 5G [11]. On the other hand, centralized control over F-RAN nodes is provided by cloud RAN (C-RAN). C-RAN captures the essence of virtualization, but by virtualizing such functions, it decouples base stations inside a cell of the mobile network from its

TABLE 3.1

Key Featured Differences Between Fog, Edge, and Cloud [15]

Features	FOG computing	Edge computing	Cloud computing
Availability – Server nodes	High	Less	Few
Type – Services	Distributed, localized (limited)	Mostly used in cellular mobile networks	Worldwide and global services
Location – Identification	Identified	Identified	Not Identified
Mobility – Features	Provided, fully supported	Provided, partially supported	With limitation
Real time – Interaction	With support	With support	With support
Real time – Response	Maximum	Maximum	Minimum
Big data – Storage and duration	Specific area target and short duration	Dependency on the services and applications	Lifetime
Big data analytic–Capacity and Quality of Computation	Short time, high-level computation	Short time, prioritized computing	Long time, categorization computing
Working – Environment and Positions	Field tracks, malls, markets, roadside, home, street	Implemented in specific area by specific service provider	Indoors
Architectural – Design	Decentralized	Decentralized	Centralized
Number of users – Facilitated	Local with related fields	Specific with related fields	General with Internet-connected users
Major – Service provided	Intel, Cisco IOx,	Cellular network industry	Microsoft Azure, IBM, Amazon, Google

baseband functions [12]. A significant number of lower cost remote radio heads are positioned arbitrarily and connected through the front-haul links to the base band unit (BBU) pool in C-RAN. For base station cell networks, both F-RAN and C-RAN are ideal and are candidates for 5G deployments. A more reliable method of network operation for energy efficiency is also provided by the use of F-RAN and C-RAN. For more details on F-RAN, we encourage the enthusiastic reader to connect to Checko et al. and Peng et al. [13,14].

3.2 Federation

1. ***Fog Federation:*** There is currently no interface or program for the fog federation that manages and federates fog resources from multiple domains of service. For the federation of fog nodes, new strategies are needed, particularly when they come from different operating realms. Resource-sharing models for fog nodes from various vendors/operators should be accounted for by the federation scheme. Likewise, new pricing models for federated fog services can be established. Finally, it is possible to suggest proposals for new fog resource-sharing schemes under the federation system.
2. ***Fog–Cloud Federation:*** Cloud and fog computing have strong differences and trade-offs, and often people choose one. However, they complement each other; the necessity of the other does not replaced the first one. Facilities that utilize smart devices can be further improved by combining cloud and fog computing. Integration of fog and cloud computing leads to improved storage capabilities, processing, and data aggregation. For example, the fog might scan, preprocess, and aggregate traffic streams from source devices in a stream processing application, whereas requests could be sent to the cloud through strong analytical processing or cataloging performance. The synchronization between cloud and fog should be manageable by an orchestrator. In general, an interoperable resource pool could be generated by a fog orchestrator, and resources could be distributed and planned for application frameworks and QoS control [16]. Via the use of Sustainable Development Solution Network (SDSN), fog service providers can have better control over how the network configures a large number of fog nodes to transfer data between the cloud, including interaction with IoT devices (Figures 3.4 and 3.5).

The above infrastructure consists of three layers:

- **End Devices:** IoT devices, laptops, smart cameras, actuators, sensors, smartphones, connected vehicles, laptops, and so on may be included. There can be multiple, heterogeneous, and mobile endpoints. The ability of end devices, in terms of CPU and memory, is usually very small. Sometimes, battery-powered terminal devices contribute to a short battery life as well as the need to reduce power consumption. Endpoints also produce information that needs to be processed and/or stored elsewhere. In other instances, data need to be received from somewhere by end devices. Therefore, terminal devices are usually attached to a certain network.

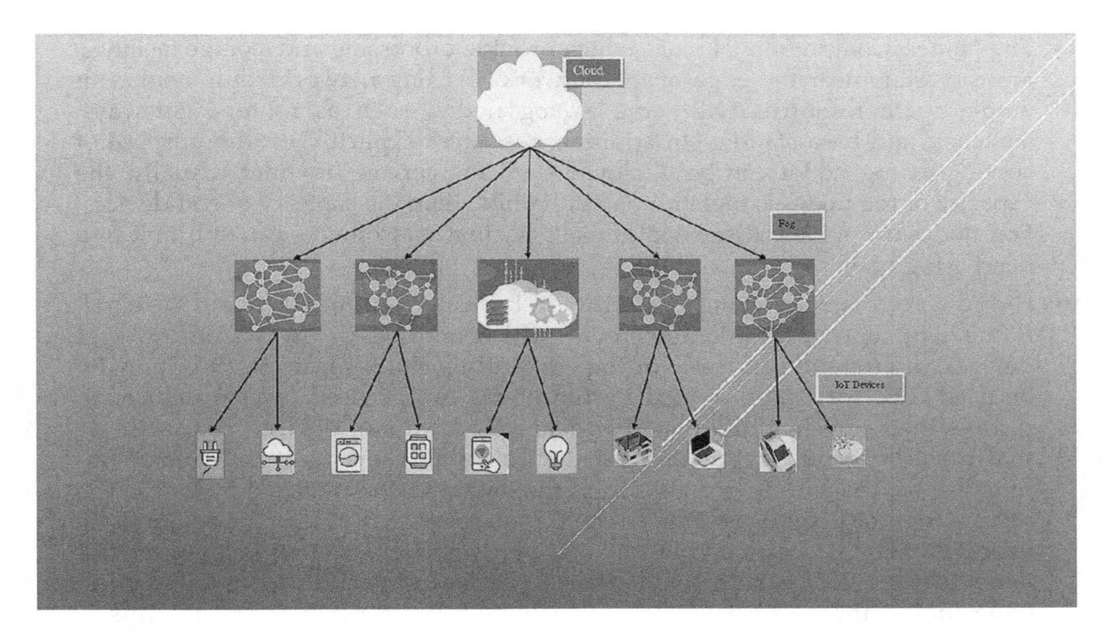

FIGURE 3.4
Data Processing of IoT Devices Using Fog and Cloud.

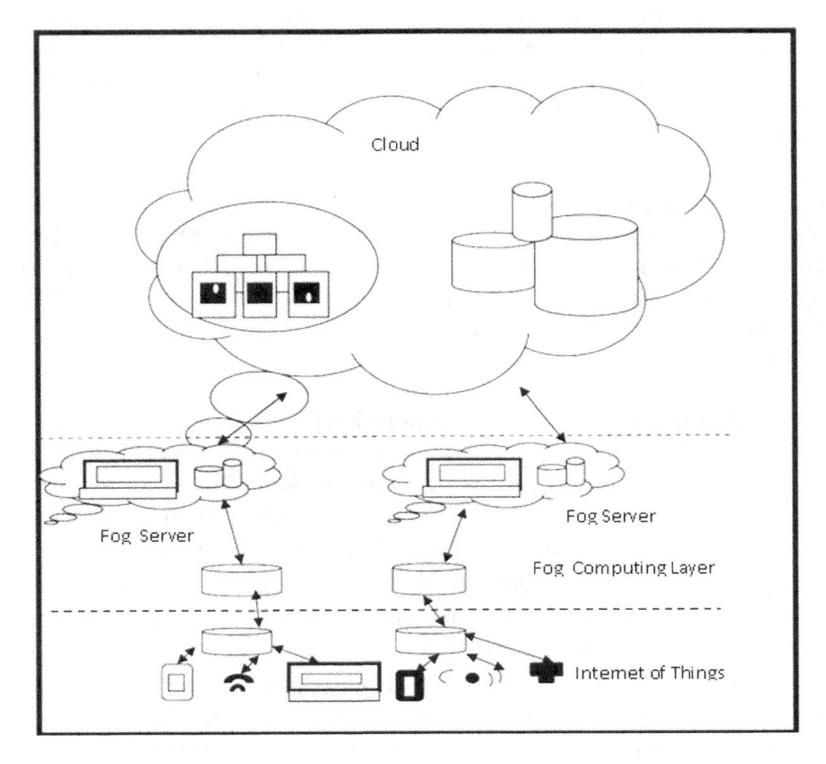

FIGURE 3.5
Basic Architecture of Fog Computing.

- **Fog Nodes:** Computational tools, which provide processing and storage facilities, are accessible near the edge of the network. Existing network equipment with ample spare resources will serve as fog nodes, such as routers, gateways, switches, and base stations. In addition, resources explicitly for the purpose of serving as fog nodes can be deployed by a fog service provider. Usually, the capacity of fog nodes is higher, but still limited, than the capacity of end devices. Fog nodes are widely dispersed globally so that end devices can still link to a nearby fog node.
- **Cloud Services:** The cloud can provide basically unlimited power. A small number of central, large data centers, which may be far away from the end devices, offer cloud services. For fog computing, cloud services can optionally be used. The services provided by fog nodes can be used by end devices to resolve their power limitations. For example, if an end device produces data that must be handled, but the end device's ability makes it difficult or impractical for the end device to process the data, the end device may link to the nearby fog node, send the data to the fog node, and discharge the fog node to process the data. Coupled with the relatively high processing power of the fog node, the low latency of network transfer between the end device and the fog node makes the entire process fast, enabling real-time interactions. This is important for many applications, such as in the fields of traffic control, virtual reality, and gaming. Off-loading to the cloud will take too long in such situations. However, cloud resources can still be used, for example, to store aggregated data for later analysis [17,18]. Some writers use other phrases, such as *edge computing*, *mist computing*, or *dew computing*, because of a lack of uniform terminology to refer to this or related concepts. In this chapter, to denote the defined interplay between end devices, fog nodes, and potentially cloud services, we often use the term *fog computing*.

The layer of software-defined resource management enforces many middleware-like services to significantly improve the use of apps on top of the rest of the cloud and fog resources. The aim of these services is to reduce the cost of using the cloud while simultaneously pushing task execution to fog nodes to achieve acceptable latency levels for application performance. This was achieved with the collaboration of a number of services, as follows.

1. **Placement of Flow and Task:** This element keeps track of the location of cloud, fog, and network resources available to determine the best applicants for incoming assignments and flows for implementation. This component interacts with the tool service provisioning to show the current number of flows and tasks, which, if considered too high, may trigger new rounds of allocations.
2. **Knowledge Base:** This element stores historical knowledge of demand for applications and demands for resources that can be leveraged by other services to help their method of decision making.
3. **Efficiency Prediction:** This service uses knowledge base service analysis to calculate the performance of the cloud resources available. The resource provisioning service uses this information to determine the number of services to be provisioned in times when a large number of tasks and flows are in use or when results are not satisfactory.

4. **Original Data Management:** This service has direct access to the origins of information, and sometimes these views can be obtained by easy querying, while more complex processing can be needed at other times. However, away from other resources, the basic process for producing the view is abstracted.
5. **Surveillance.** This service keeps track of the progress and status of applications and services and provides data to other services as needed.
6. **From Profiling.** This service constructs profiles of resources and applications based on information gained from the resources of the knowledge base and tracking.
7. **Provisioning of resources:** This service is responsible for acquiring cloud services, fog, and network infrastructure for the applications to be hosted. This apportionment is dynamic, including framework specifications, as well as the number of hosted applications, and it updates over time. It makes a decision on the number of assets using information provided by other providers and user latency specifications, as well as certificates administered by the security service. For example, once free resources are available, the component pushes tasks with low latency requirements to the edge of the network.
8. **Security:** This service provides authentication, authorization, and authentication as needed by services and applications, such as cryptography.

3.3 Benefits of Fog Computing

- Fog computing is less costly to run because data are hosted and processed on local computers rather than transmitted to any cloud system.
- Deploying the fog application per the user's need helps to facilitate and monitor business activity at par.
- Fogging gives users various options for processing their information on any physical device (Figure 3.6).

Other Benefits of Fog Computing: Fog computing is a distributed model between cloud computing and IoT [19], serving as an intermediate layer. As such, it acts as a glue among cloud, edge, and IoT computing. This is fog computing's trademark, but it also offers a range of advantages. Although the benefits of fog computing are typically summarized as CEAL [20–29], we assume that one of them is Security, so we refer to SCALE [30] as the benefits of fog computing: S-Security, C- Cognition, A-Agility, L-Latency, E-Efficiency.

1. **Security:** The fog paradigm offers a new perspective on security. In this context, security is considered a base building block of the architecture rather than an additional, and often overlooked, feature to add on top of it. As a matter of fact, the OpenFog Consortium [31] is actively working on the definition of *reference architecture* of fog computing that has security as the first pillar [32]. Particularly, the OpenFog Security Group (SWG) has drawn up the main security goals of fog computing [33–50] that we have reinterpreted and summarized as follows:

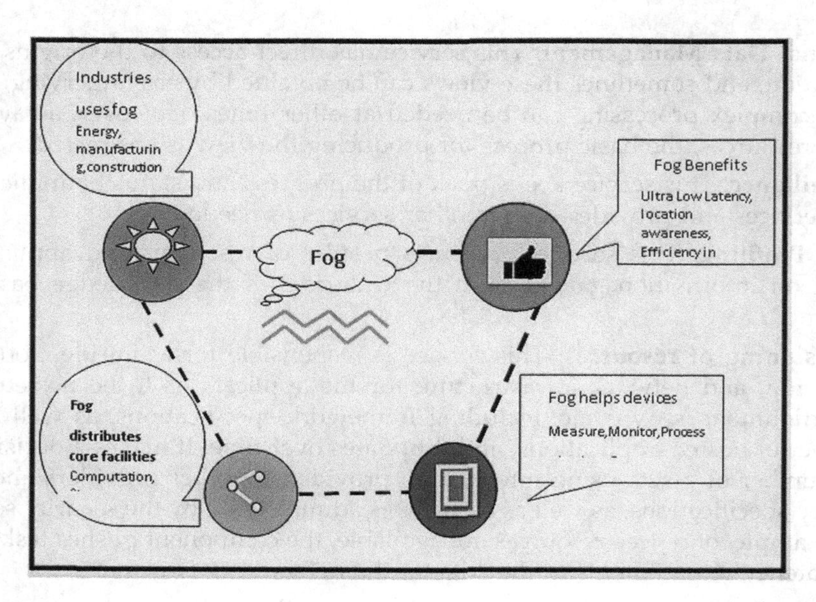

FIGURE 3.6
Pictorial Diagram of Benefits of Fog Computing.

- *Security as a Pillar (SECaaP):* Fog computing is an intrinsically secure paradigm itself that takes over the role of responsive, available, survivable, and trusted part in the cloud-to-things continuum.
- *Security as a Service (SECaaS):* Fog computing is a security service provisioned to other entities, ranging from powerful cloud servers to weak IoT devices. Thanks to the proximity of fog nodes to these entities, the fog infrastructure can both offer basic security services (e.g., protecting resource-constrained endpoints that often cannot adequately secure themselves) and improve existing security solutions (e.g., strengthening mechanisms for identity verification) [30]. This should be accomplished without interfering with the business process of the involved applications/services and respecting their domain structure.

2. **Cognition:** The fog infrastructure is aware of customer requirements and objectives; thus, it distributes more finely computing, communication, control, and storage capabilities along the cloud-to-things continuum, building applications that better meet clients' needs.
3. **Agility:** The development of a new service is usually slow and expensive due to the cost and time needed by large vendors to initiate or adopt the innovation. The fog world instead offers rapid innovation and affordable scaling, being an open marketplace where individuals and small teams can use open development tools (e.g., APIs and SDKs) and the proliferation of IoT devices to offer new services.
4. **Latency:** The fog architecture supports data processing and storage close to the user, resulting in low latency. Thus, fog computing perfectly meets the request for real-time processing, especially for time-sensitive applications [51–62].

5. **Efficiency:** The fog architecture supports the pooling of computing, communication, control, and storage functions anywhere between cloud and IoT. In this vision, the fog infrastructure "pushes" capabilities from the cloud and "pulls" capabilities from powerful IoT devices (e.g., smartphones, tablets, laptops), integrating them in the fog infrastructure, increasing the overall system performance and efficiency. In the literature, a number of other advantages and characteristics of fog computing are discussed, often with a different outline [63–69]. Nevertheless, we believe that the advantages presented in this section are generic enough to be considered the key concepts from which the other features derive.

By the aid of fog nodes, fog computing has the following properties and advantages:

- **Heterogeneity:** In various cases, fog nodes are composed of different devices, like roadside vehicle units on the Internet and smart grid service providers, among others.
- **Geographical Distribution:** Fog-based computing applications need widely distributed deployments. Moving vehicles, for instance, connect with roadside units or access points (fog nodes) deployed in distributed locations. The geographical distribution of fog nodes will broaden cloud computing coverage.
- **Low Latency, Real-time Interactions**: A large number of fog nodes are installed so that they are available to end devices. Thus, data processing and storage in fog computing are closer to end devices as opposed to the cloud. In contexts that request low latency requirements, such as Telecare Medicine Information System, virtual reality, video streaming, and gaming, among others, fog may provide high performance.
- **Data Locality:** Fog nodes will store the data they need, based on the scenario. Although the fog nodes detach from the Internet, the data can still be processed and supplied to end devices by these nodes.
- **Support for Mobility:** Due to the rapid growth of mobile devices, such as mobile phones, mobile vehicles, and mobile sensors, mobility-enhancing applications are important.
- **Save Bandwidth:** Many estimation tasks are performed by fog nodes locally. Only a subset of the data is sent to the cloud, which can effectively conserve bandwidth.

3.4 Existing Security System in Fog

Fog currently uses the Decoy system [63,70] as a malicious attacker's security service. Fog uses this tactic, as in the cloud, to trick the attacker by supplying false data as they try to extract the data. The user has to sign up then log in the decoy system. While the user is logging in, the system will ask security questions related to details provided while signing up. When an attacker attempts to login, he or she will be stuck with the problem, and the system will return a bogus file that is very close to the original file but will turn out to be fake data when the attacker tries to download it. There is a risk that the attacker will correctly guess the questions. Therefore, this system is not a very good way of securing data.

3.5 IoT Challenges and Their Solutions with Fog Computing

The IoT offers effective solutions for a variety of applications, such as the smart traffic light system, industrial control, emergency services, the waste management system, and the logistics control system. The two most appealing fields of the IoT are intelligent healthcare systems and wearable sensors. Pursuant to Chiang and Zhang [71], fog computing, as described in Table 3.2, can solve many problems.

3.6 Some of the Major Issues with Fog Computing

1. *Authentication and Trust Issues* (Table 3.3)
   - Authentication is among the most critical problems of fog computing as these facilities are available on a wide scale.

TABLE 3.2

Categories and Sub-Categories of Optimization Problems in Fog Computing

Category	Sub-category	Decision-making leading from the optimization
1 Optimization with all three architectural layers included	1.1 Offload computing activities from end systems to cloud and fog nodes	When to offload computing activities to the fog nodes from the end systems When to further offload computing activities to the cloud from the fog nodes
2 Optimization including fog nodes as well as end devices	2.1 Offloading computation tasks to fog nodes through end devices	Whether to discharge computing tasks into fog nodes from either of the end devices Which assignments should be offloaded to which fog nodes
3i Optimization that involves only fog nodes	3.1 Scheduling inside a node of fog 3.2 Fog nodes clustering 3.3 Migration among nodes of fog 3.4 Physical resource allocation 3.5 Applications/data distribution among fog nodes	How and when to set priorities within a fog node for incoming computing tasks How and when to delegate tasks within a fog node to computing resources How well the size of a cluster of fog nodes can be calculated to manage the related requests To either migrate or to leave data/applications among fog nodes where they are How physical resources should be arranged The computational tools with which fog nodes should be fitted Which node of fog should host which apps and which data
4. Including fog as well as cloud	4.1 Application/data sharing between fog nodes and the cloud	Whether to position data/applications on or in the cloud on the individual fog nodes

- Various parties such as cloud service providers, Internet service providers, and end users may be fog service providers.
- The entire structure and confidence situation of fog is complicated by this versatility. A rouge fog node is a fog device that pretends to be legal and connects to it by coaxing end users.
- It can control the signals coming to and from the user to the cloud once a user connects to it and can launch attacks easily.

2. *Privacy*
   - When there are multiple networks involved, privacy issues are still present. Since wireless technology is focused on fog computing, there is a major concern about network privacy.
   - There are several fog nodes that each end user has available to them, and data transfers from end users to fog nodes because of this more sensitive information.
3. *Security*
   - Security problems with fog computing occur when there are several devices connected to fog nodes at various gateways.
   - Each computer has a different IP address, and your IP address can be faked by any hacker to gain access to personal information stored on the specific fog node.
4. *Fog Servers*
   - To be able to offer its full operation, users must choose the correct location for fog servers.
   - Before selecting a location, the business should evaluate the demand and work performed by the fog node, helping to reduce the cost of maintenance.
5. *Energy Consumption*
   - In fog computing, energy consumption is very high as the number of fog nodes present in the fog environment is high, and they require energy to operate.

3.7 Other Applications

1. ***Fog Computing and Hydraulic Presses***

Fog computing will need to process vast quantities of information to sensorize and analyze the data produced by a hydraulic press to predict its failure or simply understand why it functions sub-optimally when pressing parts. In these cases, the best approach is to process and interpret all the data on the same machine as soon as possible, showing the machine operators what steps they can take. This is a typical scenario for fog computing.

TABLE 3.3

IoT Challenges and Techniques to Resolve Them

SR NO	IOT CHALLENGES	HOW FOG CAN SOLVE THE PROBLEMS
1	Security	A fog system is able to 1. Implement the scanning role for malware and monitor the safety status of neighboring nodes. 2. Act as a proxy in a timely manner to upgrade software credentials and identify threats.
2	Latency	The fog undergoes different computational tasks that are the perfect method for time-sensitive details.
3	Network – bandwidth	Fog computing, which allows data transfer from the cloud to IoT devices, will allow hierarchical data processing. If applications, network, and computing resources are accessible on demand, data processing will occur.
4	Continuous Services	Even with a network link problem, fog computing can operate autonomously to confirm uninterrupted services.
5	Resource-constrained devices	When certain operations cannot be submitted to the cloud, fog computing can decrease system complexity, power consumption, and life cost.

2. ***Fog Computing in Smart Cities***

There are a number of challenges facing large cities, including public safety, sanitation, traffic congestion, high energy use, and municipal services. By installing a network of fog nodes, the solution to these challenges resides in a single IoT network.

The lack of broadband bandwidth and connectivity is a major problem when establishing smart cities. Multiple cellular networks and ample coverage are available in modern cities. Such networks have limits on capacity and peak bandwidth that are not sufficient to meet the needs of existing customers. The deployment of fog computing architectures, however, enables fog nodes to provide local storage and processing. Hence, optimizing the use of the network. Smart cities are also struggling hard with security and safety problems. Time-critical performance needs advanced real-time analytics to tackle this problem. Along with life-critical systems, the network may transmit traffic and data. Security problems, data encryption, and distributed analytics requirements are resolved by fog computing.

3. ***Fog-computing–based Smart Health Monitoring System Deploying LoRa Wireless Communication***

A smart health monitoring system was created, and its performance was assessed. The test conducted demonstrated LoRa's promising performance [65] with the smart health monitoring system. Using fog communication as connectivity alongside LoRa helped us achieve remote monitoring of the biometric information of users miles away. Remote monitoring requires no Internet connectivity, and the transmission of data requires less power. So, in places where there is no Internet access, this system can provide remote health monitoring services. The proposed scheme also benefits individuals living in places where health services are lacking. It will not only help patients who need to regularly check their body vitals, but also those who want to track their body vitals in a timely manner. The proposed system also benefits countries with lower physician-to-patient ratios. In addition, the proposed architecture can help minimize the burden on the cloud and help to create an energy-efficient health monitoring system to move forward.

4. ***IoTPulse***

Dhillon et al. [72] proposed an enterprise health information system called IoTPulse based on IoT to predict alcohol addiction by providing real-time data in the environment of fog computing using machine learning. As a case study, we used information from 300 alcohol addicts from Punjab (India) to train machine-learning models. IoTPulse's performance was compared with current work using various parameters, including accuracy, sensitivity, specificity, and accuracy, showing 7%, 4%, 12%, and 12% enhancement, respectively. Finally, in the FogBus-based real fog environment, IoTPulse was validated using QoS parameters, including latency, network bandwidth, energy, and performance-enhancing response time.

5. ***Fuzzy Rule-based fog–cloud Computing for Solar Panel Disturbance Investigation***

The [30] SoC Wi-Fi microcontroller online fuzzy rule-based system has proved to be efficient in monitoring efficiency and diagnosing the condition of a PV panel with the aid of physical parameter input and electrical power output sensors on solar panels. In general, the diagnostic conditions are based on the correlation of the parameters for humidity, light intensity, and temperature with the electrical power output of the PV panel. The system developed was able to monitor the efficiency of the PV panel continuously and any disturbances found in the field. The rules were established on the basis of observational field conditions and were empirically validated from sensors that were stored in the RTU computer database. The outcome of the rule-based truth table test was discrepant. This indicates that the computer system developed some undefined rule-based values. This means that comprehensive conditions of all possibilities in the field are required in order to create a rule-based method. For optimal outcomes of PV panel diagnosis against irregularities, additional sensors may be needed. By referring to the improvement of solar panel efficiency from the conventional method, the purpose of this research was achieved. The use of an online rule-based diagnostic system will accelerate the handling of interference to boost the overall efficiency and continuity of solar panel electricity output.

6. ***Fog–cloud-based Architecture for Healthcare***

In the healthcare sector, the mortality rate can be reduced using fog–cloud architecture. In Gupta et al., several security protocols were introduced for better performance of this architecture [73].

7. ***System for Drought Prediction and Forecasting Dependent on Fog–cloud***

Kaur and Sood [74] proposed an IoT-based framework for drought prediction and forecasting. By determining the required drought severity level for the case, the framework successfully controlled the drought conditions. IoT sensors were used to collect data for enormous amounts of variables triggering climatic, hydrological, and environmental drAoughts. Key points of the architecture included latency minimization that reduced the data dimensions using PCA before sending to the cloud layer, and optimization of ANN that used GA to predict the magnitude of drought. The proposed framework implemented on Amazon EC2 will effectively predict the severity of drought over different time scales. Experimental findings indicated the efficacy of the proposed system in representing improvements in drought-inducing variables relative to other drought indices. In addition, it showed that in terms of f-measure, accuracy, sensitivity, precision, and specificity, ANN-GN performed better than other classifiers. Finally, the system's findings served as useful feedback for government agencies to establish informed strategies and take prompt action to mitigate the harm.

3.8 Case Studies

1. *Fog Computing in Healthcare Internet of Things*

In the case study by Gia et al. [75], fog computing was a gateway for increasing health monitoring systems. The authors implemented fog computing services, including location awareness, a real-time notification system, distributed databases, interoperability, and an access management graphical user interface. They also introduced a flexible, lightweight template for the extraction of ECG features (e.g., heart rate, P wave, T wave). The demonstration and findings illustrate the milestones made by the smart gateway. The wavelet transform-based template can be used to detect various parts of ECG waveforms to extract different ECG features (e.g., PR interval, QT interval) or to extract EMG and EEG features.

2. *Smart City IoT Application*

A smart city is one of the main use cases of the IoT, a mixture of a number of use cases that range from smart traffic management to building energy management. Next, this chapter presents a case study on smart traffic management and shows that using fog computing increases the application's efficiency in terms of response time and bandwidth usage. A collection of stream queries that run on data generated by sensors deployed throughout the city can be read by a smart traffic management system. Real-time congestion estimation (for route planning) or traffic event detection are common examples of such queries. Contrast the performance of a DETECT TRAFFIC INCIDENT query on fog infrastructure versus the standard implementation of the cloud in this case study. The sensors installed on the roads send the speed of each passing vehicle to the query-processing engine during the query. The "Average Speed Calculation" operator calculates the average speed of vehicles over a given time frame from the sensor readings and sends this information to the next operator. The "Congestion Calculation" operator calculates the level of congestion in each lane based on the average speed of vehicles in that lane. Based on the average level of congestion, the "Incident Detection" operator detects whether an incident has occurred. This query was simulated in the processing engine of both fog-based and cloud-based stream queries.

3. ***Smart Factory IoT Application***

A small data center and a logistics office are part of a factory. The central office is situated outside the site. In general, two machines are present on the factory floor. One is used for manufacturing, and another is used for packaging and then for shipping. A camera is placed on the production machine to ensure faultless product, taking a picture of the product produced and checking if there are any defects. This will ensure that the product is faultless when it reaches the packaging machine. One of the challenging jobs of the packaging machine is to maintain the production machine's speed in order to reduce the waiting time. The packaging machine generally operates in the atmospheric temperature range, so it is important to shut the packaging machine off whenever necessary. In order to control the speed at which the machine operates, the controller is embedded in each computer. Communication is carried out through a common wireless gateway between these controllers. Logistics personnel decide when shipments of products should be made. In addition, the central office has a data center, and the factory has a data center, so these two data centers communicate via WAN. Information on machine productivity is also required by the central office at an offsite location. Additional outsourcing via the cloud is required to outsource certain computer tasks.

References

1. Varshney, P. and Simmhan, Y. (May 2017). "Demystifying Fog Computing: Characterizing Architectures." *Applications and Abstractions*, arXiv:1702.06331 [cs]: 115124.
2. Shekhar, S. and Gokhale, A. (2017) Dynamic Resource Management Across Cloud-Edge Resources for Performance-Sensitive Applications. in Proceedings of the 17th IEEE/ACM International Symposium on Cluster, Cloud and Grid Computing, Piscataway, NJ, USA: 707710.
3. Deng, R., Lu, R., Lai, C., Luan, T.H., and Liang, H. (Dec 2016) "Optimal Workload Allocation in Fog-Cloud Computing Toward Balanced Delay and Power Consumption." *IEEE Internet of Things Journal* 3(6): 11711181.
4. Cardellini, V. et al. (Jun 2016) "A Game-Theoretic Approach to Computation Offloading in Mobile Cloud Computing." *Mathematical Programming* 157(2) : 421449.
5. Khan, S., Parkinson, S., and Qin, Y. (2017). "Fog Computing Security: A Review of Current Applications and Security Solutions." *Journal of Cloud Computing* 6(1): 19.
6. OpenFog Consortium. Openfog reference architecture for fog computing, 2017. [Online]. Available: https://www.openfogconsortium.org/ra/, February 2017.
7. Bonomi, F., Milito, R., Zhu, J., and Addepalli, S. (2012) Fog computing and its role in the internet of things. In Proceedings of the first edition of the MCC workshop on Mobile cloud computing (pp. 13–16). ACM.
8. Acharya, J. and Gaur, S. (2017). Edge compression of gps data for mobile iot. In *Fog World Congress (FWC), 2017 IEEE* (pp. 1–6). IEEE.
9. Zhang, T. Fog computing brings new business opportunities and disruptions [Online]. Available: http://internetofthingsagenda.techtarget.com/blog/IoTAgenda/Fog-computing-brings-new-business-opportunities-and-disruptions, Blog, TechTarget.
10. Armbrust, M. , Fox, A., Griffith, R., Joseph, A. D., Katz, R., Konwinski, A., Lee, G., Patterson, D., Rabkin, A., Ion, S., et al. (2010) "A View of Cloud Computing." *Communications of the ACM* 53(4): 50–58.

11. Wen, Z., Yang, R., Garraghan, P., Lin, T., Xu, J., and Rovatsos, M. (2017) "Fog Orchestration for Internet of Things Services." *IEEE Internet Computing* 21(2): 16–24.
12. Hung S.-C., Hsu H., Lien S.-Y., and Chen K-C. (2015) "Architecture Harmonization between Cloud Radio Access Networks and Fog Networks." *IEEE Access* 3: 3019–3034.
13. Checko, A., Christiansen, H. L., Yan, Y., Scolari, L., Kardaras, G., Berger, M. S., and Dittmann, L. (2015) "Cloud Ran for Mobile NetworksâĂŤa Technology Overview." *IEEE Communications Surveys & Tutorials* 17(1): 405–426.
14. M. Peng, S. Yan, K. Zhang, and C. Wang. (2016) "Fog-Computing Based Radio Access Networks: Issues and Challenges." *IEEE Network* 30(4): 46–53.
15. Anawar M. R., Wang S., Zia M. A., Jadoon A. K., Akram U., Raza S. (2018) Fog Computing: An Overview of Big IoT Data Analytics. *Hindawi Wireless Communications and Mobile Computing Volume*, Article ID 7157192: 22 pages 10.1155/2018/7157192
16. Jalali F., Hinton K., Ayre R., Alpcan T., and Tucker R. S. (2016) "Fog Computing May Help to Save Energy in Cloud Computing." *IEEE Journal on Selected Areas in Communications* 34(5): 1728–1739.
17. Mann, Z.Á., Metzger, A., Prade, J., and Seidl, R. (2019) Optimized application deployment in the fog. In: 17th International Conference on Service-Oriented Computing: 283–298.
18. Mann, Z.Á. (2011) *Optimization in Computer Engineering – Theory and Applications*. Scientific Research Publishing, Incorporated.
19. Open Fog Consortium Architecture Working Group and others, "OpenFog reference architecture for fog computing," OpenFog Consortium, White Paper OPFRA001.20817, Feb. 2017.
20. Mahmud, R., Kotagiri, R., and Buyya, R. (2018) *Fog Computing: A Taxonomy, Survey Future Directions*. (pp. 103–130). Singapore: Springer.
21. OpenFog Consortium. Accessed: Jul. 25, 2018. [Online]. Available: https://www.openfogconsortium.org/
22. Martin, B.A., Michaud, F., Banks, D., Mosenia, A., Zolfonoon, R., Irwan, S., Schrecker, S., and Zao, J.K. (Oct./Nov. 2017) Openfog security requirements andapproaches. in Proc. IEEE Fog World Congr. (FWC): 1–6.
23. Chiang, M., Ha, S., Risso, C.-L.I.F., and Zhang, T. (Apr. 2017) "Clarifying Fog Computing and Networking: 10 Questions and Answers." *IEEE Communications Magazine* 55(4) : 18–20.
24. Bonomi, F., Milito, R., Zhu, J., and Addepalli, S. (2012) Fog computing and its role in the Internet of Things. in Proc. 1st Ed. MCC Workshop Mobile Cloud Comput. (MCC), New York, NY, USA: 13–16.
25. Hu, P., Dhelim, S., Ning, H., and Qiu, T. (Nov 2017) "Survey on fog computing: Architecture, key technologies, applications and open issues." *Journal of Network and Computer Applications* 98: 27–42.
26. Chenand, W. and Zhang, T. (Mar 2017) "Fogcomputing." *IEEE Internet Computing* 21(2): 4–6.
27. Yi S., Hao Z., Qin Z., and Li Q. (Nov 2015) "Fog computing: Platform and Applications." in Proc. 3rd IEEE Workshop Hot Topics Web Syst. Technol. (HotWeb): 73–78.
28. Ni, J., Zhang, K., Lin, X., and Shen, X. (2017) "Securing Fog Computing for Internet of Things Applications: Challenges and Solutions." *IEEE Communications Surveys & Tuturials* 20(1): 601–628, 1st Quart.
29. Yousefpour, A., Fung, C., Nguyen, T., and Kadiyala, K. (13 Feb 2019) "All One Needs to Know about Fog Computing and Related Edge Computing Paradigms A Complete Survey." arXiv:1808.05283v3 [cs.NI].
30. Suryono, S., Khuriati, A. and Mantoro, T. (2019) "A Fuzzy Rule-Based Fog– Cloud Computing for Solar Panel Disturbance Investigation." *Cogent Engineering* 6: 1624287 10.1080/23311916.2019.1624287.
31. Naha, R.K., Garg, S., Georgekopolous, D., Jayaraman, P.P., Gao, L., Xiang, Y., and R. Ranjan. (3 Jul 2018) "Fog Computing: Survey of Trends, Architectures, Requirements, and Research Directions." arXiv:submit/2317175 [cs.DC], Volume 10.
32. IBM. (Sep 2016) "What is fog computing?". [Online] https://www.ibm.com/blogs/cloud-computing/2014/08/25/fog-computing.

33. Brogi, A. and Forti S. (2017) "Qos-Aware Deployment of IoT Applications through the Fog." *IEEE Internet of Things Journal* 4(5): 1185–1192.
34. Gupta, H., Nath, S. B., Chakraborty, S., and Ghosh, S. K. (2016) "Sdfog: A Software Defined Computing Architecture for Qos Aware Service Orchestration over Edge Devices". arXiv preprint arXiv:1609.01190, .
35. Jemaa, F. B., Pujolle, G., and Pariente, M. (2016) "Qos-Aware vnf Placement Optimization in Edge-Central Carrier Cloud Architecture. In IEEE (pp. 1–7). Global Communications Conference (GLOBECOM). IEEE.
36. Xiao, Y. and Krunz, M. (2017) "Qoe and Power Efficiency Tradeoff for Fog Computing Networks with Fog Node Cooperation. In INFOCOM 2017-IEEE Conference on Computer Communications (pp. 1–9). IEEE.
37. Z. Chang, Z. Zhou, T. Ristaniemi, and Z. Niu. (2017) Energy efficient optimization for computation offloading in fog computing system. In GLOBECOM 2017-2017 IEEE Global Communications Conference (pp. 1–6). IEEE.
38. Hao P., Bai Y., Zhang X., and Zhang Y. (2017) Edgecourier: an edge- hosted personal service for low-bandwidth document synchronization in mobile cloud storage services. In Proceedings of the Second ACM/IEEE Symposium on Edge Computing (p. 7). ACM.
39. Yousefpour, A., Ishigaki, G., Gour, R., and Jue, J.P. (April 2018) "On Reducing IoT Service Delay via Fog Offloading." *IEEE Internet of Things Journal* 5(2): 998–1010.
40. Fricker, C., Guillemin, F., Robert, P., and Thompson, G. (2016) "Analysis of an Offloading Scheme for Data Centers in the Framework of Fog Computing." *ACM Transactions on Modeling and Performance Evaluation of Computing Systems (TOMPECS)* 1(4):16. [313] Lixing Chen and Jie Xu. Socially trusted collaborative edge computing in ultra dense networks. In Proceedings of the Second ACM/IEEE Symposium on Edge Computing, page 9. ACM, 2017.
41. Zhang, H., Qiu, Y., Chu, X., Long, K., and Leung, V.C.M.(Dec 2017). "Fog Radio Access Networks: Mobility Management, Interference Mitigation, and Resource Optimization". *IEEE Wireless Communications*, 24(6): 120–127.
42. Abderrahim M., Ouzzif M., Guillouard K., Francois J., and Lèbre A. (2017) A holistic monitoring service for fog/edge infrastructures: a foresight study. In The IEEE 5th International Conference on Future Internet of Things and Cloud (FiCloud 2017).
43. Al Faruque, M. A. and Vatanparvar, K. (2016) "Energy Managementas-a-Service over Fog Computing Platform. *IEEE Internet of Things Journal* 3(2): 161–169.
44. Baktir, A. C., Ozgovde, A., and Ersoy, C. (2017) "How Can Edge Computing Benefit from Software-Defined Networking: A Survey, Use Cases, and Future Directions." *IEEE Communications Surveys & Tutorials* 19(4): 2359–2391.
45. Grewe, D., Wagner, M., Arumaithurai, M., Psaras, I., and Kutscher, D. (2017) Information-centric mobile edge computing for connected vehicle environments: Challenges and research directions. In Proceedings of the Workshop on Mobile Edge Communications (pp. 7–12). ACM.
46. Zhang, P., Liu, J. K., Yu, F. R., Sookhak, M., Au, M. H., and Luo, X. (2018) "A Survey on Access Control in Fog Computing." *IEEE Communications Magazine* 56(2): 144–149.
47. Martin, B. A., Michaud, F., Banks, D., Mosenia, A., Zolfonoon, R., Irwan, S., Schrecker, S., and Zao, J. K. (2017) Openfog security requirements and approaches. In Fog World Congress (FWC), 2017 IEEE (pp. 1–6). IEEE.
48. Darki, A., Duff A., Qian, Z., Naik, G., Mancoridis, S., and Faloutsos, M. (2016) (poster) don't trust your router: Detecting compromised router. In The IEEE Proceedings of the 12th International Conference on Emerging Networking Experiments and Technologies CoNEXT, volume 16.
49. Schäfer, D., Edinger, J., VanSyckel, S., Paluska, J. M., and Becker, C. (2016) Tasklets: Overcoming heterogeneity in distributed computing systems. In Distributed Computing Systems Workshops (ICDCSW), 2016 IEEE 36th International Conference on (pp 156–161). IEEE.

50. Pahl, C., Helmer, S., Miori, L., Sanin, J., and Lee, B. (2016) A containerbased edge cloud paas architecture based on raspberry pi clusters. In Future Internet of Things and Cloud Workshops (FiCloudW), IEEE International Conference on (pp.117–124). IEEE.
51. De Donno, M., Tange, K., and Dragoni, N. (2019) "Foundations and Evolution of Modern Computing Paradigms: Cloud, IoT, Edge, and Fog." IEEE Access 7: 2019.
52. Murtaza, F., Akhunzada A., ul Islam S., Boudjadar J., and Buyya R. (2020) "QoS-Aware Service Provisioning in Fog Computing." *Journal of Network and Computer Applications* 165: 102674.
53. Mahmud, R., Srirama, S.N., Ramamohanarao, K., Buyya, R. (2019) "Quality of Experience (Qoe)-Aware Placement of Applications in Fog Computing Environments. *Journal of Parallel and Distributed Computing*. DOI: 10.1016/j.jpdc.2018.03.004.
54. Bittencourt, L.F., Diaz-Montes, J., Buyya, R., Rana, O.F., and Parashar, M. (2017) "Mobility-Aware Application Scheduling in Fog Computing." *IEEE Cloud Computing* 4: 26–35.
55. Bitam, S., Zeadally, S., and Mellouk, A. (2018) "Fog Computing Job Scheduling Optimization Based on Bees Swarm." *Enterprise Information Systems* 12: 373–397.
56. Intharawijitr, K., Iida, K., and Koga, H. (2016) "Analysis of Fog Model Considering Computing and Communication Latency in 5G Cellular Networks." 2016 IEEE International Conference on Pervasive Computing and Communication Workshops. PerCom Workshops 2016: 5–8.
57. Zeng, D., Gu, L., Guo, S., and Cheng, Z. (2016) "Joint Optimization of Task Scheduling and Image Placement in Fog Computing Supported Software-Defined Embedded System." *IEEE Transactions on Computers* 65: 3702–3712.
58. Deng, R., Lu, R., Lai, C., Luan, T.H., and Liang, H. (2016) "Optimal Workload Allocation in Fog-Cloud Computing toward Balanced Delay and Power Consumption." *IEEE Internet Things of Journal* 3: 1171–1181.
59. Song, N., Gong, C., An, X., and Zhan, Q. (2016) "Fog Computing Dynamic Load Balancing Mechanism Based on Graph Repartitioning." *China Communications* 13: 156–164.
60. Oueis, J., Strinati, E.C., and Barbarossa, S. (2015) "The Fog Balancing: Load Distribution for Small Cell Cloud Computing." In: 2015 IEEE 81st Vehicular Technology Conference (VTC Spring): 1–6.
61. Aazam, M. (2015) Dynamic Resource Provisioning through Fog Micro Datacenter. Pervasive Computing and Communication Workshops. PerCom Workshops: 105–110.
62. Aazam, M., & Huh, E.N. (2015) Fog computing micro datacenter based dynamic resource estimation and pricing model for IoT. In: Proceedings – International Conference on Advanced Information Networking and Applications, AINA 2015- April: 687–694.
63. (ICDIPC), 2016 Sixth International Conference on, pp. 75-79. IEEE, 2016. 11. Wang, Qixu, Dajiang Chen, Ning Zhang, Zhe Ding, and Zhiguang Qin. PCP: A PrivacyPreserving Content-Based Publish–Subscribe Scheme With Differential Privacy in Fog Computing. *IEEE Access* 5 (2017): 1796217974.
64. Cardellini, V., Grassi, V., Presti, F.L., Nardelli, M. (2016). On QoS-Aware scheduling of data stream applications over fog computing infrastructures. In: Proceedings – IEEE Symposium on Computers and Communications 2016-February, 271–276.
65. Kharel, J., Reda, H.T., and Shin, S.Y. Fog Computing-Based Smart Health Monitoring System Deploying LoRa Wireless Communication. *IETE Technical Review*, ISSN: 0256-4602 (Print) 0974-5971 (Online) Journal homepage: http://www.tandfonline.com/loi/titr20.
66. F. Bonomi, R. Milito, J. Zhu, and S. Addepalli. (2012) Fog computing and its role in the internet of things. In Proceedings of the first edition of the MCC workshop on Mobile cloud computing (pp. 13–16). ACM.
67. Cisco Systems. (2016) Fog Computing and the Internet of Things: Extend the Cloud to Where the Things Are. www.Cisco.Com (p. 6) [Online]. Available: http://www.cisco.com/c/dam/en{_}us/solutions/trends/iot/docs/computing-overview.pdf

68. Vaquero, L.M. and Rodero-Merino, L. (2014) "Finding Your Way in the Fog: Towards a Comprehensive Definition of Fog Computing." *ACM SIGCOMM Computer Communication Review* 44(5): 27–32.
69. Banafa, A. (August 25, 2014) What is Fog Computing. https://www.ibm.com/blogs/cloud-computing/2014/08/fog-computing
70. Shrestha, N.M., Alsadoon, A., Prasad, P.W.C., Hourany, L., and Elchouemi, A. Enhanced e-health framework for security and privacy in healthcare system. In *Digital Information Processing and Communications*. DOI: 10.1109/ICDIPC.2016.7470795.
71. Chiang, M and Zhang, T. (2016) "Fog and IoT: An Overview of Research Opportunities." *IEEE Internet of Things Journal* 3(6): 854–864.
72. Dhillon, A., Singh, A., Vohra, H., Ellis, C., Varghese, B., and Gill, S.S. (2020) "IoTPulse: Machine Learning-Based Enterprise Health Information System to Predict Alcohol Addiction in Punjab (India) using IoT and Fog Computing. ISSN: (Print) (Online) Journal homepage: https://www.tandfonline.com/loi/teis20
73. Gupta, V., Gill, H.S., Singh, P., and Kaur, R. (2018) "An Energy Efficient Fog-Cloud Based Architecture for Healthcare." *Journal of Statistics and Management Systems*, ISSN: 0972-0510 (Print) 2169-0014 (Online) Journal homepage: http://www.tandfonline.com/loi/tsms20.
74. Kaur, A. and Sood, S.K. Cloud-Fog based framework for drought prediction and forecasting using artificial neural network and genetic algorithm. ISSN: 0952-813X (Print) 1362-3079 (Online) Journal homepage: https://www.tandfonline.com/loi/teta20.
75. Gia, T.N., Jiang, M., Rahmani, A.-M., Westerlund, T., Liljeberg, P., and Tenhunen, H. (2015) Fog Computing in Healthcare Internet of Things: A Case Study on ECG Feature Extraction. 2015 IEEE International Conference on Computer and Information Technology; Ubiquitous Computing and Communications; Dependable, Autonomic and Secure Computing; Pervasive Intelligence and Computing, 978-1-5090-0154-5.

4

Mobile Web Service Architecture in the Cloud Environment

Anil A. Dudhe[1] and Dr. Swati S. Sherekar[2]
[1]*Assistant Professor, Department of Computer Science, P.N. Mahavidyala, Pusad*
[2]*Professor, Dept of Computer Science and Engineering, SGBA University, Amravati*

4.1 Introduction

Mobile devices are increasingly becoming an integral part of human life because they are the most efficient and expedient communication service not bounded by time and place. Mobile users benefit from the rich experience provided by mobile applications that run on mobile handsets and/or on remote servers via wireless networks. Because of the tremendous growth in mobile cloud computing (MCC), it has become a dominant area in the information technology (IT) field as well as the commercial sector. MCC is the most emerging and well-accepted technology, and it has experienced significant growth [1].

At present, you may say almost every family worldwide has a minimum of one or two smartphones. Hence, they are directly or indirectly connected to the Internet. To entice these users, service providers are coming up with a variety of plans with very cheap rates. Huge competition is going on among the service providers. A telecommunication company, Reliance, designed Long-Term Evolution (LTE), its a high-speed wireless communication technology for mobile phones and data terminals, based on Jio 4G Network. 4G network allows their smartphone users to do their personal, official, and business-related tasks at their fingertips [1] wherever they want.

In spite of the recent developments in MCC and the enhancement of the capabilities of mobile devices in terms of memory size, battery life, screen size, and many other features, there are still a lot of challenges both service providers and mobile users are facing. The challenges are speed, network connectivity, and response from the server. Mobile network infrastructure is constantly improving, and data transmission is becoming more affordable and available to users. That makes mobile phones a popular client to exploit web resources, especially web services like Whatsapp, Facebook, Twitter, Instagram, Amazon, Flipkart, and Snapdeal.

Web services suffer from the limitations of mobile devices and broadband wireless networks, making the standard architectures in mobile environments inefficient [2,3]. The limitation in resources significantly hampers service quality. This chapter focuses on the

DOI: 10.1201/9781003166702-4

study of various mobile web service architectures that enhance the performance of cloud-based mobile web services and provide reliable services to users.

4.2 Mobile Cloud Computing (MCC)

According to Khan et al. and Abolfazli et al. [4,5], integration of cloud computing, mobile computing, and wireless networks forms MCC, which provides comprehensive computational resources to mobile clients, network operators, and cloud service providers. MCC also provides a big and complex level of mobile application services to its client, with great user experience [6]. Apart from that, it also provides equal business opportunities to mobile network service providers as well as cloud service providers [7,8].

Past literature and experience state that most mobile applications require intensive computing power and software platform support for their execution (Figure 4.1). In such situations, low-end but browser-enabled mobile phones are unable to support such applications. MCC provides the computing power, storage, and platform support required to execute these applications through the cloud.

MCC is the future of the IT industry because it gives double benefits to both mobile computing and cloud computing. It means providing the best possible services for mobile users by securing the utmost advantages of cloud computing, which is developing rapidly. Resources in MCC are virtualized, and servers are distributed across geographical areas rather than in traditional local computers or servers. They can be accessed through mobile devices like smartphones, tablets, and so on.

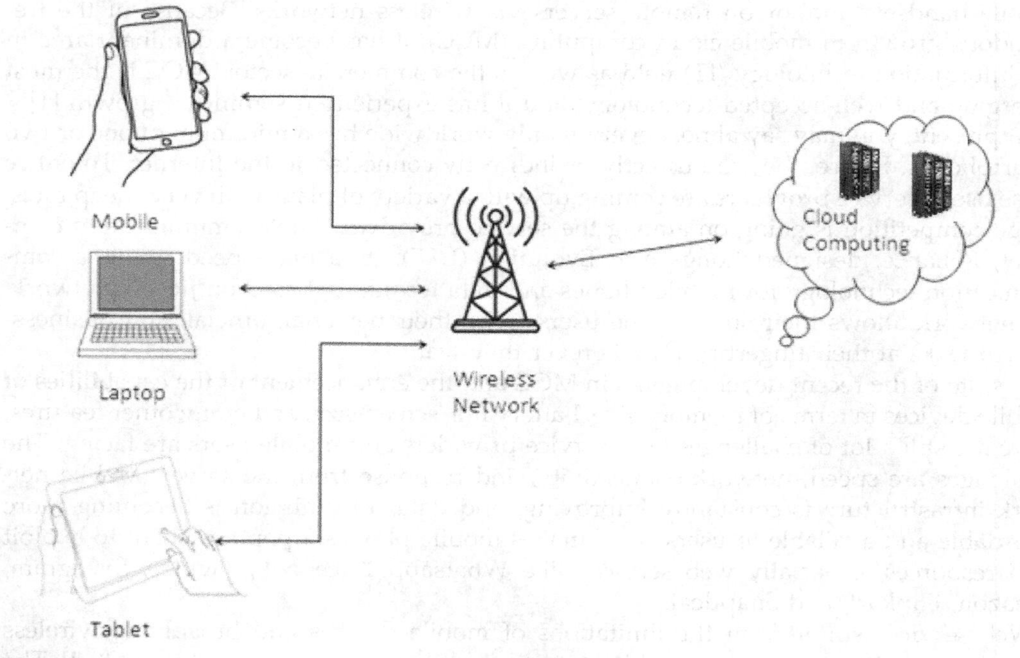

FIGURE 4.1
Mobile Cloud Computing.

4.3 Web Services

There are many definitions of *web services*, but we prefer the definition given by W3C:

> *A Web service is a software system designed to support interoperable machine-to-machine interaction over a network* [9].

There are two types of web services:

1. SOAP web services
2. RESTful web services

4.3.1 SOAP Web Services

SOAP is an XML-based protocol that exchanges data as SOAP messages. The components of web services are *Web Service Description Language (WSDL)* [10], an abstract definition of communication between two applications, and *Universal Description, Discovery and Integration (UDDI)* [10], a specification that helps providers register their service, by which they advertise their product as well as store and share their service information.

4.3.2 RESTful Web Services

REpresentational State Transfer is a stateless protocol; it has no restrictions of specifications like SOAP. It is fast in execution as it consumes less bandwidth and optimizes resources [10]. It is lightweight and has no standard. You can use JSON, XML, text, or any type of data to send a request or get a response back.

IBM was the first company that launched web services on mobile devices, through a shopping kiosk application [11]. The important thing about web services is that they allow different applications from different platforms to communicate with each other with little or no configuration change [12], as shown in Figure 4.2.

Web services communicate through XML only, so they are not tightly coupled to any other programming language. For example, Java cannot talk to .NET technologies or Windows operating system applications cannot interact with UNIX applications.

4.3.3 Technological Impact of Mobile Application Development

In this era of digitalization, the mobile application development industry and its use in all spheres of life is becoming a crucial part of the digital world. A lot of investments in mobile application development occurred in the past year. Even small-scale industry is implementing and incorporating mobile applications in their businesses.

As per a report published in 2019, the Google Play store has 2.46 million apps, and 1.96 million are predicted to be added in 2021. Technologies, such as Internet of Things (IoT), artificial intelligence, machine learning, virtual reality, or Blockchain, are making impressions in the market through mobile web services through cloud computing.

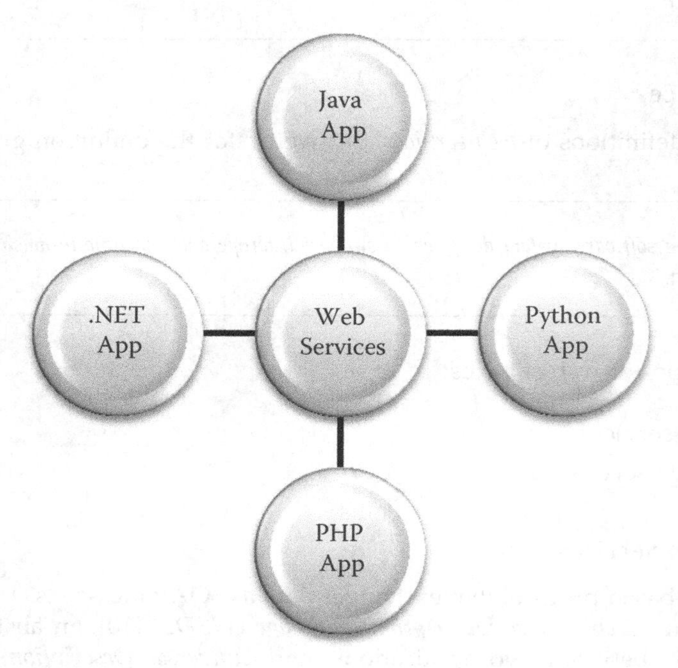

FIGURE 4.2
Web Services Heterogeneity.

4.3.4 IoT

IoT is a beyond-measure network of interconnected digital, mechanical, and computing devices that enable the exchange and transfer of data and information. The use of IoT in mobile apps helps remotely control smart gadgets via smartphone. Moreover, mobile apps powered by IoT can connect bands, wristwatches, and other wearables to smartphones [13].

4.3.5 Virtual Reality

Mobile app development changed the market with apps like ARKit by Apple, ARCore by Google, iOS, and Android app. Nowadays, demand for virtual reality and augmented reality applications occurs in most industries: retail, healthcare, education, travel, etc. Mobile apps make the use of these technologies incredible experiences for users and help businesses increase sales [14].

4.3.6 M-Commerce

With more and more people converging toward mobile purchasing, the future of m-commerce looks promising. Apple Pay and Google Wallet have encouraged customers to shop via smartphones instead of their debit or credit cards. Retail and eCommerce businesses prefer apps that let their customers shop easily and make transactions without cash or physical cards [15].

4.3.7 Cross-Platform Mobile App Development

Cross-platform mobile app development allows applications to run on heterogeneous platforms. It diminishes development costs for organizations and saves developers time, allowing them to survive business competition [16]. This is the latest trend in the field.

4.3.8 Blockchain

Blockchain is changing the entire mobile application development industry and providing a best feature to mobile apps by adding security, tracking, and quality services. This technology has already been adopted for secure money transactions [17]. Blockchain is also the future of the mobile and business industries (Figure 4.3).

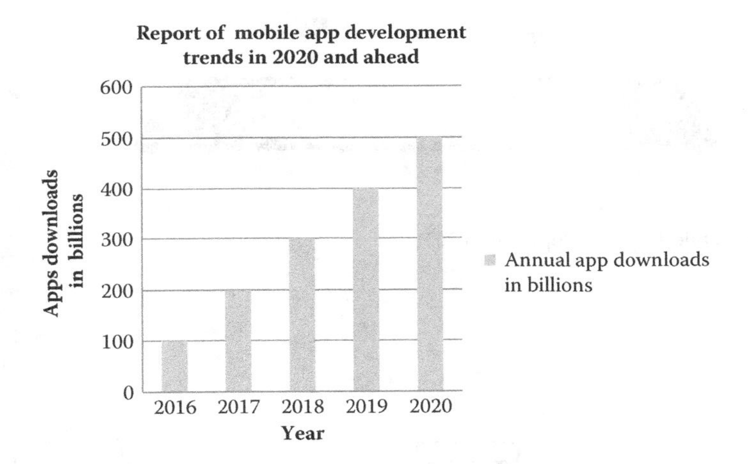

CHART 1
Chart Shows Mobile App Development Trends in 2020 and Ahead.

4.4 MCC Architectures

4.4.1 Reliable Web Service Architecture

Mobile devices are ubiquitous in the world and connect with the Internet through intermittent wireless connectivity to access web services. Reliability in low network range areas and retrieving appropriate responses from the web service are very important. Many researchers have used the middleware approach to study web service reliability.

In Nguyen et al. [18], reliable web service architecture using middleware achieved reliable web service consumption. The architecture uses: request size, response size, and consuming time as major factors. The above architecture focuses on surety of executing the request under low network range area and service temporal unavailability situation (Figures 4.4 and 4.5).

This architecture uses the middleware approach instead of a direct connection to the cloud service, using reliable web service architecture with the middleware protocol to ensure consumption reliability.

FIGURE 4.3
Prediction Report Mobile App Development in 2021.

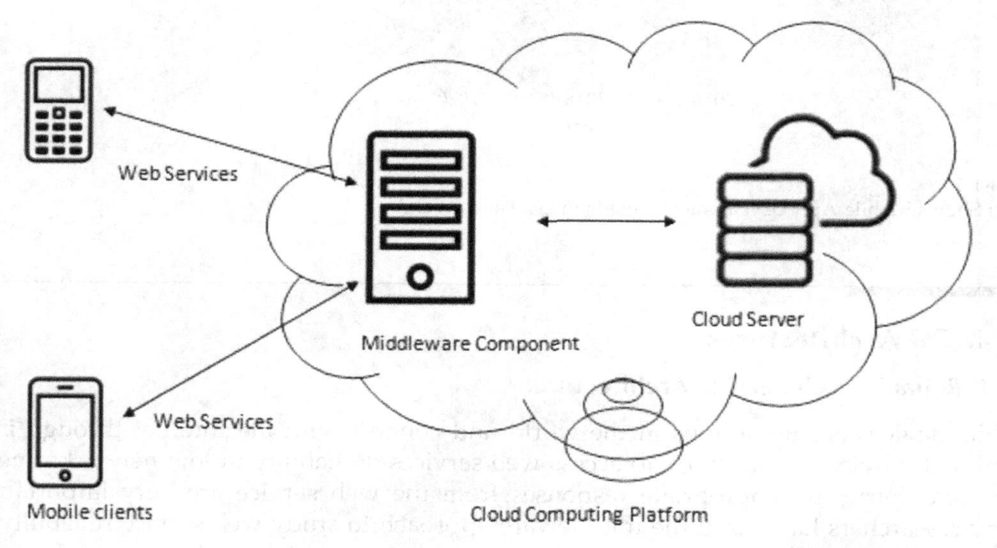

FIGURE 4.4
Middleware Architecture.

1. **Mobile Cloud Service Client**
   This mobile client component is responsible for:
   - Composing the request.
   - Making communication possible through the cloud service directly or through the service middleware component.

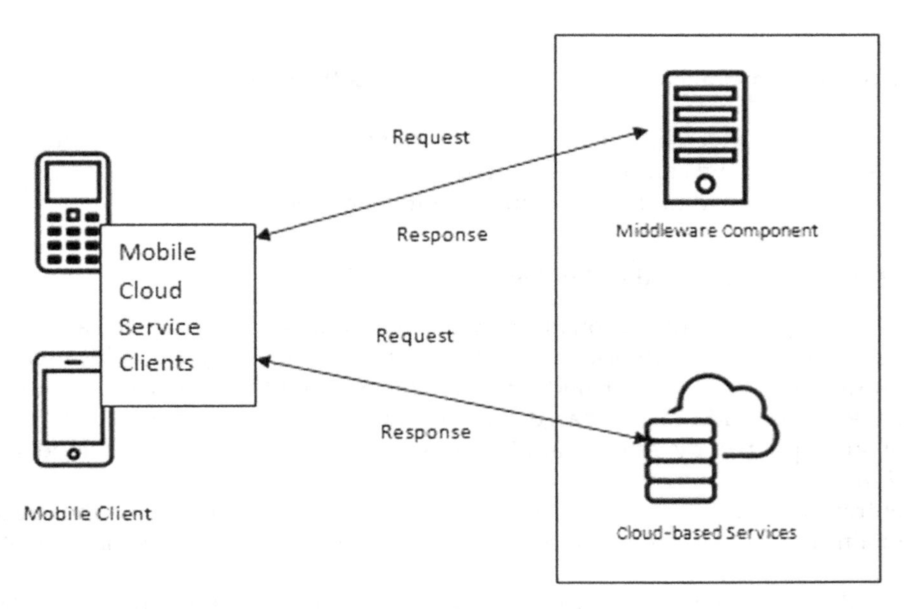

FIGURE 4.5
Reliable Service Architecture using Middleware.

- Getting back the response with the appropriate state.

2. **Middleware Component**
   - Accepting the request from the mobile cloud service client, and generating the respective response.
   - Storing the request and response in caching for later use.

4.4.1.1 Reliable Web Service Through Middleware Component

The middleware is a communication gateway that has accountability for getting back a response from the web service. The reliable web service architecture using middleware architecture is an entry gate between the mobile client and the web service and is responsible for handling heavy load communication with the service. The middleware improves the communication between mobile clients and web services.

The following factors are most effective regarding the reliable web service architecture using middleware architecture:

Request Size: The request size is increased due to the addition of attributes necessarily for the middleware to ensure a reliable request.

Response Size: The size of the response will not change because there are no additional properties added from the middleware for any reason.

Consuming Time: Due to the use of middleware, the consuming time may be longer than the direct cloud service connection. It can be calculated for both architectures: the middleware and the direct cloud service connection.

$$Tm = Tmm + Tm$$
$$TC = Tmc$$

where, TM is the time taken by the middleware. TC is the time taken by the direct cloud service.

For cloud services accessed directly from the server, results time out each time a large response size with heavy computations is required. Also, the request size is constant, and the response size is alike to the traditional architecture [19].

4.4.2 A Framework of Mobile Cloudlet

This framework [20] is a ubiquitous framework based on smartphones for accessing remote cloud services. The network of cloudlets is spread within a geographic area and is connected to the main server for finding its online availability and resource availability. This framework provides cloud-based network and software services, which produce data that are updated by remote applications or cached from remote cloud centers (Figure 4.6).

The central directory, also referred to as *root server,* reduces the load of the broadcast and maintains track of the online presence of mobile cloudlets and the availability of their resources.

To serve the requested service, root servers first check the online presence of the requested cloudlets. Then, the mobile cloudlet sends the particular requested services to its

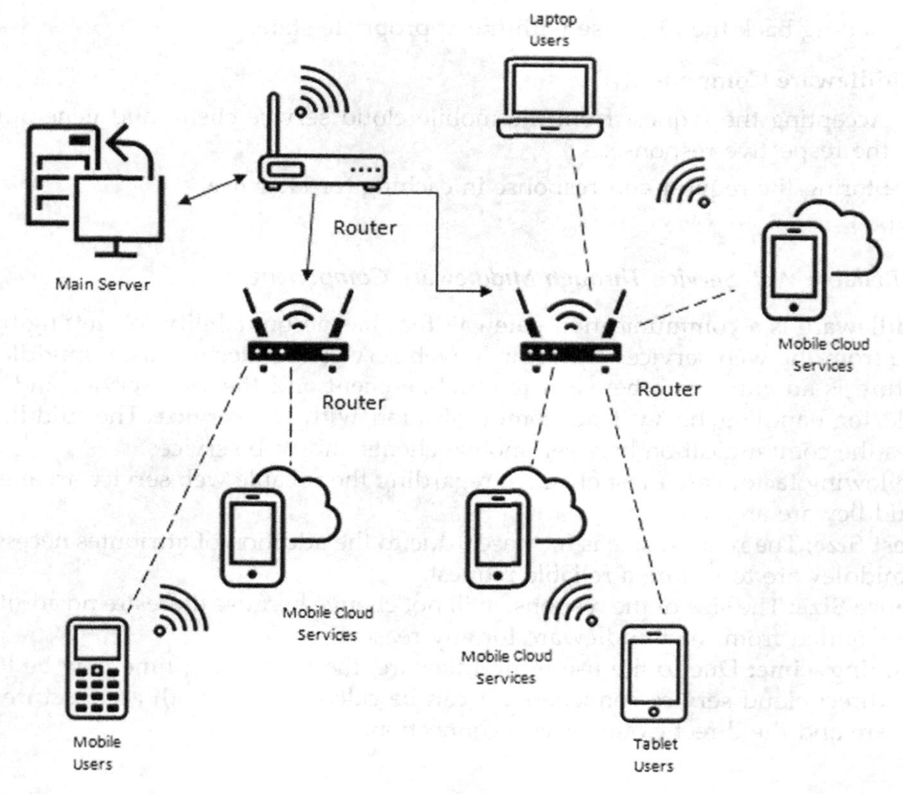

FIGURE 4.6
A Framework of Mobile Cloudlet.

users. This framework is most useful where mobile users are interested in network and software services.

4.4.3 Virtual Caching Service Architecture

The concept of Caching-as-a-Service (CaaS) [21] is related to cache virtualization. As you know, a mobile user expects rich multimedia services over mobile networks. However, traditional dedicated networking equipment in mobile network operators is unable to accommodate the phenomenal growth of the traffic load and user demand dynamics, and it also consumes energy resources inefficiently.

The growing techniques for mobile content caching and delivery are more demanding because the popular content can be cached inside mobile front-haul and back-haul networks (Figure 4.7). The most demanded content from users in propinquity can be easily accessed without redundant transmissions from the main server that abolish the duplication of traffic flow. This can be achieved with the concept of CaaS.

As the demand for multimedia applications increases, it is tempting to develop cloud applications for mobile devices. But due to various issues associated with the wireless network and other technical issues, mobile users suffer from low latency and Quality of Experience (QOE) [22]. At the same time, cloud providers are also facing challenges like computation complexity, expected bandwidth, and scalability. In Regunathan et al. [23], a system called User Preference Profile (UPP), which has two profiles, Reactive and

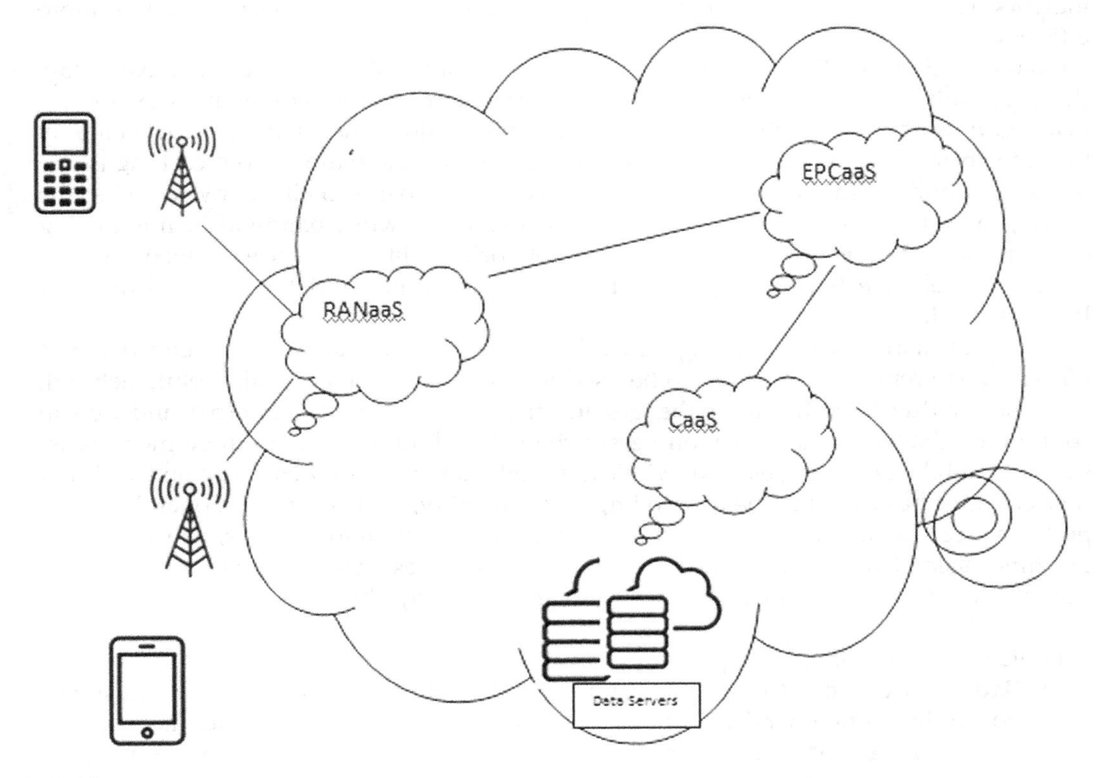

FIGURE 4.7
Virtual Caching Service Architecture.

Proactive UPP, is defined. R-UPP (Reactive) is an active user's profile that requests videos from Internet Content Delivery Netwrok (CDN) if they are not present in cache. As and when cache gets full, eviction is done based on Least Recently Used (LRU) replacement policy. P-UPP (Proactive) profile preloads the cache with videos that are most visited or requested by R-UPP users, but it utilizes high computational complexity and more bandwidth, so a hybrid solution performs updates in cache only if the expected cache hit ratio of improvement due to replacement exceeds a preset threshold. This system uses a caching technique to change the video encoding and transmission rate constantly. This leads to improvement in QOE and fast and effective streaming through the use of caching.

If the distance to the cloud server that provides the services is long, mobile devices require high latency to access content. To treat such issues, various powerful desktop-based models of local cloudlets have been proposed to provide services to nearby users. Dhule and Shrawankar [24] designed and implemented a solution for mobile users that can access cloud services from neighboring devices, and its major objective is to pull out the computational competence of mobile devices by creating a group of cloudlet networks distributed within particular areas. It is connected to a main server that finds online availability and resource availability. This framework is applicable to those cloud clients who are more concerned about network and software services. The framework performance was tested on CloudSim simulator. This simulator is extensively used for modeling and simulating cloud computing resources. CloudSim contains interfaces for virtual machines, data center brokers, cloudlets, storage, and bandwidth. The result of the framework is compared with another design; observed resource consumption is more efficient.

Quality of Service [25] is the key concern for users of the World Wide Web. Every day, the exponential growth in size of the network results in overcrowding and server overloading. In contrast, with the hasty growth in content and its users, the Internet can only provide "best-effort" content delivery to its users. In such situations, web caching is one of the best options that can be used to enhance web delivery to clients by reducing the latency perceived by clients, minimizing the use of network bandwidth, minimizing consumption of power, and enhancing the network quality of service. In web caching, technique web objects (e.g., hypertext documents, multimedia, database) are stored for later retrieval.

The performance of the caching depends on the implementation of an effective and efficient page replacement policy to be used for cache replacement [26]. Mobile network users suffer due to technical challenges in efficient access to the Internet and due to continuous change in their location versus the network range. Due to these mobility issues in mobile technologies and wireless mobile phone networks, new and efficient services must be offered to the user to improve reliability, availability, and overall system performance. Caching the most-used items to keep data close to the user, which reduces the time needed to retrieve the desired document, saves network bandwidth, lessens traffic congestion, and reduces response time observed by clients.

Benefits of Mobile Cloud Computing:

1. **Battery Lifetime:** MCC supports storage of data and processing in the cloud and not in the mobile handset, which certainly enhances the battery lifetime of the mobile device. Mobile battery consumption is more for heavy and complex computation. Data offloading and migration on the cloud is an effective solution that extends the life of a mobile battery.

2. **Unlimited Storage:** As mentioned in the above point, data are stored on the cloud; users need not worry about handset storage capacity. MCC provides a huge space of storage. Google Drive and Dropbox are examples of clouds that provide storage to their users.
3. **On-demand Service**: As MCC provides on-demand, seamless service from the cloud, users do not need to install hardware or software to their mobile handset. For example, the iPhone App Store and the Android Google Play Store allow you to install software any time with or without payment.
4. **Cost-efficient:** MCC is one of the most cost-efficient methods as it charges its users per use.
5. **Backup and Recovery:** Data backup and restoration is very convenient because all data are stored on the cloud itself. The digital copy of records is also maintained in the cloud by a disaster management committee as a security measure.
6. **High Reliability**: In MCC, data and applications are spread on multiple computers, so there is less chance of data loss. Also, the cloud supports security services like automatic virus scanning, authentication, malicious code detection, etc., so it is reliable.

Limitations of Mobile Cloud Computing:

1. **Data Security:** Data security is a major concern, so selection of a reliable cloud server is important.
2. **Network Connectivity:** Because MCC is an Internet-based activity, connectivity is the most important thing.

4.5 Conclusion

In this chapter, we discussed various MCC architectures, benefits, and limitations. MCC provides the computing power, storage, and platform support required to execute mobile applications through the cloud. Studied architectures have supported web caching to obtain reliable services, improving availability of web resources and reducing the network bandwidth. But there are certain limitations related to security and other network issues. In the future, the existing approaches can be modified to address these limitations [27].

References

1. Somula, R., and Sasikala R. (2018) "A Survey on Mobile Cloud Computing: Mobile Computing + Cloud Computing (Mcc = Mc + Cc)." *Scalable Computing: Practice and Experience* 19(4): 309–337.
2. Huaming, W. (2018) "Multi-Objective Decision-Making for Mobile Cloud Offloading: A Survey." *IEEE Access* 6: 3962-2976.

3. Khalid, A., Tarek M., Tamerm K., Elias, Y., Loay, S., Mohamed H., and Mohsen, G. (2019) "A Survey on Mobile Crowd-Sensing and Its Applications in the IoT Era." *IEEE Access* 7: 3855–3881.
4. Khan, A.U.R., Othman, M., Madani S. and Khan S. (2014) "A Survey of Mobile Cloud Computing Application Models." *IEEE Communications Surveys & Tutorials* 16(1): 393–413.
5. Abolfazli, S., Sanaei, Z., Ahmed, E., Gani, A. and Buyya, R. (2014) "Cloud-Based Augmentation for Mobile Devices: Motivation, Taxonomies, and Open Challenges." *IEEE Communications Surveys & Tutorial* 16(1): 337–368.
6. Khadija, A., Micheal, G., and Hamid, H. (2018) "Mobile Cloud Computing for Computation Offloading: Issues and Challenges." *Applied Computing and Informatics* 14(1): 1–16.
7. Ghahramani, M., Zhou, M., and Hon, C. (2017) "Toward Cloud Computing QoS Architecture: Analysis of Cloud Systems and Cloud Services." *IEEE/CAA Journal of Automatica Sinica*4(1): 6–18.
8. Khan, A., Othman, M., Xia, F., and Khan, A. (2015) "Context-Aware Mobile Cloud Computing and Its Challenges." *IEEE Cloud Computing* 2: 42–49.
9. Dinh, H.T. (2013) "A Survey of Mobile Cloud Computing: Architecture, Applications, and Approaches." *Wireless Communications and Mobile Computing* 13: 1587–1611.
10. Link Web Service at W3 https://www.w3.org/
11. AlShahwan, F., Faisal, M., and Ansa, G. (2016) "Security Framework for RESTful Mobile Cloud Computing Web Services." *Journal of Ambient Intelligence and Humanized Computing* 7: 649–659.
12. Boz, F., Finley, B., Oulasvirta, A., Kilkki, K., and Manner, J. (2019) "Mobile QoE Prediction in the Field." Pervasive and Mobile Computing (Elsevier) 59.
13. Wesling, P. "Heterogenous Integration Roadmap," 2019 Edition Chapter 3: Internet of Things (IoT). http://eps.ieee.org/hir
14. Parekh, P., Patel, S., Patel, N. et al. (2020) "Systematic Review and Meta-Analysis of Augmented Reality in Medicine, Retail, and Games." Visual Computing for Industry, Biomedicine, and Art3: 21.
15. Huang, J. (2020) "How Mobile Payment Is Changing The World." *Student Theses, Papers and Projects (Computer Science)* 5.
16. https://digitalcommons.wou.edu/computerscience_studentpubs/5.
17. Wafaa, S., El-Kassas, Bassem, A., Ahmed, H., et al. (2017) "Taxonomy of Cross-Platform Mobile Applications Development Approaches." *Ain Shams Engineering Journal* 8(2): 163–190.
18. Nguyen, D., Pathirana, P., Ding, M., and Seneviratne, A. (2019) "Integration of Blockchain and Cloud of Things: Architecture, Applications and Challenges." *IEEE Communications Surveys & Tutorials* 22(4): 2521–2549.
19. Amr S., Tamer, A., EI-Sayed, M., and EI-Horbaty. (2018) "RSAM: An Enhanced Architecture for Achieving Web Services Reliability in Mobile Cloud Computing." *Journal of King Saud University – Computer and Information Sciences* 30: 164–174.
20. John Kolb, P., Chaudhary, A., Schillinger, A., Chandra and Weissman, J. (2015) "Cloud-Based, User-Centric Mobile Application Optimization." *2015 IEEE International Conference on Cloud Engineering*, Tempe, AZ, 26–35.
21. Artail, A., Frenn, K., Safa, H. and Artail, H. (2015) "A Framework of Mobile Cloudlet Centers Based on the Use of Mobile Devices as Cloudlets." *2015 IEEE 29th International Conference on Advanced Information Networking and Applications*, Gwangiu. 777–784.
22. Li, X., Wang, X., Zhu, C., Cai W., and Leung, C. (2015) "Caching-as-a-Service: Virtual caching Framework in the Cloud-Based Mobile Networks." *2015 IEEE Conference on Computer Communications Workshops (INFOCOM WKSHPS)*. Hong Kong. 372–377.
23. Regunathan, R., Aramudhan, M., and Lavanya, M. (2018) "Neural Based QoS aware Mobile Cloud Service and Its Application to Preeminent Service Selection using Back Propagation." *Procedia Computer Science* 132: 1113–1122.

24. Dhule C. and Shrawankar, U. (2019) "Energy Efficient Green Consolidator for Cloud Data Centers," *2019 6th International Conference on Computing for Sustainable Global Development (INDIACom)*, New Delhi, India, , 405–409.
25. Talaat, F.M., Ali, S.H., and Saleh, A.I. (2020) "Effective Cache Replacement Strategy (ECRS) for Real-Time Fog Computing Environment." *Cluster Comput, Springers Link* 23: 3309–3333.
26. Ghahraman, M., MengChu, Z., and Chi, T. (2017) "Toward Cloud Computing QoS Architecture: Analysis of Cloud Systems and Cloud Services." *IEEE/CAA Journal of Automatica Sinica* 4(1).
27. Talaat, F., Ali, S., and Saleh, A. (2020) Effective Cache Replacement Strategy (ECRS) for Real-Time Fog Computing Environment." *Cluster Computing* 23: 3309–3333.

Section II

Internet of Things

5

Internet of Things – Essential IoT Business Guide with Different Case Studies

Nutan Hemant Deshmukh
Savitribai Phule Pune University

5.1 Introduction to the Internet of Things

There was an era when people used to get rid of things by throwing them in the thrash. Finding them again used to be a big task. But as soon as technology evolved, these dumped things became important to us. The Internet of Things (IoT) is not only about dumped things with radio-frequency identification (RFID) placed on them just to find them, but much more than what people know about it.

Kevin Ashton is the inventor of the term *Internet of Things*. There are multiple definitions that describe IoT in their own words. Many definitions on the web talk about proposed systems for the IoT, but the reality is that the systems are REAL systems that are in use by many. The simplest definition is "Multiple intelligent devices (sensors, actuators, etc.) with unique identifiers for each device are linked together using wireless networks for sharing information without human-to-human interface or human-to-machine interface in between the devices and using the shared data for further processing."

In 1966, the first embedded system was designed using a microprocessor for the first time to control an electronic fuel-injection system by Volkswagen. This embedded system was developed for vehicles and had a big remarkable contribution towards the Evolution of Embedded Systems. Then, slow and steady, the concept of evolving embedded systems into intelligence began. There are many alike conceptualizations, but the IoT has got more credits than the others as people have taken up the term and also supported the concept. The term came into existence in 1999, to assist the RFID concept. But the popularity of the IoT system boosted after 2014. Until then, no advertising reached people related to new concepts that could make life simpler. Machine-to-machine or Industrial Internet is a concept that supports IoT, i.e., they are also called sub-segments of the IoT.

But there are certain important parameters playing a major role in the system, like intelligent things, network or Internet, devices, resources, and protocols. Every parameter plays a vital role in the system.

- **Things:** Things in the IoT can be any intelligent device that has the ability to boot itself, read the information from the outside world, and transfer it to the devices

DOI: 10.1201/9781003166702-5

for further processing. Things can be a sensor, transducer, or any other component that can fulfill the given task.

- **Network/Internet:** An Internet is a global system of interconnected devices that forms a system in performing specific tasks. It connects devices all over the world. Through the Internet, people can communicate and share information easily.
- **Devices:** The IoT is a package of electronic intelligent gadgets that may include sensors; resources, also called as software program to make the devices work; actuators; and computing devices. All these devices are interfaced to a computing device that can communicate to them easily using the Internet, allowing data to transfer easily, or performing human monotonic tasks without a human interference. A couple of matters are related to the gadgets that help in imparting large information, and this fact is used for evaluation or further processing to get the desired tasks executed.
- **Resources:** Resources in the IoT are the software that supports every device or thing to perform. It is as good as a device driver that supports the hardware to run.
- **Protocols:** Protocols are a medium of communication in between the devices, connected in a system. There are various protocols used at various layers [1–7].

5.2 Phrases of the Internet of Things

- **Machine to Machine (M2M):** M2M was introduced approximately in 2012. At that time, nobody was dependent on such communication systems. M2M is very famous in the telecoms sector. M2M communiqué, to begin with, was a pair-to-pair connection where one system will communicate with the other with the aid of linking. Now, the generation has grown up so rapidly due to cellular connectivity. Humans can easily share facts from one location to another using a network with a huge range and multiple gadgets. M2M now does not simply connect gadgets or machines. It also connects humans to the device.
- **Industrial Internet (of Things):** The term Industrial Internet was brought into existence by General Electric Corporation in the United States. It refers to integrating and linking huge data, referred to as big data; various analytical tools; and wireless networks with physical and industrial equipment. Major industrial equipment is linked with tools required for various analyses.
- **Internet:** The Internet is a magical box that connects various devices/components and /or humans to each other.
- **Web of Things (WoT):** The WoT is a subset of the IoT as it has a narrower scope and focuses on systems developed using software for controlling.
- **Internet of Everything (IoE):** The IoT relates to the interconnectivity of physical objects and data input and output, while the IoE relates to interconnectivity of various technologies, processes, and people.

- **Enterprise 4.0:** The time period Enterprise 4.0 specializes in manufacturing environments. However, it has a wider scope for all standards. Enterprise 4.0 describes a fixed set of ideas to force the next business revolution. This consists of all styles of connectivity principles inside the industrial context. However, it also consists of real changes to the physical world around us, inclusive of 3D printing technologies or the introduction of recent augmented reality hardware [1–3].

5.3 Working of the Internet of Things

The IoT surroundings consist of smart devices that support Internet connectivity. These devices are designed using embedded system specifications, which include hardware components like transducers, controllers, and verbal exchange hardware to help gather data, pass on huge data, and act on the data that have been gathered from surroundings. The data that are read by intelligent devices are shared to other devices with the help of a gateway or router placed in between the devices for smooth transfer of information via Internet connectivity. The data shared in different domains use software; every record is sent on to the cloud for processing or is analyzed locally. All these devices have strong communication with each other. They react based on the facts that they receive via their connected devices. These smart devices do many tasks without humans present. Humans can interact with the components if required – for example, construct, supply them, dictate to them, or access their records. Many IoT systems deploy use of web-enabled software for operations where integration, mingle, and communiqué protocols are used with those web-based devices in large domains. The important thing is the IoT platform. The devices and objects with built-in sensors are connected to this platform. This IoT platform pinpoints the important information from all the received data and safely ignores the remaining data. The required data segregated by IoT platform can be used for analysis, characteristic learning, pattern recognition, etc.

The environment is read by the sensors, which are placed in intelligent hardware devices that are part of the IoT system. Devices make use of the Internet for completing their activities. After the operations are performed on the read data from the environment, the output is displayed on actuators. Actuators, if required, may use the Internet to display the information or generate output in any form, like moving the arm of a robot. Actuators interface with people, who are more concerned with the output after the intelligent work done by the devices or machines. People make use of the Internet to complete tasks and use Internet-enabled devices. The IoT also can employ various domains that can support artificial intelligence and/or machine learning to make statistics-accumulating tactics less difficult and more dynamic [5,6,8].

5.4 Protocols Used in the Internet of Things

Protocols are the mode of communication between the devices connected on the network. There are various protocols that are used in the IoT. They are as follows:

1. **Message Queuing Telemetry Transport (MQTT):** Two network entities are defined by this protocol: a message broker and the number of clients connected to it. The message broker is a server that receives all the messages from the clients and then routes the messages to the appropriate destination clients. A client can be any device that needs to connect to the MQTT broker by running an MQTT library over a network. An example of an MQTT is if there is a huge network of sensors connected to each other in a parking zone to identify the number of vacant and free slots. This network is monitored or managed via the Internet using the MQTT protocol.
2. **Extensible Messaging and Presence Protocol (XMPP):** This protocol is based on using Extended Markup Language (XML), which aims to present information and focuses on message-oriented middleware at an instance. An example is a smart thermostat that can be accessed from a smartphone via a web server.
3. **Data Distribution Services (DDS):** DDS protocol uses broker-less architecture. It is an IoT protocol developed for M2M. The standard for DDS is decided by the Object Management Group (OMG). It enables data exchange via publish-subscribe methodology. DDS makes use of broker-less architecture, unlike MQTT and constrained application protocol (CoAP). It uses multicasting to bring high-quality QoS to the applications. DDS protocol can be deployed from low footprint devices to the cloud. Examples are subscribing to a YouTube channel or to a publisher on a certain topic, e.g., recipes, information on birds.
4. **Advanced Message Queuing Protocol (AMQP):** AMQP is the asynchronous open-source standard for communication. It is used between organizations and various applications as it helps with encryption of data and ensures interoperability of messages. This protocol is used in client/server messaging and in IoT device management. An example is the stock market, where shares of publicly listed companies are bought and sold. These transactions are done on a large scale and on a regular basis. Accordingly, messages are sent to multiple shareholders, and transactions are updated.
5. **Representational State Transfer (REST):** It is the most widely used application protocol interface today. Restful API is generally used with http so that data transfer is easy and the modern web can easily implement it with JSON or XML. An example is a lady who ordered a suit for Diwali from Amazon. The purchasing and payment process was done online. The activity – searching, selecting/choosing, and ordering – was also done online.
6. **Constrained Application Protocol (CoAP):** CoAP is an Internet utility communication protocol for constrained gadgets that have low bandwidth availability. It is designed to enable simple, constrained devices to join the IoT network. The protocol is designed for an IoT system that is generally based on HTTP protocol. This protocol is primarily used for M2M communication. An example is where a smart home system of electronic devices, like fridge, doorbell, washing machine, etc, are monitored and controlled. The system can identify scheduled tasks about to happen in a sequence and send alerts/messages. Electronic devices can also be controlled by remote control [9].

5.5 Criticality of the Internet of Things

The IoT allows human beings to live their lives and be smart workers, as well as benefit from lifelong management. In addition to supplying smart gadgets to control domestic appliances, the IoT is important for corporations. The IoT provides agencies with a synchronous inspection of how their systems genuinely operate, handing over insights into the whole thing, including the overall execution of machines, moving a product from supplier to consumer, and strategic operations. The IoT allows companies to robotize technology and reduce the price of manual labor. It additionally improves service delivery by cutting down on waste, which makes it cost effective to manufacture and deliver items, in addition to providing clarity to client transactions. When compared to the other technologies in human beings' lives, the IoT is one of the critical technologies. It will continue evolving as more groups recognize the capacity of linked gadgets to keep them competitive. The IoT has encouraged many companies to reexamine their current technology for their organizations in terms of making or implementing smart workplaces that will automatically help employees, which in turn will enhance their commercial organization technology. Usually, the IoT is used in Industrial IoT, smart cities, smart supply chains, etc., using software programs and the hardware support of components like sensors and other IoT devices; however, it has also been used for corporations inside the agriculture field, for the infrastructure and control of domestic appliances, and for virtual reality. The IoT may help farmers in agriculture to make their jobs simpler. Sensors can acquire information on temperature, pressure, moisture level, rainfall, humidity, and soil content, as well as many other parameters that could assist in simplifying and automatizing agricultural technology.

Screening operations surrounding infrastructure are another aspect of the IoT that can be helpful. For example, sensors are used to display events or changes to the interior structure of houses, bridges, and different infrastructures. The advantages gained with it include having a paperless workflow, cutting costs, saving time, and making high-quality workflow changes. A domestic controller system that is designed for commercial organization can make use of the IoT to explore and build electronically powered structures to make tasks easier. On a huge scale, smart cities/towns can help residents reduce waste and electricity intake by designing intelligent IoT products that can be used at homes or offices. The IoT affects every industry, including healthcare, finance, retail, and manufacturing [3,4,7].

5.6 Benefits and Detriments

A number of the benefits of the IoT:

1. Capacity to get access to information from anywhere at any instance and on any smart device
2. Increased communication between related digital devices/smart components
3. Moving huge packets of data over a coupled network that responds by saving time and capital
4. Self-operating tasks to enhance a commercial enterprise's services and minimize the need for human presence

A few detriments:

1. As the number of associated devices grow(s) and more information on facts is shared between intelligent things, security plays a major role as the chances of leakage of confidential data or stolen data increases, which can create issues for the system.
2. Organizations may ultimately have to survive with large numbers – perhaps even hundreds of thousands – of IoT intelligent things, and acquiring and dealing with the huge amount of data from all the things will be tough.
3. Multiple devices are connected in a network; if one device gets a malicious program, it may harm the system. This can result in corruption of data.
4. There is no worldwide standard in terms of specifications, design, or compatibility for IoT components, so it is tough for devices from different manufacturers to speak with each different [1–3,7].

5.7 Applications of the Internet of Things

There are various domains that use the Internet, starting from consumer IoT and business IoT, to production and Industrial IoT (IIoT). IoT applications span numerous domains, including automotive, telecom, and power. In the consumer segment, for example, smart houses might be equipped with smart thermostats, smart appliances, and related heating, lights, and digital gadgets managed remotely through computers and smartphones.

Wearable devices like smart watches with sensors and software can collect data like heart rate, oxygen, etc., and examine user facts, sending messages to different technology. Wearable devices are also used for public safety – for example, they can improve first responders' reaction times during emergencies by providing optimized routes to an area or can monitor firefighters' vital signs.

The IoT in the healthcare domain provides many benefits, such as screening patients carefully to generate a report of the data. Hospitals frequently use IoT systems to finish tasks, along with managing stock for prescription drugs and clinical gadgets.

Smart buildings can, for example, reduce energy consumption by using sensors that estimate the number of people in a room. Depending on the occupants in the room, temperature can be adjusted accordingly – for example, the air conditioner can be turned on and off depending on sensor values. In agriculture, the IoT is primarily based in smart farming systems that monitor temperature, humidity, soil, and water level for plant growth; crop fields use connected sensors. Automated irrigation systems are also implemented using IoT technology. In a smart city, IoT sensors and devices, including smart streetlights and smart meters, can help manage visitors, preserve electricity, reveal and address environmental worries, and enhance sanitation [5,6,8,10–12].

5.8 Design Methodology for the Internet of Things

There are four important pillars of the IoT, as follows:

Components: A few important concepts that you should know before you start working on the design methodology of any IoT system.

1. **Transducers:** A digital device that transforms electricity from one shape to another is called a transducer. Examples include microphones, loudspeakers, thermometers, role and strain sensors, and antenna.
2. **Sensors:** Every transducer can be a sensor, but every sensor is not necessarily a transducer. Sensor is the next name given to sensor. In some cases, the task of a sensor is less than a transducer. A microphone takes in sound energy and gives out sound waves based on the strength of the signal received.
3. **Actuators:** Every transducer in an IoT structure is an actuator. In simple words, an actuator operates on the opposite path of a sensor. It takes in electricity and turns it into physical action. Example are an electric motor, a hydraulic gadget, and a pneumatic system. All are one-of-a-kind actuators.
4. **Controller:** In a standard IoT system, a sensor may gather statistics and direct them to middleware. There, previously defined logic dictates the selection. As a result, a corresponding command controls an actuator in reaction to that information. Hence, sensors and actuators in the IoT work together from contrary ends [5,6].

5.9 Various Types of Sensors

Sensors play an important role in the IoT for sensing information present in the surroundings. All sensors are electronic devices that help measure environmental conditions and share the information with devices for further processing. A few sensors are mentioned here that are commonly used in many IoT systems (Table 5.1).

5.10 Design Methodologies

There are 10 different steps that can be used for IoT system designing.

1. **Purpose and Requirement Specification**

Step 1 in the methodology defines the cause and need for the device. The machine to be used should have a reason; take the necessary training and read any instructions so that you have a clear idea about the system. This step covers all sub-factors like cause, conduct, system management requirements, records analysis requirements, and application deployment requirements [14].

2. **Process Specification**

The next stage defines the specific plan of action. How will the device be used? Write down requirement specifications to provide clarity about the system [5,6,15].

TABLE 5.1

Different Sensors Used for Projects [13]

Sr.no	Sensor Name	Details of the sensor
1	Temperature sensor	It measures the heat energy in the surroundings. It detects changes in temperature and converts these changes to data.
2	Infrared sensor	Infrared radiation is measured and detected from the surroundings. There is one panel that senses infrared radiation emitted by the other panel. If it is interrupted, then an object is present in between. If it is continuously received, that means there is no disturbance.
3	Passive infrared sensor (PIR)	It is used to sense motion. It detects whether a human has moved in or out of the sensor's range.
4	Soil sensor	They are also called "volumetric water content sensors". They are used to measure the water content in soil. They estimate the quantity of stored water in a profile, or how much irrigation is required to reach a preferred amount of saturation. They maintain the water content in soil and make it more beneficial to crops.
5	Water sensor	They are used in a location where water should not be present. If water is present, the sensor collects the information and shares the location.
6	Humidity sensor	It measures the quantity of water vapor inside the atmosphere of air or other gases. It is generally seen in hospitals and meteorology stations to document and forecast weather.
7	Pressure sensor	It senses changes in gases and liquids. When the pressure changes, the sensor detects these changes and communicates them to connected systems. Generally, they are used for checking gas leakage at home or in industrial domains.
8	Gas sensor	These sensors monitor and detect changes in air quality, including the presence of toxic, combustible, or hazardous gases. Generally, they are used in industries for mining, oil and gas, chemical research, etc.
9	Proximity sensors	These sensors are used for non-contact detection of items close to the sensor. These sorts of sensors regularly emit electromagnetic fields or infrared beams.
10	Optical sensors	They convert light rays into electrical signals. They are commonly used inside the auto industry, where cars use these sensors to diagnose signs and symptoms, boundaries, and other things that might help drivers while driving, riding, or parking. Driverless vehicles are a top instance. There are many different domain names where they are used.

3. **Domain Specification**

The next stage in the design method is defining the domain model. Provide a description of the important fundamentals, entities, and items within the area of the IoT machine to be designed. Attributes of the device and their relationship among each other are defined by this model. It draws a clear picture of the physical items and virtual items that will use various resources to connect to smart devices via software or an application through which it is controlled. With the area version, the IoT device designers can get an explanation of the IoT domain for which the gadget is to be designed. It covers minute details like physical entity, digital entity, tool, sources, and service. Range of digital entities can be equal to the number of physical entities. The major components in the domain specifications are mentioned below:

Physical Entity: It is a discrete and identifiable entity in the physical environment. The physical entities in gadgets are monitored and are present in the surroundings.

Virtual Entity: It is a representation of the physical entity in the virtual world. For every physical entity, there is a virtual entity.

Device: These provide a medium of interaction between physical and virtual entities. In this system, the device is a mini computer that has various sensors attached to it.

Resource: This is network and device resources. The on-device resource is the Operating System on the mini computer.

Service: This provides an interface for interacting with the physical entities [5,6,15].

4. **Information Specification**

The fourth stage is to define the information model. The statistics model defines the shape of all the data inside the IoT machine, for example, attributes of virtual entities, relations, etc. The records version does describes the specifics of how the information is represented or stored. To outline the statistics model, first list the digital entities described inside the domain version. The records version provides more info to the virtual entities by defining their attributes and relationships [5,6,15].

5. **Service Specification**

The fifth stage talks about outlining the service specifications. Carrier specifications outline the offerings in the IoT gadget, service kinds, service inputs/output, service endpoints, carrier schedules, provider preconditions, and carrier effects [5,6,15].

6. **IoT-level Specification**

The IoT system comprises components like devices, resources, controller services, databases, web services, etc.

There are several levels in the IoT – Level 1 to Level 6. Every level is different and has its own importance for designing an IoT system. Following are the levels briefly explained and when they are used:

1. **IoT Level 1:** It has a single node or device that performs sensing and/or actuation, stores facts, makes evaluations, and hosts software.
2. **IoT Level 2:** Asingle node performs sensing and/or actuation, and neighborhood analysis is accomplished through the node itself. Data are stored on the cloud.
3. **IoT Level 3:** A single node monitors or senses the information, and analysis and storage are done on the cloud.
4. **IoT Level 4:** There are multiple nodes sensing the environment with the observer node, and data analysis is done at local site but stored on the cloud.
5. **IoT Level 5:** There are multiple nodes observing and/or sensing the environment along with the observer node. The data sensed by all the nodes individually goes to a coordinator node, and then the coordinator is responsible for sending the data to the cloud for storage and analysis purposes.
6. **IoT Level 6:** Multiple nodes are monitoring the information with the observer node, and they send their information or data to the centralized controller, which

is placed on the cloud (i.e., not local to the nodes). This centralized node is responsible for sharing the data required for analysis and storage purposes [5,6,15].

7. **Functional Specification**

This step defines the purposeful view. The useful view (functional view) defines the capabilities of the IoT structures grouped into functional categories. Each functional category both presents the process for interacting, with times described inside the area model, or offers facts associated with those processes [5,6,15].

8. **Operational View Specification**

The eighth stage in the IoT design method is to outline the operational view specifications. In this step, various alternatives relating the IoT system deployment and operation are described, which include provider hosting alternatives, storage options, device options, utility hosting options, etc. [5,6,15].

9. **Device and Component Integration**

The next stage inside the IoT layout technology is the integration of the gadgets and supplements.

10. **Application Development**

The very last footstep in the IoT design technology is to implement and deploy the IoT software.

5.11 Case Studies

5.11.1 Case Study 1: Single Axis Solar Tracker Using Arduino

Step 1: Purpose and Requirements Specification

Purpose: To make use of maximum solar electricity by means of the panel. The sun panel tracks the sun from east to west daily from maximum depth of light (solar energy).

Behavior: This solar tracking system contains a solar panel fixed on a structure that moves, captures energy, and stores energy according to the position of the sun.

System Management Requirement: Provides the control function [16–18].

Step 2: Process Specification

Figure 5.1 describes the complete flow of the system, starting from initialization of pins till the activation of the respective output pin based on certain threshold values until the time the value is low.

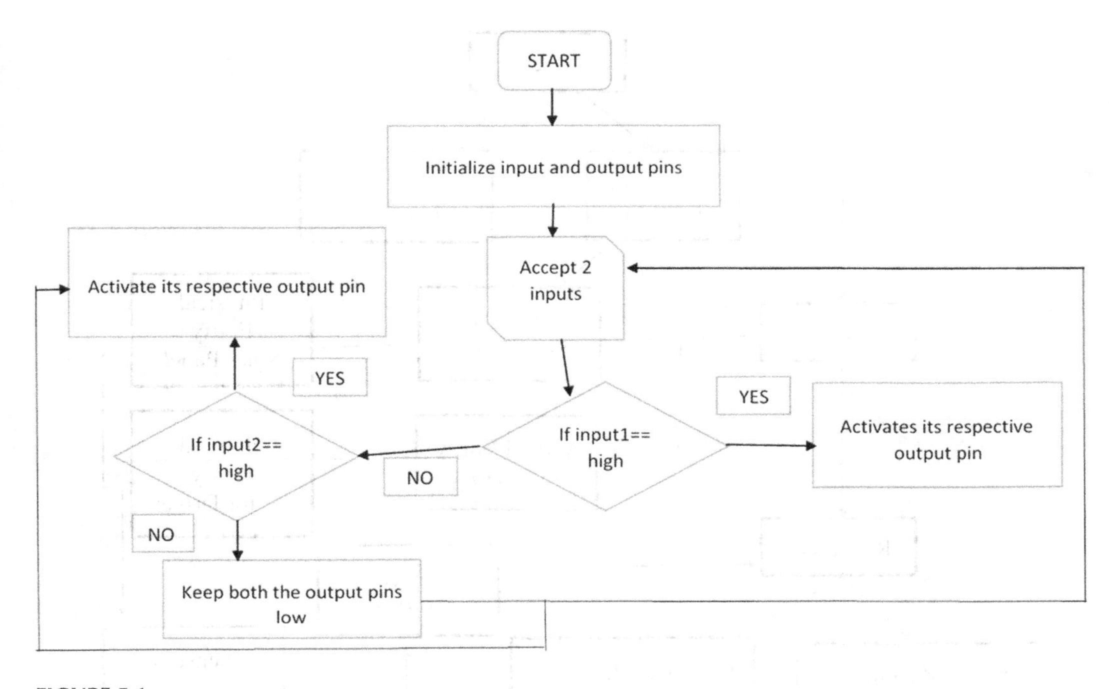

FIGURE 5.1
Process Model for Solar Tracker.

Step 3: Domain Model Specification

This step describes a number of physical entities, virtual entities, the resources used to achieve the system, the application mode and services used for the system by the devices as shown in Figure 5.2. User interface to application can be manual or automated. Resources used can be in network or on individual devices. The virtual entities are equal to the number of physical entities as the hardware, i.e., the physical entity is activated or addressed using the virtual entity, which works as a software logic pin for the hardware connected to the system. The domain model specifications talk about the internal connectivity in between the hardware components and their logical addresses.

Step 4: Information Specification

Information specification step talks about the attributes that the virtual entities hold, as shown in Figure 5.3. Defining attributes play an important role as it talks about the characteristics that it poses. As in this case, the attributes of motor and panel are mentioned. Motor has a movement called rotation on either side, i.e., clockwise or counterclockwise.

Step 5: Service Specification

Figure 5.4 describes the service named as light and rotation, their related service that it will work for when certain input is provided, and the obtained output for that input.

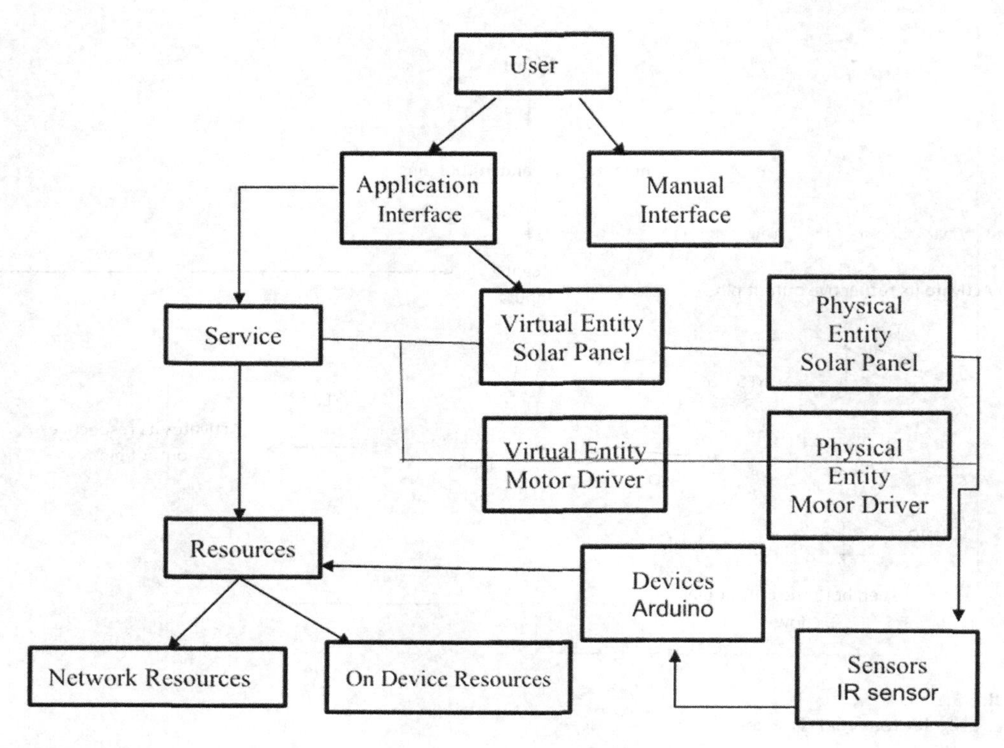

FIGURE 5.2
Domain Model for Solar Tracker.

Step 6: IOT-level Specification

IoT level that is applicable for the given problem is IoT Level 1. The solar tracking works only for the local area where devices are connected to each other and the information read is stored in a global database with the help of a controller service that is receiving data from devices connected at the environment side. With the help of REST /Web sockets Application-based Protocol, the data is sent on the application created for easier access. Figure 5.5 describes this process.

Step 7: Functional Specification

Functional specification talks about interface of the IoT level with the IoT architecture layers. The IoT architecture layers focus on:

1. **Devices:** Devices is another layer that deals with the sensors embedded into the devices, which helps in reading the environmental data. As it deals with devices, it also makes sure to work with the resources used by the devices and keep them updated according to the requirements of the devices or components. Devices can be actuators or sensors or any computing devices used as an individual entity or group of entities.
2. **Management:** It handles the task of managing the global database used to save information read by the components connected in the environment or also used

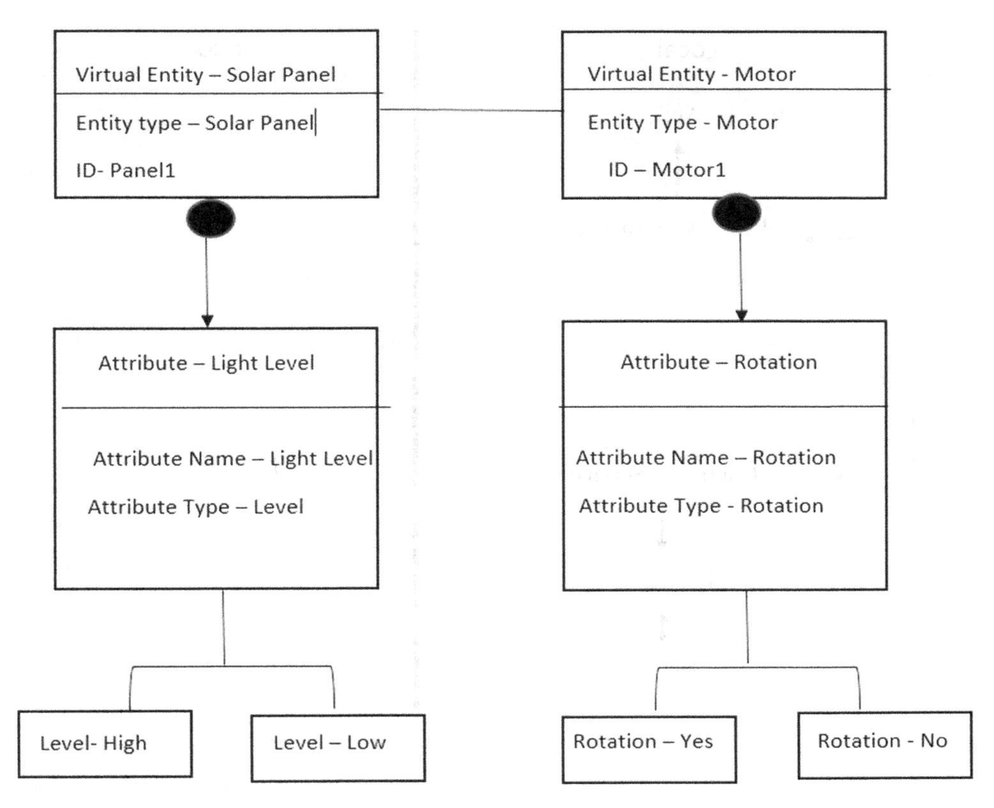

FIGURE 5.3
Information Model for Solar Tracker.

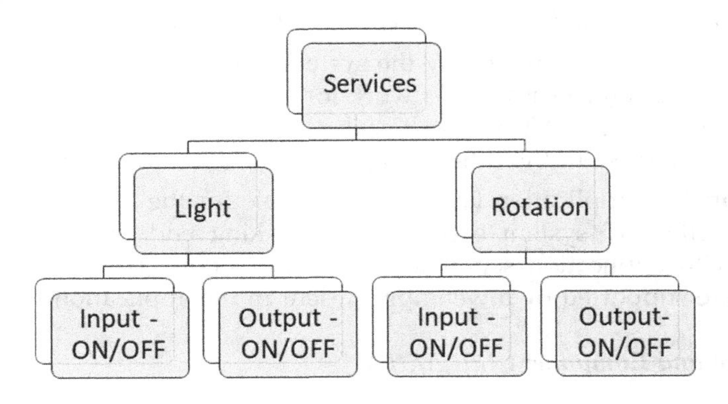

FIGURE 5.4
Service Model for Solar Tracker with service light and motor.

by the services to perform certain tasks. Devices connected to the IoT system are also handled by a device manager present in the management architecture clock.

3. **Communication:** Communication of various components is performed via the device present in the IoT level. As IoT makes use of various components

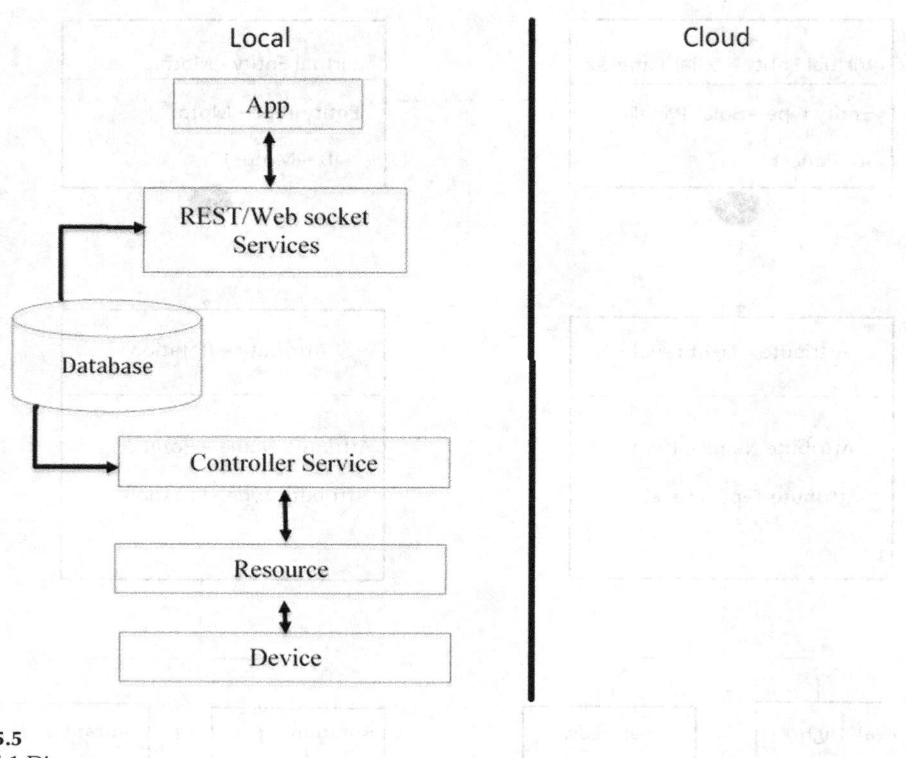

FIGURE 5.5
IoT Level 1 Diagram.

similarly, it also makes use of various resources for communication between the components and the system. These resources work as software, which can help devices or components for communication with the outer world.

4. **Services:** Services are provided by the system with the help of a controller service. Mainly the controller services work for web services. Application protocol interfaces (APIs) like REST and web sockets are also used for data transferring from global database to the different applications used by the customers.
5. **Application:** The application layer is responsible for the applications designed for the specific IoT system execution for making end user jobs simpler. It supports with application server and database server. The main applications designed are supported by a web app present in the application layer.

Step 9: ***Device and Component Integration***

Step 8: Operational View Specification

1. Application
   - Application Server: Google App engine
2. Offerings/services
   - Local: controller carrier
   - Web: relaxation

FIGURE 5.6
Hardware Connections for Single Axis Solar Tracker Using Arduino

3. Communication: Communication APIs: REST SPIs
4. Communication Protocol:
   i. Protocol that could be used for link layer is 802.11.
   ii. Protocol that could be used for network layer is IPV6.
   iii. Protocol that could be used for transport layer is TCP.
   iv. Protocol that could be used for application layer is HTTP.
5. Management: Device Management: Arduino device management
6. Security: Login Management

The device used in Arduino makes use of a microcontroller that is easy to program. It can handle at maximum one or two tasks. Wherever there are limited features, the Arduino can be used. The diagram talks about integrating components with the Arduino device.

Figure 5.6 shows the actual hardware connection for Solar Tracker with required devices and components.

5.11.2 Case Study 2: Plant Monitoring System using Raspberry-Pi

Step 1: Purpose and Requirement Specification

Purpose: A plant monitoring system monitors the current temperature (in Celsius and Fahrenheit) and moisture content of the plant soil through an application.

Behavior: The system measures the current temperature and checks if there is adequate moisture in the soil after every five minutes. When the soil moisture is low, an email notification is sent to the user, which specifies the current temperature, that low moisture has been detected, and that the user might need to water the plant.

System Management Requirements: The system provides on-site control functions.

Data Analysis Requirement: The system does not provide a data analysis feature.

Application Deployment Requirement: The application should be deployed on local devices.

Security Requirement: The email is to be sent to an authorized user only.

Data Collection Requirement: The system keeps collecting data related to moisture adequacy in soil and temperature measurements until the user manually interrupts the application [19].

Step 2: Process Specification

This step, draws use cases for the IoT system based on the specifications derived in step 1 of the design methodology. The system uses two sensors: temperature and moisture sensor. As shown in Figure 5.7, the temperature sensor senses the current temperature, and the moisture sensor senses the soil moisture level (low or adequate). If the soil moisture level is low, then email notification is sent to the plant owner along with the current temperature.

Step 3: Domain Model Specification

This step represents the main concepts like resources available and used by devices, the related entities to the external components present in the system, and various devices present within the domain of the IoT device. In the case of the IoT system for plant monitoring, Figure 5.8 describes the domain model with physical entities, virtual entities, resources used by devices, devices and sensors connected to the system, and the services provided by the system.

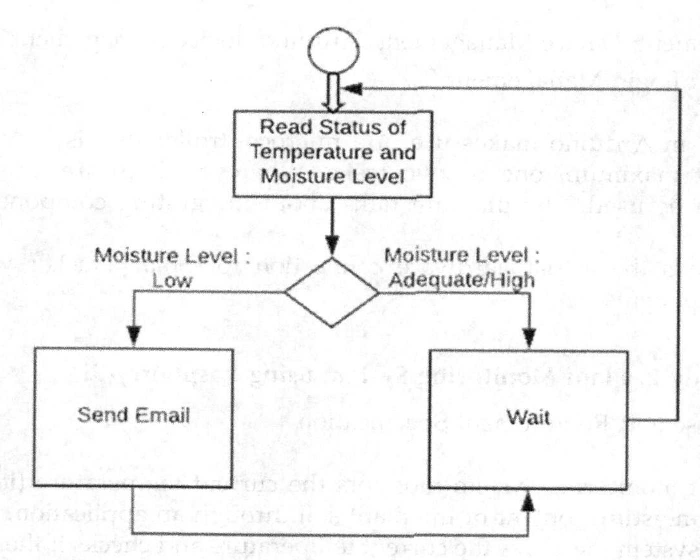

FIGURE 5.7
Process Model for Plant Monitoring System.

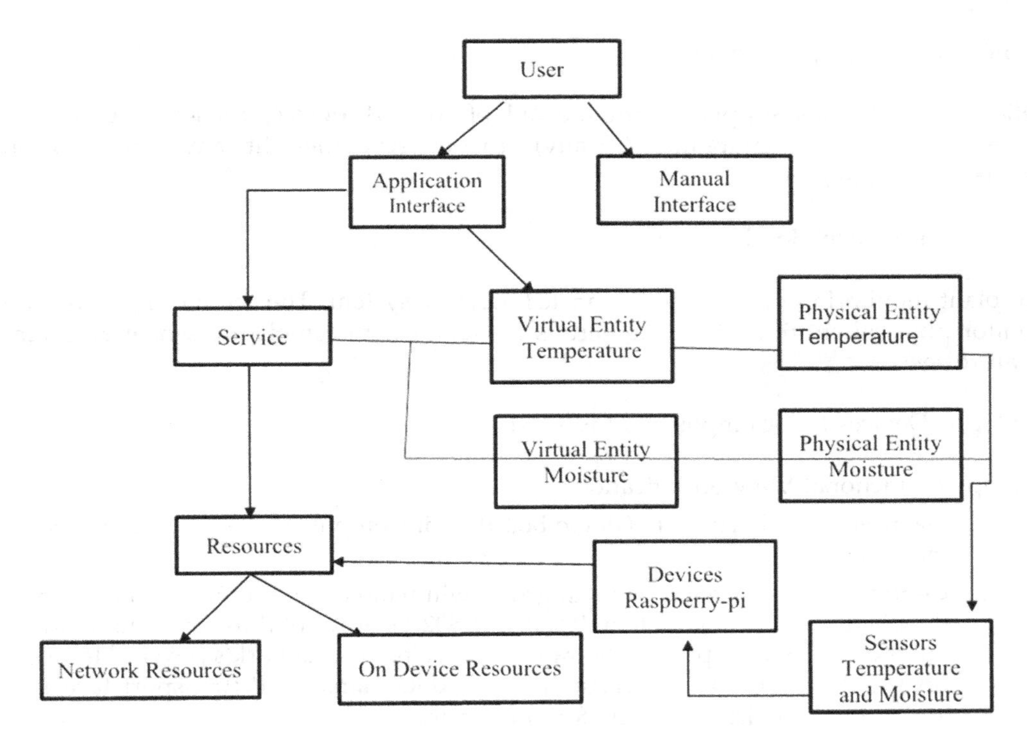

FIGURE 5.8
Domain Model for Plant Monitoring System.

Step 4: Information Model Specification

Figure 5.9 defines the shape of all of the facts within the IoT gadget and adds more information to the digital entities by defining their attributes and relations.

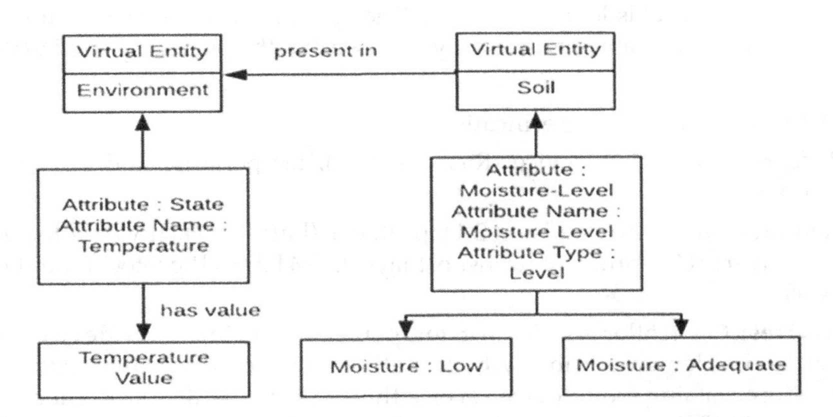

FIGURE 5.9
Information Model for Plant Monitoring System.

Step 5: Service Specification

Defines the offerings supplied with the aid of the IoT device, carrier type, service input-output, service endpoints (if any), carrier schedules (if any), and so on (Figure 5.10(a)(b)(c)).

Step 6: IoT-Level Specification

The plant monitoring system will be an IoT Level 1 system. The monitoring node can monitor physical entities. The node here is equipped with a moisture sensor and temperature sensor.

Step 7: Devices and Components Involved

Step 8: Functional View Specification

1. **Devices:** Raspberry Pi (single-board minicomputer), temperature and moisture sensors.
2. **Communication:** This functional group will handle communication between the plants monitoring system. It will use 802.11 protocol (link layer) to figure out the flow of data physically transferred using the network's physical layer, IPv4 (network layer) for transmitting IP datagrams, TCP (transport layer), and SMTP (application layer) for email purposes.
3. **Services:** Services include device monitoring, device control services, data publishing services, and services for data discovery. Here, we have a controller service that is a native type of service, FTP-type service that uploads the captured image of the intruder onto the server, and an SMTP-type service that sends mail of the captured image.
4. **Management:** Application management and device management are needs of the system.
5. **Security:** Only an authorized user is sent an email notification.
6. **Application:** The interface allows the user to monitor plant details to check if the moisture level is low or adequate. The system alerts in case moisture levels are low and sends an email to notify the user that the plant may need watering.

- Step 9: Operational View Specification

1. **Devices:** Computing device (Raspberry Pi), temperature and moisture detectors (sensor).
2. **Communication Protocols:** 802.11 protocol (link layer), IPv4 protocol (network layer), TCP protocol (transport layer), SMTP/FTP protocol (application layer)
3. **Services:** Controller service for temperature sensing – on device service that is implemented in Python; native service. Controller service for moisture sensing – on device service that is implemented in Python; native service. SMTP service – used to send an email to the user, if the moisture level is low. This service uses Simple Mail Transfer Protocol (SMTP) to send the mail.

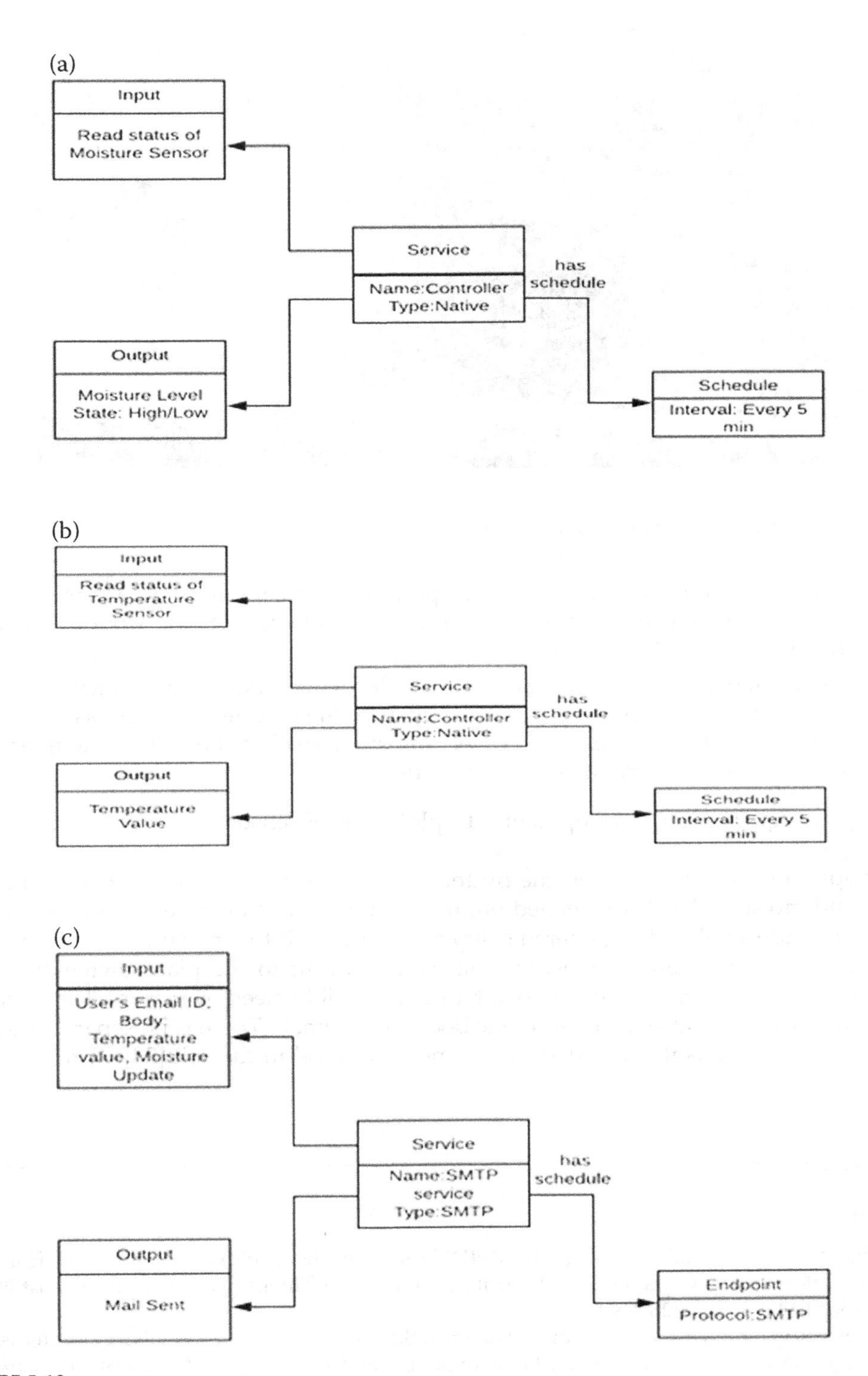

FIGURE 5.10
(a): Service Specification for Moisture Sensor (b): Service Specification for Temperature Sensing (c): Service Specification for Simple Mail Transfer.

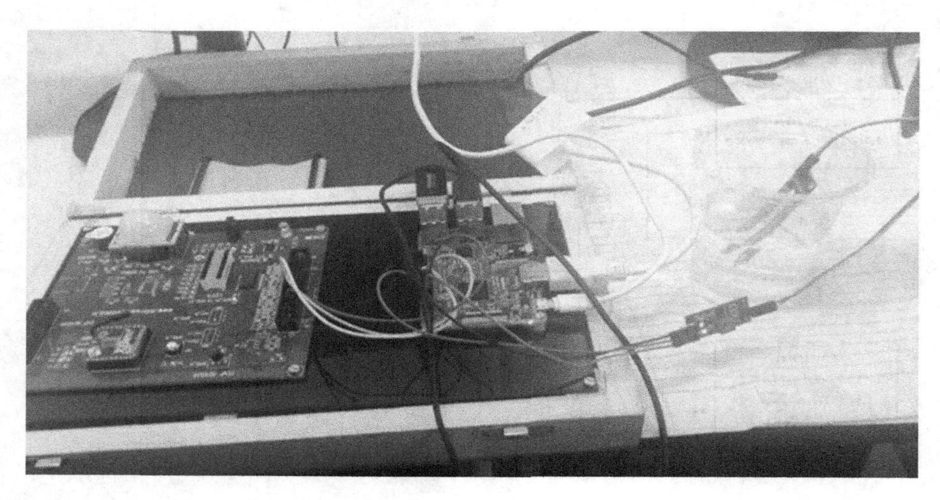

FIGURE 5.11
Hardware Connections for Plant Monitoring System.

4. **Security:** Authorization – Interface application, email sent only to authorized users. Authentication – Interface application authenticates the sender's email ID by accepting the password.
5. **Management:** Device management – Raspberry Pi device management. Application management – Raspberry Pi minicomputer is used as microprocessor. Moisture sensor to detect moisture level in the soil and temperature sensor to detect current temperature.

- Step 10: Application Development – Deployment of sensors

In the application interface, when the button for starting the gathering of temperature reading and moisture level is switched on, the temperature is noted every five minutes. The soil moisture level is also gathered every five minutes. If the moisture level is low for a set number of consecutive times, then an email is sent to the plant owner that the moisture level is low and that the plant might possibly need watering. The current temperature reading is also included in the body of the email. There is no separate button for sending an email as it is decided by the moisture level in the soil [20] (Figure 5.11).

References

1. Aggarwal, R. and Lal Das, M. (2012) RFID Security in the Context of "Internet of Things." First International Conference on Security of Internet of Things, Kerala, 17-19 August 2012: 51–56. 10.1145/2490428.2490435
2. Khekare, G. and Verma, P., Dhanre, U., Raut, S., and Yenurkar, G. "Analysis of Internet of Things Based on Characteristics, Functionalities, and Challenges." *IGI Global International Journal of Hyperconnectivity and the Internet of Things (IJHIoT)* 5(1): 44–62 accessed (January 22, 2021). doi: 10.4018/IJHIoT.2021010103

3. Gershenfeld, N., Krikorian, R., and Cohen, D. (2004) "The Internet of Things." *Scientific American* 291: 76–81. doi: 10.1038/scientificamerican1004-76
4. Feki, M.A., Kawsar, F., Boussard, M., and Trappeniers, L. (February 2013) "The Internet of Things: The Next Technological Revolution." *Computer* 46(2): 24–25 [Online]. http://ieeexplore.ieee.org/xpl/articleDetails.jsp?arnumber=6457383
5. Bagha, A. and Madisetti, V. (2014) *Internet of Things – A Hands-on-Approach*. Universities Press.
6. Srinivasa, K.G., Siddesh G.M., and Hanumantha Raju, R. *Internet of Things*. Cengage Publication.
7. https://www.zdnet.com/article/what-is-the-internet-of-things-everything-you-need-to-know-about-the-iot-right-now/
8. Vermesan, O. and Friess, P. (2013) *Internet of Things – Converging Technologies for Smart Environments and Integrated Ecosystems*. River Publishers.
9. Hanes, D. and Barton, R. *IoT Fundamentals- Networking Technologies, Protocols, and Use cases for the Internet of Things*. CISCO publication.
10. Gigli, M. and Koo, S. 2011"Internet of Things, Services and Applications Categorization." *Advances in Internet of Things* 1: 27–31. 10.4236/ait.2011.12004
11. Zanella, A., Bui, N., Castellani, A., Vangelista, L., and Zorzi, M. (Feb 2014) "Internet of Things for Smart Cities." *IEEE Internet of Things Journal* 1(1): 22–32.
12. IoT application areas. https://iot-analytics.com/top-10-iot-project-application-areas-q3-2016/.
13. Behr Technologies, Numberg. https://behrtech.com/blog/top-10-iot-sensor-types/
14. Uchihira Naoshi. "Innovation Design Method for the Internet of Things: Requirements and Perspective." 2019 Proceedings of PICMET '19: Technology Management in the World of Intelligent Systems, 978-1-890843-39-7.
15. Mexghani, E., Exposito, E., and Drira, K. (2017) "A Model-Driven Methodology for the design of Autonomic and Cognitive IoT-Based Systems: Application to Health Care." *IEEE Transactions on Emerging topics in Computational Intelligence* 1(3).
16. Wang, Y., Luo, P., Zeng, X., and Peng, D. (2019) "A New Design Method for Solar Energy Harvesting based on Neural Network." *IEEE International Symposium on Circuits and Systems*, 978-1-7281-0397-6/19, IEEE.
17. Subhasri, G. and Jeyalakshmi, C. (2018) "A study of IoT based Solar Panel Tracking System." *Advances in Computational Sciences and Technology*11 (7): 537–545.
18. Sawant, A., Bondre, D., Joshi, A., Deshmukh, A., and Tambavekar, P. (2019) "Design and Analysis of Automated Dual Axis Solar Tracker Based on Light Sensors", published in 2018 2nd International Conference on I-SMAC (IoT in Social, Mobile, Analytics and Cloud), INSPEC Accession No- 18493264, IEEE Publisher.
19. Ayaz, M., Sharif, Z., Mansour, A., Ammad-Uddin, M., and Aggoune, EL-Hadi M., (July 7, 2019) "Internet of Things(IoT) -Based Smart Agriculture: Toward Making the Fields Talk." (*Special section on new technologies for smart farming 4.0: Research challenges and Opportunities*). *IEEE Access.*
20. Jacob, M. Simple explanation of Internet of Things. https://www.forbes.com/sites/jacobmorgan/2014/05/13/simple-explanation-internet-things-that-anyone-can-understand/?sh=7465b4d51d09

6

Research Issues in IoT

Sayali A. Sapkal[1] and Sandhya Arora[2]
[1]*Research Scholar, SKN COE, Pune*
[2]*Professor, Cummins College of Engg for women, Pune*

6.1 Introduction

The Internet of Things (IoT) is a system made up of objects that are able to gather and transmit data over the Internet without human interference. IoT abilities are boundless for personal as well as business work. IoT devices have their own software and hardware in order to work. They integrate with application software, which supplies instructions to the device and analyzes the collected data. Various operating systems, hardware, software, and firmware are available for IoT systems so it is impossible to assess the collective group of hardware and software that makes up IoT.

The main IoT challenges are classified as follows:

- There are too many IoT platforms to choose from.
- The IoT has too many communication protocols.
- There are increasing threats to IoT network security.
- Diversity is a big challenge for IoT apps.
- Real-time data increases the load on the network.

IoT devices offer many business prospects but the IoT is difficult to maintain. IoT developers should know the newest techniques in order to overcome IoT challenges to provide the best results to customers. There are continuous developments to IoT tools and the network structure to make the system work better.

6.2 Internet of Things (IoT)

6.2.1 Introduction

The IoT is a group of billions of physical devices all over the world dependent on the Internet for gathering and processing data [1]. It is easy to convert anything into

DOI: 10.1201/9781003166702-6

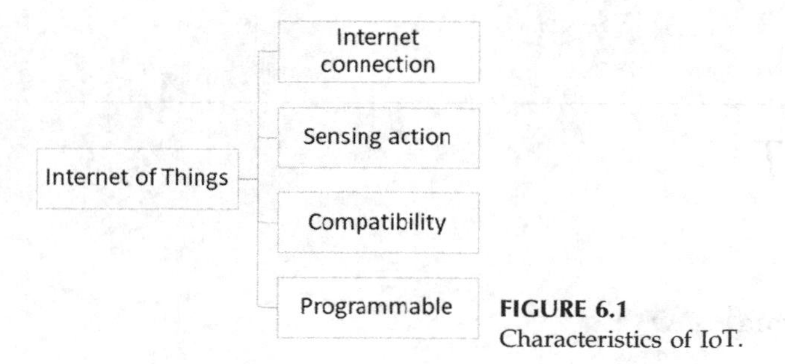

FIGURE 6.1
Characteristics of IoT.

an IoT device due to wireless networks and cheap computer chips; something as small as a pill to something as big as an airplane can become part of the IoT. Connecting these objects and adding sensors to them gives them a digital form to become IoT devices, which allows them to communicate with real-time data without a human.

Characteristics of the IoT are shown in Figure 6.1. The IoT is making the world around us more responsive and smarter while merging the physical and digital universes.

6.2.2 Examples of IoT

- PCs as well as smartphones are not IoT devices even though they have sensors.
- A fitness band or smartwatch can be an IoT device.
- The first IoT application was RFID tags applied to expensive pieces of equipment for tracking their location.

6.2.3 IoT Challenges

The IoT mainly evolved for business and manufacturing, but the current use of IoT is filling offices and homes with smart devices. There are many IoT applications on the market that are quite assorted in nature and are part of day-to-day life. Organizations, committees, or individuals can use these applications.

Some of applications are as follows:

- Smart living
- Smart environment
- Smart cities
- Healthcare
- Smart agriculture and water management
- Retail and logistics

As there are various domains that work in the IoT, a plan is needed to support the above-mentioned applications along with their functionality [2]. Because there are so many methods available, the IoT has yet to choose an effective method for all its applications. New research is needed to overcome the problems that are arising.

Some of challenges are as follows:

- **The IoT has too many platforms.**
  As there are various types of hardware and software available on the market, deciding which one to choose for a particular application is difficult. If a tester has proper knowledge about how to pair a device and operating system, he or she can run a small test to check if the device can pair with the operating system.
- **The IoT has too many communication protocols.**
  IoT devices and controllers are connected using communication protocols. For automated testing, testers have to design the automated tests dependent on which protocols will be used. To be effective, the testing tool should support the protocols and APIs.
- **There is an increase in IoT security threats.**
  Currently, 80% of IoT devices suffer from security issues, but finding the main reason for the issues is tedious. Testers should focus on the password policy of IoT devices so that fewer passwords are needed on devices.
- **IoT applications are too diverse.**
  IoT applications are so various that they demand very durable test capabilities. Testers should have a good plan for conducting tests so that they have the proper architecture and software as well as the correct version of the software.
- **Real-time data increases the load on the network.**

IoT devices show better results when they have fast communication, which is dependent on the status of the network. Overloaded WiFi networks increase problems in the structure of the network, such as unreliable hardware as well as poor Internet connections. IoT applications must be verified for different situations to ensure that they give required output efficiently by providing lossless data. IoT devices contain the reflexive property, so testers should know if all devices are working.

Research challenges for the IoT associated with technical issues include managing and designing a national-level technology structure for industry purposes. Other research categories are design, scientific, engineering, and operations for business, social, or political parameters. Challenges can fall into more than one category. For example, reliability is a design challenge, while robustness is an operational challenge.

Challenges in the IoT related to technical and business parameters may be solved with Operations Research (OR) and mathematical tools [3]. This chapter focuses on OR techniques. System thinking can be included in OR; it assumes that the IoT system has a complex nature but is self-organizing. Although the system has a large number of devices connected to it, it contains a framework along with required features. Research from 2000 until now has shown that the IoT should focus on socio-organizational issues rather than purely technological issues. Oxford discussed some issues faced by the IoT related to social and legal parameters [4]:

- Protecting data and privacy
- Fixing misleading global systems
- Addressing public opinions and actions
- Creating strongly united systems
- Ensuring Quality of Service
- Addressing risks to various forms

IoT devices that are used to process data, sense data, collect information, and store and manage data are heterogeneous in nature [5], which increases the challenges in different research areas.

A. **Privacy and Security** [6,7]

Due to increased use of the Internet, security and privacy are important to address in the IoT. Researchers know the weak points of IoT devices. As the IoT has the same structure as Wireless Sensor Networks (WSN), it possesses the same issues as WSNs. Solutions such as cryptography can be applied, but more research should be done to explore these techniques. Cryptographic services should also be able to operate on constrained IoT devices.

B. **Data Processing and Monitoring** [6,7]

Currently, most IoT systems use a centralized method to unload data and carry on complex computational tasks on the cloud. It is difficult to maintain the simultaneous constraints related to time as well as computational load because the cloud stores a massive amount of data. Most systems are dependent on the edge processing concept. Information Centric Networking (ICN) is used in the IoT for data management, but extending the ICN paradigm to the fixed network edge creates difficulties for using the ICN. Even though data analysis plays an important role in the IoT, it has challenges.

C. **Data Sensing and Data Monitoring** [6,7]

Monitoring and sensing of IoT devices uses a lot of energy. Sensors are constantly active as the IoT works on real-time data. New nanotechnology and miniaturization will help actuators as well as sensors to be developed on the nanoscale and be more energy efficient.

D. **IoT Machine-to-Machine (M2M) Communication and its Protocol** [6,7]

Message Queuing Telemetry Transport (MQTT) and Constrained Application Protocol (CoAP) are communication protocols mainly used for IoT systems. IEEE 802.15.4, Bluetooth, and Lora can also be used for wireless connections in the IoT, but they have a hard time covering the range of the network. In order to increase efficiency at low traffic, MAC protocols are being used, such as Frequency Division Multiple Access (FDMA), Time Division Multiple Access (TDMA), and Carrier Sense Multiple Access (CSMA).

E. **Blockchain of Thing** [6,7]

Blockchain is technology for Bitcoin cryptocurrency. The IoT works on the concept of the WSN, so it has security and privacy issues. Blockchain's implementation was actually developed on parameters such as trust, security, and immutability, so combing the IoT and Blockchain can create a more efficient, secured, and trustworthy system.

F. **Interoperability** [6,7]

Interoperability means the Internet connection between the devices must be similar in their encodings and protocols. Due to the large amount of data, it is difficult to maintain interoperability of IoT devices while handling system limitations. Nowadays, many industries have their own standards to support their applications. Hence, designing IoT systems in order to be interoperable is a crucial task.

6.3 Tools and Techniques for IoT Challenges

6.3.1 Data Analytics

Data analytics is the science of using data to predict conclusions. As billions of devices connect to the IoT, large amounts of data are collected on environmental, historical, identification, positional, and descriptive information. Created IoT data will be useful for issues related to managing and analysis of data. As discussed in Haghighi et al. [8], characteristics of the IoT, such as real-time data, assortment, imprecision, and unspoken semantics, result in various data management issues. Figure 6.2 shows the relationship between IoT challenges and OR tools.

6.3.2 Decision Analysis and Support Systems

Athreya et al. [9] created a framework that includes interviews and literature surveys for business purposes. The model is built on different channels, key activities, customer relations, key resources, value propositions, key partners, and cost structures. Value propositions are more important in the model while customer relationships and key partnerships are less significant.

Westerlund et al. [10] discussed challenges related to IoT business model development, including the types of objects, IoT immaturity, and the formless nature of the IoT system. They [10] developed an adaption of an analytic hierarchy process model for main IoT applications such as healthcare, logistics, and the market environment. They used decision support to assess probable IoT solutions.

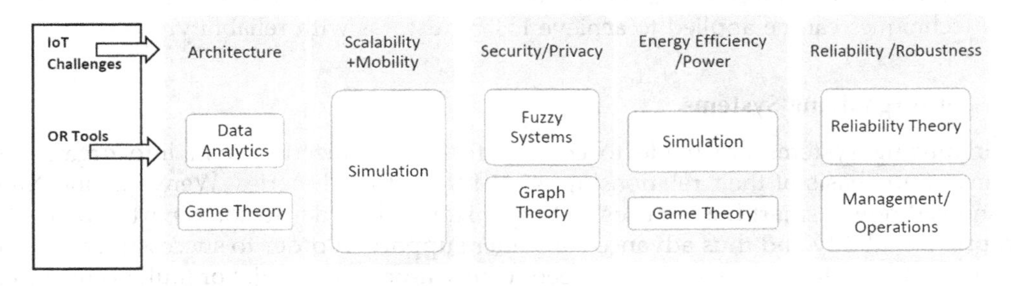

FIGURE 6.2
Relationship between IoT Challenges and OR Tools/Techniques.

6.3.3 Game Theory

Game theory is defined as the technique of thoughtful communication among groups in which each group has a different point of view [8]. Wang et al. [11] discussed a method to enhance the distribution of tasks and consumption of energy in IoT systems.

Dyk et al. [12] discussed a game theory approach for IoT that uses an OPNET simulation tool to help distribute data and conduct parallel tasks.

6.3.4 Simulation

Discrete-event simulation is a process that stimulates the working system in an isolated order of events while each event pattern tries to make a change to the state of the system. No modification is acceptable in between the system state. IoT network issues can be studied with the help of discrete simulation. Gubbi et al. [13] used discrete-event simulation to create a framework that connects smart devices to a diverse sensor network. The model works on weather and crisis situation phenomena while maintaining the state of the network.

6.4 Probable Solutions to Address IoT Research Challenges

6.4.1 Simulation for IoT Scalability

Stankovic [14] focuses on IoT scalability using simulation. Existing approaches may fail and be inadequate for the expected huge number and variety of IoT objects. IoT networks collect the same number of actuators and sensors that connect all nodes to each other. Discrete-event simulation is easily applicable for problems to overcome network traffic flow through IoT nodes.

6.4.2 Reliability Theory for IoT Robustness

Robustness and reliability are part of the word *scalability* only. Metcalfe's Law explains that large interconnectedness between self-defining devices as well as sub-systems increases the complexity of the IoT system, which results in system failure [15]. Kempf [16] discusses that in large networks, bit errors result in problems that cannot be solved. Reliability theory of OR techniques can be applied to achieve IoT robustness with reliability.

6.4.3 Self-organizing Systems

Self-organizing systems are made to convert the unorganized system into organized systems on the basis of their relationship with different machineries. Wenyang and Xue [17] showed how a smart home needs self-organization in order to sort smart appliances to improve usability and thus advance consumer support. In order to successfully run an IoT system as per design, the networks need to be self-organized [9]. For fault occurrence, the IoT device must know its working environment. Then, it can connect to neighboring devices to start communication paths that may overcome local faults and be able to return to normal operations.

6.4.4 Context Awareness

Context-aware computer systems will play a vital role in research for at least the next 20 years. Looking after personal health and fitness and regulating traffic in cities relies on smart processors. Systems Thinking using complex adaptive systems can be helpful in research as it is the study of living organisms. For designing context-aware information systems in the IoT, the Soft Systems Thinking approach can be used.

6.5 Case Study

"An SSM model for a home security system based on a context-aware application using context modeling" [4]

Figure 6.3 shows an abstract SSM model that has basic activities with a root definition. In context-aware applications, services depend on the type of context. Context modeling provides different options for selecting how to conduct the services. If the correct SSM model is prepared, the model will monitor and control the various services in order to achieve the efficiency of the system.

Figure 6.4 represents an SSM conceptual model for a home security system service based on the following root definition:

> "The system provides services such that, when in an active state, it should be able to collect sensor data such as entering and exiting of people, determining any movement within the property, sensing security threats, and taking appropriate actions."

The conceptual model shows only necessary sets of activities as stated in the root definition. The root definition only tells "what" is to be done but the conceptual model tells "how" it is to be done. In Figure 6.4, the conceptual model uses the context model for the following activities:

- Collecting sensor data
- Making decisions about security threats
- Deciding appropriate responses to a threat

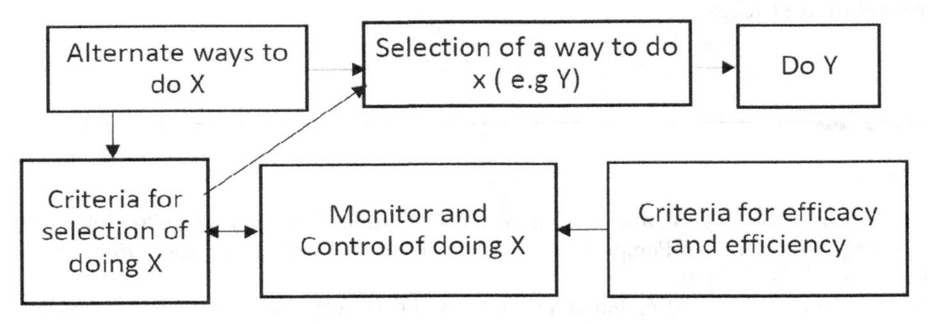

FIGURE 6.3
SSM Abstract Model.

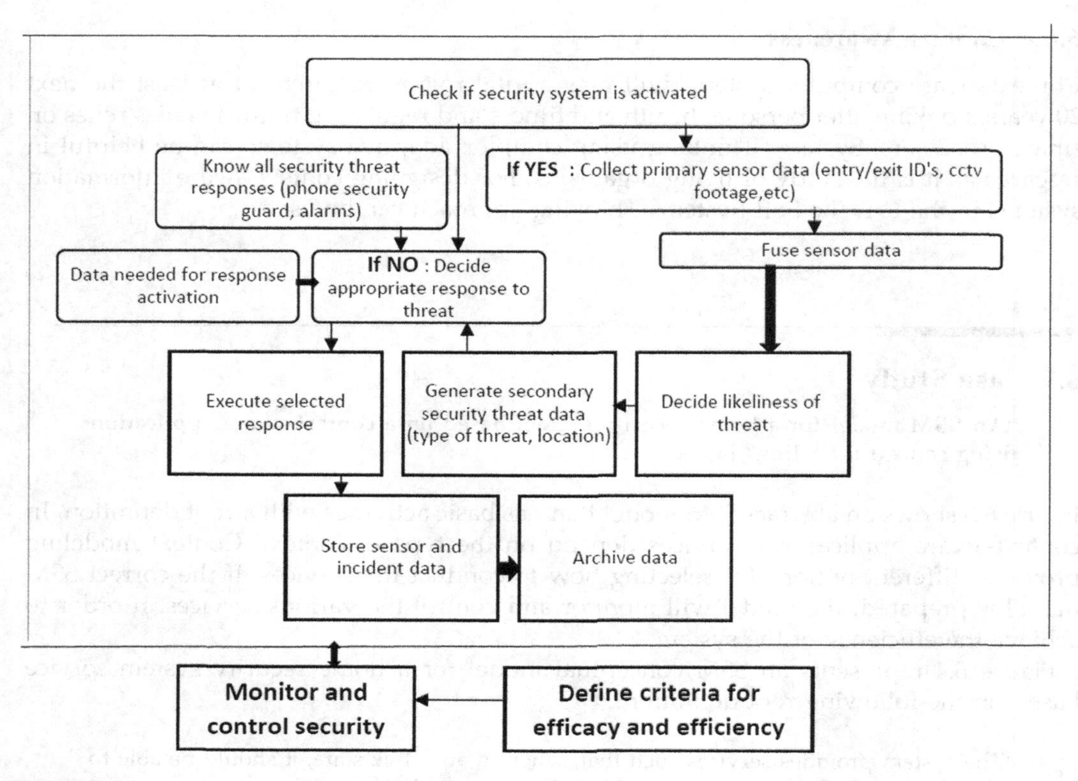

FIGURE 6.4
SSM Conceptual Model.

6.6 Conclusion

Major research challenges of the IoT use OR techniques for management of data in the system while OR data analytics is used for the analysis of "big data". OR researchers have been of great help with business models for investment purposes. Some case studies show that efficiency is being improved, but due to the IoT's complex nature and large real-world data, it is difficult to evaluate the performance of the IoT system. The daily IoT device count is increasing, which results in a larger scope for IoT researchers, who need to do more detailed studies.

References

1. Muralidharan, S., Roy, A., and Saxena, N. (2016, 20 June) "An Exhaustive Review on Internet of Things from Korea's Perspective." *Springer Wireless Personal Communication International Journal* 90(3): 1463–1486.
2. Kim, S. and Kim, S. (2016, January) "Multi-Criteria Approach toward Discovering Killer IoT Application in Korea." *ScienceDirect Technological Forecasting and Social Change Journal* 102: 143–155.

3. Mortenson, M.J., Doherty, N.F., and Robinson, S. (2015, 13 Jan) "Operational research from Taylorism to Terabytes: A Research Agenda for the Analytics Age." *European Journal of Operational Research* 241(3): 583–595.
4. Ryan, P.J. and Watson, R.B. (14 March 2017) "Research Challenges for the Internet of Things: What Role Can OR Play? *Multidisciplinary Digital Publishing Institute (MDPI) Journal* 5(24): 1–32. doi:10.3390/systems5010024
5. Chen, S., Xu, H., Liu, D., Hu, B., and Wang, H. (Aug 2014) "Vision of IoT: Applications, Challenges, and Opportunities with China Perspective." *IEEE Internet of Things Journal* 1(4): 349–359.
6. Hussein, A.R.H. (2019) "Internet of Things (IOT): Research Challenges and Future Applications." *International Journal of Advanced Computer Science and Applications (IJACSA)* 10(6): 77–82.
7. Rehman, H.U., Asif, M., and Ahmad, M. (Dec 2017) "Future Applications and Research Challenges of IOT." International Conference on Information and Communication Technologies (ICICT): 68–74.
8. Haghighi, M., Maraslis, K., Tryfonas, T., Oikonomou, G., Burrows, A., Woznowski, P., and Piechocki, R. (Dec 2015) "Game Theoretic Approach towards Optimal Multi-Tasking and Data-Distribution in IoT." *IEEE 2nd World Forum on Internet of Things (WF-IoT)*: 406–411.
9. Athreya, A.P. and Tague, P. (June 2013) Network self-organization in the internet of things. 10th Annual IEEE Communications Society Conference on Sensor, Mesh and Ad Hoc Communications and Networks (SECON): 24–27.
10. Westerlund, M., Leminen, S., and Rajahonka, M. (July 7, 2014) "Designing Business Models for the Internet of Things." *Technology Innovation Management Review (TIM Review)* 4: 5–14.
11. Wang, J., Liao, J., Li, T., and Wang, J. (April 2016) "Game-Theoretic Model of Asymmetrical Multipath Selection in Pervasive Computing Environment." *ScienceDirect Pervasive and Mobile Computing Journal*27: 37–57.
12. Dyk, M., Najgebauer, A., Pierzchał, D., and Sim, S. (14–16 December 2015) "An Agent-Based and Discrete Event Simulator for Wireless Sensor Networks and the Internet of Things." *IEEE 2nd World Forum on Internet of Things (WF-IoT)*, Milan, Italy: 345–350.
13. Gubbi, J., Buyya, R., Marusic, S., and Palaniswami, M. (September 2013) Internet of Things (IoT): A Vision, Architectural Elements, and Future Directions. *ScienceDirect Future Generation Computer Systems* 29(7): 1645–1660.
14. Stankovic, J.A. (Feb 2014) "Research Directions for the Internet of Things." *IEEE Internet of Things Journal* 1(1): 3–9.
15. Brown, B.R. (2016) "What the Internet of Thing Needs: Systems Engineering." *International Council on Systems Engineering News*.
16. Kempf, J., Arkko, J., Beheshti, N., and Yedavalli, K. (25–26 March 2011) "Thoughts on Reliability in the Internet of Things." *Interconnecting Smart Objects with the Internet Workshop*: 1–4.
17. Lui, W. and Li, X. (2013). "A Study of the Application of Self-Organizing Networks in Designing Appliances of Internet of Things." 3rd International Conference on Consumer Electronics Communications and Networks (CECNet) (20–22 November 2013): 45–48.

7

Intrusion Detection Systems in IoT: Techniques, Datasets, and Challenges

Priya R. Maidamwar[1,2], Dr. Mahip M. Bartere[3], and Dr. Prasad P. Lokulwar[4]

[1]*Ph.D. Scholar, Department of CSE, G H Raisoni University, Amravati, India*
[2]*Assistant Professor, Department of CSE, G H Raisoni College of Engineering, Nagpur, India*
[3]*Assistant Professor, Department of CSE, G H Raisoni University, Amravati, India*
[4]*Associate Professor, Department of CSE, G H Raisoni College of Engineering, Nagpur, India*

7.1 The IoT and Security

7.1.1 IoT Architecture

Variety and heterogeneity make Internet of Things (IoT) framework security more urgent. Security for IoT frameworks contrasts conventional frameworks for the following reasons:

- IoT frameworks are limited as far as computational ability, memory, battery life, and network capacity. Therefore, it is difficult to install existing conventional security arrangements that have constraints in terms of resources.
- IoT frameworks are vigorously disseminated and heterogeneous. In this way, traditional arrangements that are centralized in nature may not be appropriate. Besides, additional troubles may be due to the distributed nature of the IoT and limitations in IoT security.
- IoT frameworks are installed in an actual climate that is not predictable. For this reason, physical assaults have been included in the list of customary security dangers.
- IoT frameworks are associated with the Internet since every gadget approaches through its IP address. Therefore, one more danger area is the Internet.
- IoT frameworks are made out of countless constrained items that produce gigantic amounts of information. So, it is simple to attack and flood these little gadgets on one hand, and on the other hand, these networks have limited bandwidth.

DOI: 10.1201/9781003166702-7

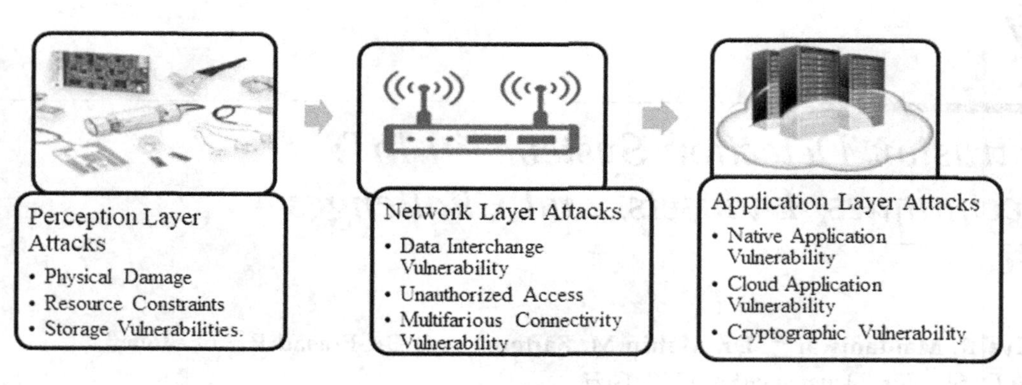

FIGURE 7.1
IoT Architecture.

- Systems in IoT architecture include a different variety of protocols and technologies. Therefore, the solutions proposed for IoT security should take into account the huge range of these protocols and technologies.

The real working environment is related to the virtual one in IoT frameworks. An architecture of an IoT framework is depicted in Figure 7.1. It comprises three layers [1]: perception/physical layer, network/transport layer, and application layer.

To start with, the perception layer is well known as the equipment layer. It includes various sensor nodes that are able to send as well as receive information utilizing distinctive communication standards, for example, 6LowPAN, Bluetooth, and RFID. The second one is the network layer, which confirms the effective routing/transmission of the perception layer that can guarantee accurate delivery of information/data. Communication protocols such as 3G, WiFi, GSM, IPv6, etc., are utilized here. The last one is the application layer, also called the product layer, that furnishes frameworks with the business rationale and presents graphical user interfaces (GUIs) to the customers.

Next, this chapter examines IoT framework threat classification. After that, conventional defense components utilized against various threats are presented.

7.1.2 Classification of IoT Threats

Since various challenges are faced by IoT frameworks, IoT attacks can be divided into two categories. The first category is related to the framework layers, whereas the second category is related to design challenges.

1. **IoT Threat Classification by Layers**

Each layer may represent multiple weaknesses, as illustrated in Figure 7.1. Since gadgets are kept at various physical locations, they might be subject to natural disasters like irregular wind/snow/rain or unexpected harm or malicious attacks. Additionally, through actual access, information stored might be stolen. Sensors are minute things. Therefore, they experience ill effects from resource requirement constraints like computational assets, memory, energy, etc. The moment commands are floated in the network layer, they can go through several network issues such as information exchange weaknesses (movement of information can be closed down due to unauthorized access, network flooding, and connectivity weaknesses) [2]. Additionally, the application layer is mainly open to programming issues, for example, insecure account credentials, account enumeration lists, and suspended accounts due to several failed attempts to guess a password. The applications that are cloud based are at high risk of getting infected by Trojan horses, worms, and so forth. Because the IoT depends on low computational capacity gadgets, sometimes transport encryption is neglected or utilized in a feeble, weak form. Accordingly, communications are easily traceable and effortlessly found [3].

2. IoT Threat Classification by Challenges

In order to summarize security threats in the IoT, we present some specialized terms related to IoT attacks. Later, we categorize threats on the basis of IoT challenges.

a. Specialized Terms of IoT Attacks

To start with, **spoofing assault** [4,5] steals authentication details to achieve unauthorized access to services. Credentials can be taken straight from a gadget, through snooping the path of communication. This threat can be classified as: IP address spoofing, ARP spoofing, and DNS server spoofing. In IP address spoofing, the contents in the IP header of the source node are misrepresented in order to cover the sender's identity to launch distributed denial of the service attack. ARP spoofing attacks refer to address resolution protocol. Here, IP addresses are resolved to MAC (media access control) addresses. Whenever a spoofed ARP message is sent by an intruder over the local area network (LAN), its MAC address will be attached to the IP address of an authentic individual in the network. Subsequently, malicious parties can steal information, alter information on the way, and deny traffic on the LAN. DNS server spoofing manipulates the data on a DNS server and maps the particular domain name to the infected server with an unapproved IP address.

The second type of attack is **routing attacks** [6], which target routing protocols wherein the information flowing over the network is modified or spoofed to create false network behavior to attract traffic. A sinkhole attack [7] is a kind of attack that attracts a large amount of traffic through a malignant node by introducing an imaginary route as an ideal route. A selective forwarding [8] attack is another harmful attack where a malicious node selectively drops the packets and thereby degrades the network performance. A black hole attack [9] disturbs an ordinary information stream inside an organization. At first, the attacker misleads at least one or several faulty nodes as an optimal path, and at this point, the nodes begin to leak information packets directed via the defective path. Next, is the wormhole attack [9] that requires a minimum of two damaged nodes connected to each other, either in wired or ad-hoc mode. These faulty nodes tunnel the packets quicker than an ordinary path. Next, a replay attack [5] occurs when an intruder interferes on a network and then re-transmits or postpones the legitimate information to obtain unauthorized access over an already ongoing session [52].

The third type of attack is a **data tampering attack** [5], which can be further divided into device tampering and data tampering. Device tampering can be effortlessly employed, particularly when an IoT gadget leaves the majority of its energy unattended. It very well may be taken in secret and thus utilized malignantly. The gadget can be taken as equipment or similarly as programming. The data tampering includes malicious adjustment of information, for instance, information stored in datasets or information traveling between two gadgets.

Fourth, a **repudiation attack** [8] is related to gadgets doing a noxious activity and afterward denying performing it. This scenario results when a gadget sends an infection throughout the network without leaving a trace to distinguish it.

Fifth, **data revelation** [5] is related to unapproved data access. An intruder accomplishes this by sneaking around gadgets, by interfering in the network path, or by getting actual admittance to a gadget; for example, probe is where attackers attempt to accumulate data about a target hub and its weaknesses by checking associations (e.g., port filtering). With data revelation comes delicate data leakage, for example, side channel assault.

The sixth type of attack is **DDoS** [10], the distributed denial of service assault, launched together by various compromised hubs located at different areas. Also, DDoS assault involves a poisonous aggressor that aims to consume by overflowing the framework with hostile traffic. In order to lead an effective DDoS assault, botnets can be used. These networks include controlled or infected web-associated gadgets. As referenced in Bou-Harb et al. [11], DDoS assaults are the most incessant assaults, particularly in IoT networks related to social activities, for example, smart cities. DDoS assaults are classified into the following types [12,13]:

1. Flooding assaults depend on overloading the network with countless packets, like UDP, ICMP, and many more, which results in bargain of network transmission capacity. These attacks can be launched easily utilizing botnets.
2. Amplification attacks occur upon spoofing of IP sources. The packets are sent by the attacker toward the reflector servers, where a source IP address is subsequently allotted to the victim's IP in a roundabout way, overpowering the casualty with the response packets. To continue, programmers misuse weaknesses in various protocols to transform little questions into an immense number of queries to crash the victim's server. DNS and SSDP are examples of amplification [13] distributed attacks.
3. Attacks related to malicious exploit of various protocols are known as protocol exploit attacks. Examples of these attacks are TCP reset, SYN flood, and water torture attack [14].
4. Attacks based on deformed network packets are known as malformed network attacks. For example, utilizing similar IP addresses both for source and destination [13].
5. Software attacks are related to application protocols. Examples of these attacks are ping of death and teardrop attack. In ping of death, the attacker sends a segmented request packet, having a size greater than an IP packet, so that the intruder cannot rearrange it. In teardrop assault, the intruder forwards two fragments so that, with the help of the offset value of the packet, it cannot be reassembled [53].

The seventh type of attack is **elevation of privilege** [5], which is concerned with obtaining privileges to use a service without having any legal right. Well-known examples of elevation of privilege assault are user-to-root (U2R) and remote-to-local (R2L) [15]. U2R is related to gaining privileges of the root node when an attacker has access to a normal user account. R2L happens when an assailant does not have a record on the victim hub and therefore abuses weaknesses to obtain local access as a client by guessing a password.

The eighth type of attack is **man-in-the-middle attack** [5,6], where the attacker interferes in communication between two entities to listen in on a discussion. Examples of this attack include harming ARP Cache, session hijacking, DNS spoofing, port stealing, ICMP redirect, and so on.

Ninth, **client security** [13] resembles data exposure. Also, a programmer does not have to approach unapproved data to get the information of a client. It can be done by traffic investigation.

Last is **cloning nodes** [14,16], which concerns reintroducing a part in a framework after catching the characteristics and credentials of the original one. This kind of assault empowers the malicious client to control the framework, embed false data, disable capacities, and so forth. When an object is heavily influenced by the intruder without the information on its proprietor (botnet), the whole network gets infected.

b. **Classification of IoT Threats by Design Challenges**

Since there are various limitations identified with IoT framework designs, engineers along with industry persons should pay attention to the possible dangers. Several researchers have studied IoT security constraints and research openings; for example, Zhang et al. [17] presented detailed IoT security difficulties, for example, object recognizable proof, authentication and IoT protection, and so on, in IoT organizations.

Based on the design challenges, classification of IoT architecture is introduced in Figure 7.2 and explained in detail below.

- Heterogeneity

At the backend of the IoT framework, various devices, such as gateways, sensors, and actuators, are utilized that were developed by different sellers with various versions. In order to achieve interoperability, the use of a dispositive between heterogeneous gadgets is required. This segment can be overloaded with counterfeit demands, which prompts DDoS assaults. Generally, in a heterogeneous climate, MITM, routing, and spoofing attacks are bound to happen in contrast with homogeneous frameworks. It is simpler for a malevolent hub to imitate a real thing, pick up unapproved admittance to information, and transfer correspondence between two hubs via message infusion. As should be obvious from Figure 7.2, the IoT is a world of numerous standards and protocols [17]. This makes IoT security arrangement increasingly mind boggling. A decent study about these advances is presented in Al-Fuqaha et al. [18].

- Connectivity

Various segments of the framework need to be available to each other either physically or through the administrator. Generally, information from peripheral gadgets is associated with an IP network that might be the reason for directing assaults, as in MITM assaults. Gadgets should be informed of changes in the accessibility of the administrator so they do not flood the framework unconsciously with dreary and non-accessible solicitations. Such a flood can prompt a DDoS assault. In addition, the Quality of Service (QoS) in IoT organizations can be vital, especailly in crisis circumstances. Hence, powerful routing of

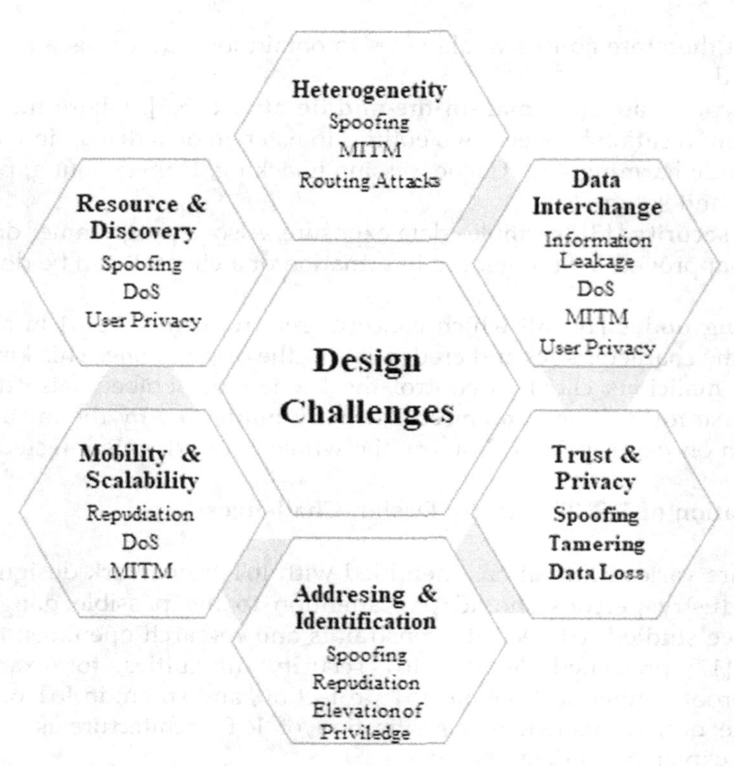

FIGURE 7.2
Classification of IoT Threats by Design Challenges.

packets and a decent QoS in information conveyance should be guaranteed, even in profoundly unique geographies [19].

- Scalability and Mobility

Gadgets within IoT frameworks are constantly moving throughout the field region; thus, they can change their associations with other devices. This causes frequent interruption due to discontinuity as well as associations with unapproved administrators. Assaults like renouncement, DDoS, sinkhole, and wormhole MITM have a higher risk of occurrence. To alleviate such dangers, security arrangements should think about cell phones as well as network segments, e.g., router and switches [20].

- Device Identification and Addressing

In IoT applications, field devices normally utilize low-range radios for smaller distance associations (under 1 km). In such scenarios, facilitator hubs assign local addresses that do not observe a typical norm, to peer gadgets. Thus, these addresses can take cover behind the FAN gateway, making it difficult to trace malicious practices. Subsequently, separation of the fake hub and identification of spoofing attacks is troublesome. Moreover, the hub can try to obtain unauthorized advantages without any verification from the external organization (rise of advantage).

- Spatio-temporal Services

Activities in the IoT framework are represented as an amplitude of the spatio-fleeting drive. Subsequently, information from IoT gadgets of the same framework ought to have sensible worldly conduct and spatial geolocation. In any case, these spatio-worldly labels should be shielded from malicious clients to prohibit replay assaults. Likewise, the client's area information should not be revealed to unapproved clients.

- Resource Constraints

The majority of IoT components are small in size; hence, they are resource constrained as far as registering power, installed memory, network transmission capacity, and energy accessibility. Altering, data loss, and hub cloning are potential assaults since the smart gadgets as well as sensors are constrained in terms of resources. This limits the installation of heavyweight cryptographic algorithms, and lightweight arrangements are primarily used. Alabady and Al-Turjman [21] tried to remove these limitations by proposing a novel mistake revision and discovery procedure named *low complexity parity check* (LCPC) to enhance the quality of cutting-edge IoT organizations.

- Interchange of Information

Any exchanged information should be encoded at the source nodes of the IoT. The method of encryption relies on the kind of equipment, its capacity limit, and its computational capability. Improper determination prompts security weaknesses, for example, data spillage (i.e., keys are being divided among various gadgets when scrambled parcels are decoded and repacked at different focuses in the correspondence chain). Moreover, hubs that encode information can be assaulted by means of denial of service or assaults for exhausting resources. Thus, end-to-end security through encryption is necessary.

- Discovery of Services and Resources

In IoT frameworks, resource mechanisms and service discovery must be employed to empower autonomy as well as self-revelation of gadgets. IoT gadgets must be secured by implementing two-route verification to prevent spoofing activities and avoid viruses that send fake requests in the network to promote denial of service attacks.

- Trust and Privacy

Sensor devices of the IoT oversee private client information (e.g., client habits, patients' information, common protection information); subsequently, privacy and information protection [22,23] is critical. Truth be told, trust and protection are crucial tasks to be fulfilled for IoT networks. Clients and IoT gadgets should be validated by means of solid administration to prevent spoofing, altering, and data spillage assaults. Trust and protection stand out enough to be noticed with advanced mobile phones, for example, Android OS.

7.1.3 Traditional Defense Mechanisms

In the wake of specifying and characterizing IoT attacks, we talk about attack detection and protection strategies that secure available IoT networks. In the literature for regular

IT security arrangements, researchers have discussed servers, organizations, and distributed storage. These arrangements can be employed for security of IoT frameworks. Depending upon the treated dangers, defense mechanisms can either be isolated or combined [24]. In the following section, conventional methods used to secure IoT are portrayed.

Initially, **packets are filtered** [24] through firewalls and proxy servers, which represents significant protection against IP spoofing assaults. Filtering can be two types: ingress filtering and egress filtering. In ingress filtering, packets coming from outside the organization with internal source addresses are restricted to secure the server from external spoofing assaults. Egress filtering is used to monitor outgoing traffic and packets so that they are allowed to leave the network only if they satisfy the rules set by an administrator.

The second defense mechanism is the **encryption mechanism,** which can be adopted through cryptographic protocols, data encryption methods, or virtual private organizations (VPNs). By using cryptographic protocols like HTTP Secure (HTTPS), Transport Layer Security (TLS), Secure Shell (SSH), etc., information or data can be encrypted prior to being sent and verified. The protection depends on computerized marks/authentications to guarantee that information was sent by a genuine device and never altered. It also ensures that information/code/updates are scrambled and cannot be perused or utilized by an unapproved person.

To secure IoT devices against altering, IP spoofing, renouncement, MITM, client security, attacks that compromise the network, and hub cloning, cryptographic network protocols can be utilized. Encryption of data storage prevents data revelation, and it also maintains client security. VPNs are a kind of protected channel of communication between source and destination. Communication is encrypted by making a virtual private connection above the current unstable network. Hence, encryption is a decent solution to maintain secrecy as well as security. In any case, IoT networks are helpless since resources are limited.

Third, networks can utilize vigorous **password verification schemes**. By allocating resources with proper rights, access is provided to only limited information. The better solution is to utilize one-time password (OTP). Data exposure, tampering, spoofing, and MITM can be forbidden by the above-mentioned mechanisms. For IoT networks, verification techniques should be lightweight. For example, Al-Turjman et al [25] proposed a lightweight framework to improve security within IoT frameworks. The authors mentioned a cloud-based mobile node confirmation, secure validation based on an elliptic-curve and key understanding (S-SAKA). Al-turjman and Alturjman [26] proposed a "hash"-based authentication mechanism for developing 5G innovation. They presented a modern system for the Industrial Internet of Things (IIoT) known as context-sensitive seamless identity provisioning system.

The fourth type of defense activity is tracing audit and logging **events** on database servers, web servers, and application servers. All these activities enable detection of anomalies. More explicitly, log key occasions, for example, exchange, login/logout, admittance to record framework, or failure to access resources, can identify malicious behavior. A good way to secure these documents is to take their backup, analyze it consistently for identification of malicious movement, and move the log records from their specified areas. Also, the log records must be secured by using a limited access control list and encrypting the exchange log. Thus, these strategies will keep IoT frameworks from repudiation and privilege rise assaults.

The fifth defense mechanism is **intrusion detection** using intrusion detection systems (IDSs). An IDS is used to monitor network activities to identify malicious events and raise alerts upon detection [27]. IDSs are, for the most part, classified by deployment and detection methodology. An IDS system can be deployed as a host-based or network-based IDS [28]. Host-based IDSs are deployed on a host machine such as a device or object. In this deployment strategy, system application documents and the host's related operation are monitored and analyzed. Host-based IDS is generally favored to prevent insider attack detection and avoidance. Network-based IDS (NIDS) is used to catch and examine packet stream inside the network. At the end of the day, this type of IDS filter sniffs parcels. NIDSs are solid against outside interruption assaults. Since our focus is toward security of IoT frameworks constrained in terms of resources, the following section focuses on solutions for NIDSs.

The sixth defense mechanism is to **prevent intrusions with the help of IPS (intrusion prevention system)**. This IDS reacts to an expected danger by trying to prevent it from happening. It reacts promptly and prevents malicious traffic from passing prior to the occurrance of events such as dropping and resetting of meetings, blocking of packets, or proxying traffic. In any case, an IDS reacts subsequent to identifying passed assaults. There are several kinds of IPS, primarily application firewalls, layer seven switches, in-line detection, and crossover switches [26].

The above introduced components can be utilized to secure IoT frameworks. Techniques like encryption and authentication are lacking in their ability to secure the IoT. Hence, IDSs play a vital and more reasonable role for this kind of framework. When other tools are broken, these systems are said to be the last line of defense. One more plus point of IDSs is that they are adaptable to requirements. Moreover, they can be partnered with learning logic, for example, artificial intelligence as well as machine learning methods along with other cutting-edge innovations. This topic will be elaborated in the following section.

7.2 Intrusion Detection Systems Based on Learning Techniques

In this section, we examine the situation after an intrusion has occurred. An IDS distinguishes the malicious behavior and limits the loss of data by immediately recognizing the attack.

7.2.1 Design Choices of Machine Learning Based IDSs

The major difference in the choices for designing IDSs relies upon the mentioned elements as shown in Figure 7.3:

- **Intrusion Detection Techniques:** The IDS can be categorized as rule-based or signature-based, anomaly-based, and hybrid-based detection.
- **Network Architecture:** Architecture is categorized as centralized and distributed architecture.
- **Data Source:** Data are generated from host-based, network-based, or hybrid-based input data.

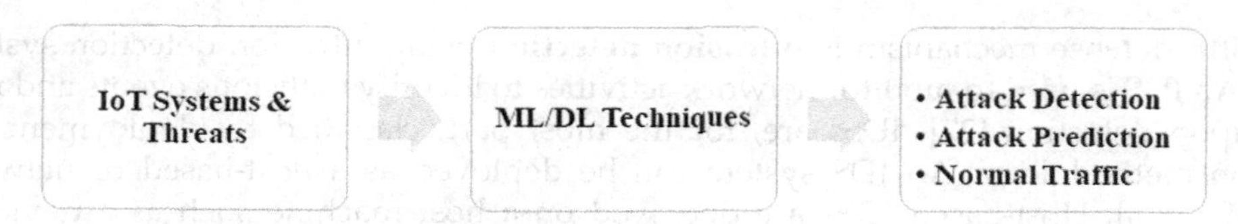

FIGURE 7.3
Role of Machine Learning Based IDS for IoT System.

- **Time of Location:** Time is either online or offline for detection.
- **Environment:** The IDS operates in wired, remote, or ad-hoc networks.

1. **Intrusion Detection Strategies of IDSs**

The intrusion detection strategies utilized for IDSs are divided into four categories [29], as depicted in Figure 7.4 and elaborated in detail as follows.

a. **Signature-based Detection Techniques**

Rule-based or signature-based identification procedures contain a database of attack signatures and analyze the ongoing network traffic or framework activities to compare them with the signature repository. When any match is discovered, an identification alert is raised. In spite of the fact that this kind of IDS is efficient against familiar attacks for which a signature is available in the database, it does not recognize zero day (new) or unknown attacks. Cyber-security arrangements incline toward this strategy of intrusion detection since it is easy to deploy and powerful for distinguishing known assaults. It has high detection accuracy along with a low false alarm rate.

b. **Anomaly-based Detection Techniques**

Anomaly-based detection matches the normal behavior of network traffic with current input data. At first, the ordinary network as well as the framework behavior are modeled. For any event, if there is a deviation from normal network behavior, then the IDS marks it as an attack. As mentioned by researchers in literature, an anomaly can be detected through statistical data analysis, data mining methods, as well as algorithmic learning approaches. This strategy is highly effective in detecting unseen or newer attacks, but it generates high false positive rates because earlier unseen practices might be sorted as irregular. Another favorable position appears when normal network behavior is customized for each application, every system, as well as every network, which makes situations hard for the intruder. It becomes tedious to identify which events are undetected [30].

c. **Specification-based Detection Techniques**

The specification-based detection technique has the same fundamental guideline as the anomaly-based technique, where the normal behavior of a framework is captured and matched against current framework activities in order to distinguish deviations that are out of range. In this technique, normal behavior of the network is learned through the

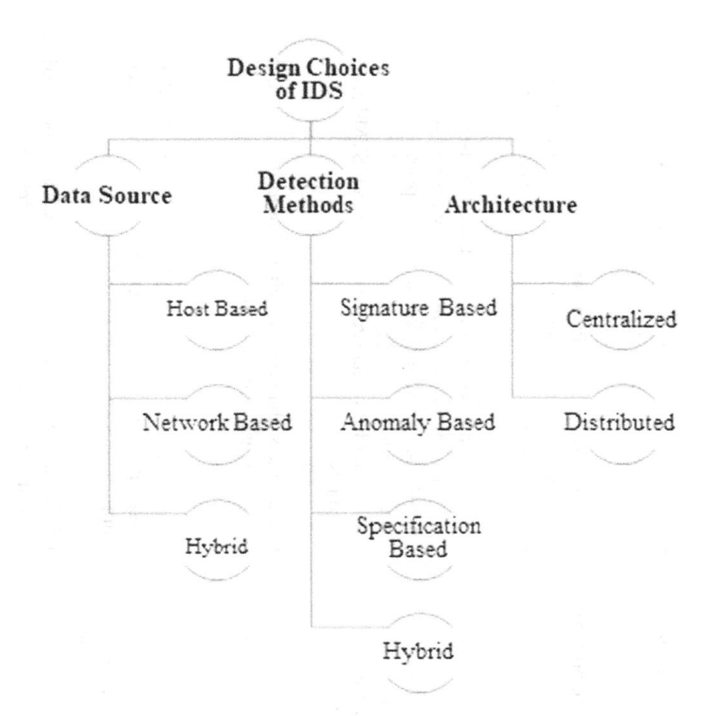

FIGURE 7.4
Flowchart for Design Choices of IDS in IoT System.

machine learning model. However, it should be physically indicated via rules of a database, and a related range of deviations should be defined by a human expert for specification-based techniques. This takes into consideration bringing down the false positive rates when contrasted with the anomaly-based detection techniques. Even though this method has merit for not needing a machine learning stage after determining a standard set of rules [31], it suffers from the ill effects of absence of versatility to different conditions, and there are high chances of specification errors.

d. **Hybrid-based Detection Techniques**

The hybrid-based detection techniques are a combination of all the previously referenced strategies to balance the weaknesses and upgrade the benefits of identifying known as well as unknown attacks. This technique thus reduces false positive events, thereby improving the accuracy. The majority of available anomaly-based detection frameworks are truly hybrid ones. These methods usually start with an anomaly detection. At that point, the technique attempts to relate it with the compare signature [32].

7.2.2 Machine Learning Techniques for IDSs

As mentioned in an earlier section of this chapter, except the specification-based detection method, the remaining types of detection strategies depend upon a wide range of machine learning algorithms for training the IDS data. A brief overview of various machine learning techniques utilized in IoT-based IDSs is introduced [57]. Table 7.1 presents a

TABLE 7.1

Summary of Machine Learning Methods for IoT Systems Security

Machine Learning Algorithm	Attack Category	Merits	Demerits
Naïve Bayes Classifier	R2L, DDoS, Probe, HTTP attacks (shell attacks, buffer overflow)	• Very few feature samples are needed for training. • Both binary and multi-label classification are possible.	• For classification problems, it fails to consider interdependencies among features; hence, its accuracy is affected.
K-nearest-neighbor classifiers	Flooding attacks, U2R, R2L	• It is efficient to classify new or unknown features. • It is simple to use.	• It is tricky to identify optimal value of k. • Identification of missing nodes is a challenging issue.
Decision Tree Classifier	U2R, R2L, DDoS	• It is an easy and simple method to implement.	• Larger storage capacity is needed. • It is complex in terms of algorithmic computations. • Outcomes are accurate and easy to predict if there are fewer decision trees.
Support Vector Machine (SVM)	Smurf attack, TCP and UDP flooding attack, port scanning, port sweeping	• It is suitable for datasets that contain a large number of features. • Due to its simplicity, SVM classifier is easily scalable and has the ability to perform intrusion detection in real time. • It requires less memory for storage. • It is vigorous to overfitting.	• In SVM, the kernel that is utilized to isolate the information when it is not directly distinct remains a test issue to accomplish required speed of classification. • It is hard to analyze and interpret SVM-based models.
Ensemble Learning Classifier	R2L, U2R, DDoS, probe attacks	• It is strong to over fitting. • It has improved performance compared to single classifier. • It decreases variance.	• Since more than one classifier is used in parallel, it has increased time complexity.
Random Forest Classifier	R2L, U2R, DDoS, probe attacks	• It generates a highly accurate outcome, which is resistant to overfitting. • It utilizes fewer input data; hence, the process of feature selection is not needed.	• Since random forest method builds a few decision trees, its utilization might be unfeasible for real-time applications requiring enormous datasets.
K-means	R2L, U2R, DDoS, probe attacks	• Labeled data are not required.	• It is not much proficient compared to supervised learning techniques, especially for detection of known attacks.
Principle Component Analysis (PCA)	Used generally for feature selection/extraction in combination with other machine learning methods	• PCA technique is useful to transform large sets of features to minimum sets of features without any information loss. Hence, PCA is suitable for large datasets. • Implementation of PCA reduces the complexity in the dataset.	• It is used to extract important features of datasets. To design a security model it must be used with some other machine learning methods.

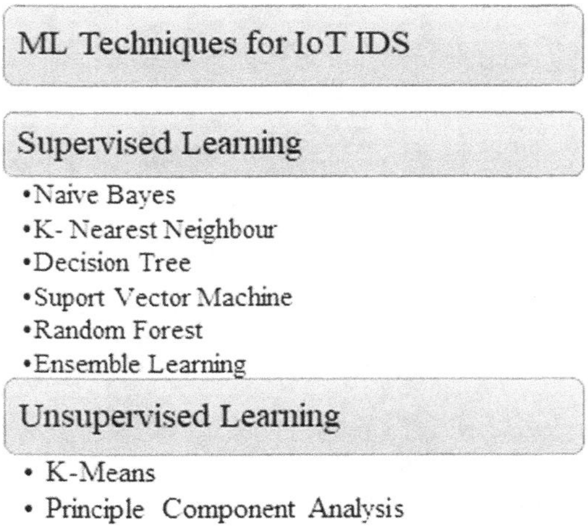

ML Techniques for IoT IDS

Supervised Learning

- Naive Bayes
- K- Nearest Neighbour
- Decision Tree
- Suport Vector Machine
- Random Forest
- Ensemble Learning

Unsupervised Learning

- K-Means
- Principle Component Analysis

FIGURE 7.5
A Hierarchy of Machine Learning Techniques for IoT-based IDSs.

concise outline of machine learning techniques, their pros and cons, and details of references related to literature work (Figure 7.5).

1. **Naive Bayes (NB) Classifier**

In this classification algorithm, which is dependent on past perceptions of network traffic, Bayes' theorem is used to predict the likelihood of an event. During machine learning situations, this classifier can be utilized for categorization of genuine and fake behaviors dependent on past perceptions during supervised learning processes. It is simple and the most widely utilized supervised classifier. This classifier measures posterior probability, and depending on that, a decision is made to split unlabeled traffic as normal or anomalous. Various features from captured traffic, such as protocol, status flag, and delay, are utilized to determine whether the traffic is normal or something else [30,33]. Due to the simplicity of this, different IDSs have utilized this classifier to recognize peculiar traffic. For the training phase, it does not require many examples and can form groups in both binary as well as multi-label classification [56].

2. **K-Nearest Neighbor (KNN)**

K-nearest neighbor (KNN) is another supervised machine learning algorithm that uses labeled input data to train a model and produces a suitable output. This classifier does not need any parameters for its operation. The distance between neighbors is measured through Euclidean distance. The basic concept behind the KNN classification algorithm is shown in Figure 7.6, which classifies an unknown data sample into already noticed classes on the basis of its overall distance to the neighboring classes. The normal behavior class is depicted through black circles, and the abnormal behavior class is depicted through the light gray diamond shape. When any unknown example is recently observed (depicted as dark grey hexagon), then it can be classified on the basis of a greater distance neighbor, which is nearer to any of the classes. Thus, this recent occurrence is classified as a normal class. Here k is defined as the number of closest neighbors used for classification [31,32].

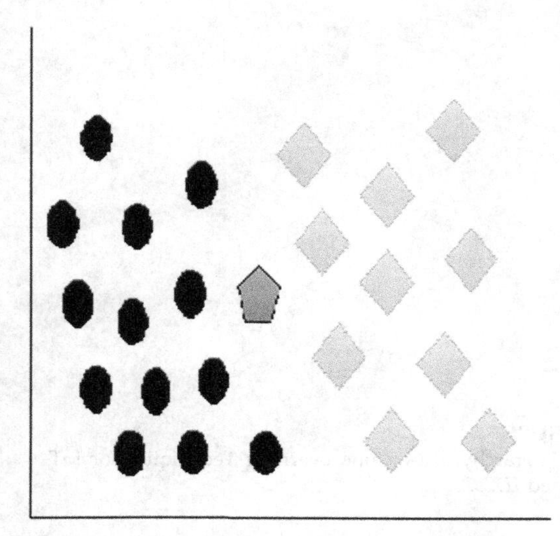

FIGURE 7.6
K-Nearest Neighbor (KNN) Classification Principle.

The classification will vary as per the value of parameter *k*. Here, light grey hexagon is classified as malicious class for *k* =1; however, it is classified as a normal class for *k* = 2 as well as *k* = 3. Thus, for the accuracy of this algorithm, getting the ideal estimation of k through testing is essential. For classification of an anomaly, many of the authors have utilized KNN-based classification, and it has achieved reasonable accuracy in identifying remote-to-local (R2L) and user-to-root (U2R) attacks. Because KNN is easy to use, distinguishing missing hubs and deciding the ideal estimation of k are tedious and time consuming as far as accuracy is concerned.

3. **Decision Trees (DTs)**

One of the predictive modeling approaches used in statistics is the decision trees (DTs) technique. This approach works by selecting important sample features from the dataset and, based on the value of a feature, creating an ordered tree. The node of the tree represents a feature of the dataset, and the branches arising from that node depict its related values. In order to optimally divide the tree into two parts, the starting node for the tree can be any feature node [33]. Also, to ideally partition the training datasets, different metrics are used for determing the starting node.

Decision tree nodes are illustrated in Figure 7.7. Two processes, namely acceptance and deduction, are involved in this algorithm, which aims to build the model and later make the classification. The acceptance process for developing a DT begins with the addition of nodes and branches. At the beginning, nodes are vacant. By gaining information through different measures, the feature selection algorithm is implemented. A feature is chosen in such a way that it is considered to be part of a dataset during the testing phase. Then, this feature is allotted the vertex of the decision tree [34].

In order to minimize the overlapping between the classes of the training dataset, the cycle keeps on choosing feature root nodes. As a result, the accuracy of the classification algorithm improves in recognizing unique samples of a class. Eventually, for each sub-DT the leaves are distinguished, and they are ordered according to their related classes. Once the DT is developed, the derivation cycle starts, where new features of the sample are highlighted and are grouped together through repetitive examination with the built DT.

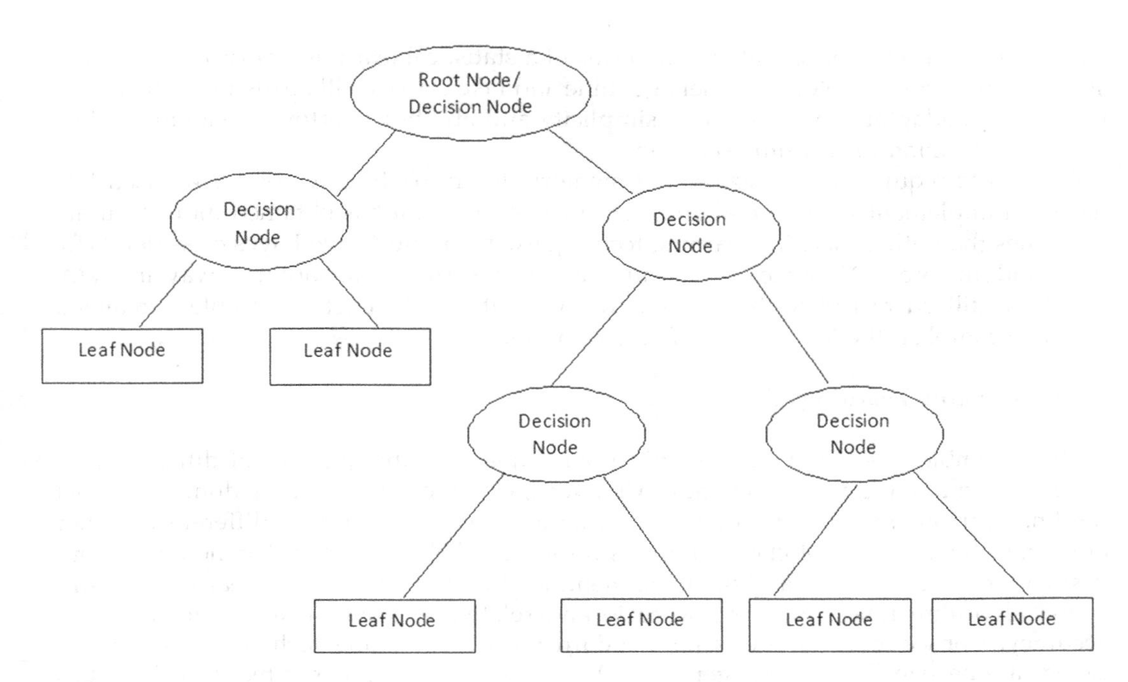

FIGURE 7.7
Depiction of Decision Tree Structure.

Again, if the similar leaf node is procured, the classification method is over for this new example. For the scenario of intrusion detection, DTs can be used as a classifier for detecting attacks. In literature, various researchers have utilized DT to identify DDoS assaults by analyzing the network traffic for identifying fake nodes (Figure 7.7)

4 **Support Vector Machines (SVMs)**

Support vector machine (SVM) is one of the widely used supervised algorithms for classification and regression. This classifier works by making a hyperplane of at least two classes using their features. The aim of SVM is to partition n-dimensional space into classes by creating the best hyperplane so that new data points can be placed in the correct category [35,36]. In Figure 7.8, the concept of SVM is depicted. This algorithm is suitable for classes having a large number of features that have to be classified on the

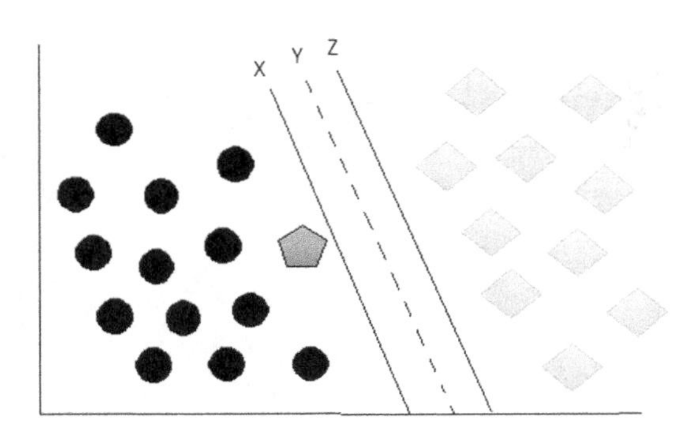

FIGURE 7.8
Hyperplane Splitting using Support Vector Machines (SVM).

basis of a fewer number of features. In terms of a statistical learning approach, SVMs are suitable for intrusion detection where genuine and fake classes will be distinguished. This technique is adaptable because of its simplicity and ability to perform assignments like anomaly detection in real-time scenarios.

Also, SVM requires less utilization of memory. In an IoT framework, SVM-based IDS has been implemented in several areas of research, where it has obtained more accurate outcomes than other machine learning techniques, including Naïve Bayes classifier, DTs, and random forest. Nonetheless, the utilization of kernel in an optimal way in SVM, which is utilized to isolate the information when it is not directly divisible, remains a test to accomplish the ideal speed of classification.

5. **Ensemble Learning (EL)**

The ensemble classifier model is built upon expanding the qualities of different classifiers, by combining their outcomes. Afterward, classification voting is done to predict the final outcome, as shown in Figure 7.9. Because of the mixture of different classifier outcomes, the accuracy of classification is improved. This algorithm depends on the investigation [33,37], where authors have mentioned that every machine learning classification algorithm relies upon the scenario and related information for achieving good accuracy. Consequently, there is no standard machine learning technique that can be called a one-size-fits-all arrangement". In some scenarios, ensemble learning (EL) methods are considered to be the most suitable way to improve accuracy by decreasing variance and avoiding overfitting. Since various classifiers are used in parallel, the precision of EL results in increased time complexity. Researchers have tested the feasibility of EL for intrusion detection by proposing a generalized application called the lightweight EL system for detecting anomalies in IoT networks. This investigation demonstrated that by using such an EL algorithm, better results are produced than individual classifiers.

6. **Random Forest**

Random forest is a popular supervised machine learning algorithm. It comprises numerous decision trees to produce error resistant and more accurate results. This classifier collects predictions from each decision and, depending on majority votes for prediction, generates a final output. In spite of the fact that DTs are considered to be parts of random forest, there are two separate classification algorithms because in DTs, a rule set is

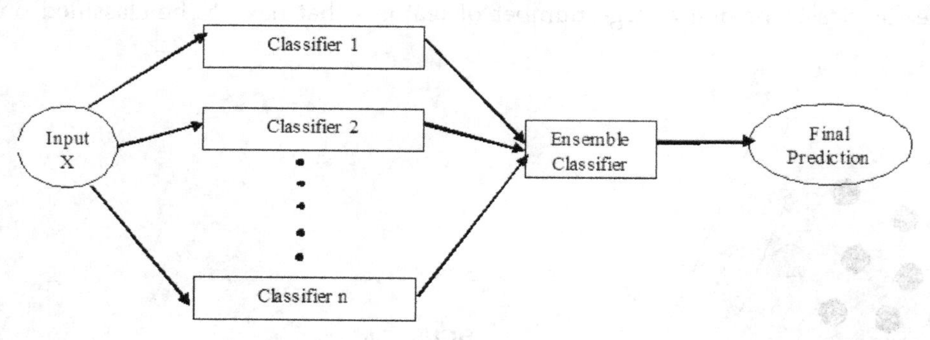

FIGURE 7.9
Working of an Ensemble Classifier.

prepared during the training phase for newer examples whereas random forest generates a standard subset of rules utilizing all part of DTs [33,38]. This produces a more robust and exact output, which does not result in overfitting, requires less input sources, and does not need the step of feature selection. A few researchers have examined if the random forest technique is reasonable for detection of threats in the IoT. Additionally, another investigation demonstrated that this method is superior to K-nearest neighbors (KNN), artificial neural network (ANN), and SVM for DDoS identification in IoT networks since it needs less input data as well as can bypass more weighted calculations needed for feature selection in real-time intrusion detection.

7 **k-Means Clustering**

K-means clustering is one more supervised learning algorithm that is used to solve clustering problems in machine learning as well as data science. Each case of test information is allotted to a specific cluster depending on its features. Each cluster is assigned a centroid; hence, it is called a centroid-based algorithm. The primary goal of this technique is to limit the sum of the distances between the data points and their neighboring clusters. The algorithm takes as input the unlabeled dataset, partitions this dataset into k-number of clusters, and repeats the cycle until the best cluster is found. In this calculation, the estimation of k has to be predetermined [33,39]. This clustering method basically performs two tasks: (1) determines the optimal value for K centroids through an iterative cycle, and (2) allocates every data point to its nearest centroid. The points that are nearer to the specific k-center combine to form a cluster. Therefore, each cluster has data points with shared characteristics, and it is away from different clusters. Researchers have recommended that a combination of DT and k-means clustering algorithms can be used to identify anomalies in IoT networks; hence, improved performance is achieved (Figure 7.10)

8. **Principle Component Analysis (PCA)**

Principle component analysis (PCA) is a widely used unsupervised algorithm that is normally utilized for feature reduction or feature selection in machine learning. Chosen feature sets are utilized with some other machine learning classifiers to identify inconsistencies in an IoT environment. This method modifies a huge set of features into a minimized set of features without losing a significant part of the data. In literature, many researchers have utilized PCA in combination with different classifiers to identify abnormalities in the IoT environment [38].

7.2.3 Metrics for IDS Evaluation

To quantify the effectiveness and efficiency of IDS, several metrics have been designed. As presented in the literature, metrics are categorized into three classes: threshold, ranking, and probability metrics.

In threshold metrics, F-measure (FM), classification rate (CR), cost per model (CPE), and so forth are included. To determine how near the predicted value is to a threshold value does not matter. The important point is that it should be above or below the threshold. The range of threshold metrics should be from 0 to 1. Ranking metrics include parameters like detection rate (DR), precision (PR), false positive rate (FPR), area under ROC curve (AUC), and capability of intrusion detection (CID). The ranking metrics range

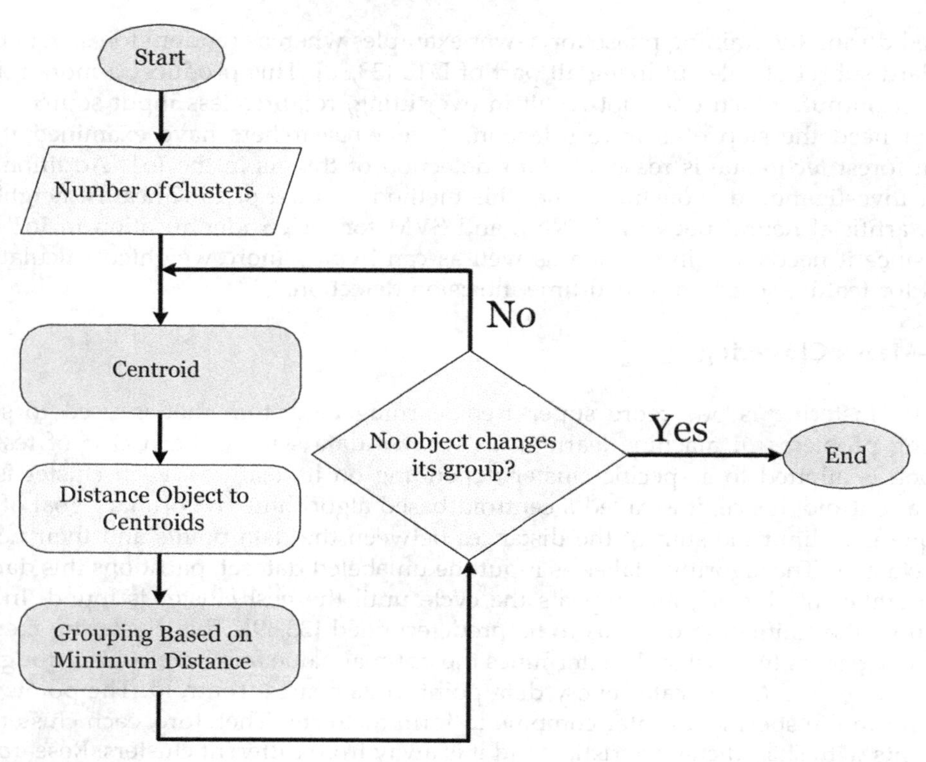

FIGURE 7.10
Illustration of k-Means Clustering.

is from 0 to 1. With the help of these metrics, we can measure how the attack scenarios are arranged prior to normal cases, and every single threshold value can be considered a brief summary for the implementing model. The most widely used probability metric is root mean square mistake (RMSE). It is used to output a probability value from 0 to 1. The value of this metric is limited when there is overlapping between the predicted value of an attack class with the genuine conditional probability of that class being a normal class. Besides, CID is a data hypothesis based metric that gives a superior correlation of different IDSs than the other mainstream metrics like AUC. The estimation of CID lies somewhere in the range of 0 to 1. The greater the estimated value of the CID, the better performance of IDS is obtained. Usually, confusion metrics are used to calculate these measurements. In order to present the classification results of IDS, the confusion matrix is the most ideal approach [40].

1. **Confusion Matrix**

This matrix highlights the results of classification in terms of true and false. The potential outcomes to classify the events are portrayed in Table 7.2:

- **True Positive (TP):** Intrusions are detected successfully by the IDS.
- **False Positive (FP):** Normal/non-intrusive behavior is classified incorrectly as malicious by the IDS.

TABLE 7.2
Confusion Matrix

Actual	Predicted Attack	Predicted Normal
Attack	TP	FN
Normal	FP	TN

- **True Negative TN):** Normal/non-intrusive behavior is classified correctly as normal by the IDS.
- **False Negative (FN):** Intrusive behavior is missed by the IDS and is classified as normal.

2. **Performance Metrics**

Regardless of the powerful classification results representation ability of the confusion matrix, it is not yet considered to be an efficient tool for the examination of the different IDSs. In order to take care of this issue, various performance metrics are characterized in terms of confusion matrix variables. Some numeric values are produced by these metrics, which can be compared easily and are summarized in brief in the resulting passages [40].

a. **Classification Rate (CR):** This metric is the ratio of correctly classified instances and the total number of instances.

$$CR = \frac{\text{Correctly Classified Instances}}{\text{Total No. of Instances}}$$
$$= \frac{\text{TP} + \text{TN}}{\text{TP} + \text{TN} + \text{FP} + \text{FN}}$$

b. **Detection Rate (DR):** This metric is the ratio between the number of correctly detected attacks and the total number of attacks.

$$DR = \frac{\text{Correctly Detected Attacks}}{\text{Total No. of Attacks}}$$
$$= \frac{\text{TP}}{\text{TP} + \text{FN}}$$

c. **False Positive Rate (FPR):** This metric is the ratio between the number of normal instances detected as an attack and the total number of normal instances.

$$FPR = \frac{\text{Number of normal instances detected as attacks}}{\text{Total No. of Normal Instances}}$$
$$= \frac{\text{FP}}{\text{FP} + \text{TN}}$$

d. **Precision (PR):** This metric is the fraction of data instances predicted as positive that are actually positive.

$$PR = \frac{\text{TP}}{\text{TP} + \text{FP}}$$

e. **Recall:** The missing part from precision, such as the percentage of the real attack instances classified by the classifier, is measured in this metric. It is good to obtain a higher recall value for a classifier. It is similar to the DR metric.

$$FM = \frac{2}{\frac{1}{PR} + \frac{1}{Recall}}$$

f. **Area under ROC Curve (ROC):** In ROC, for all possible thresholds, a graph of sensitivity versus 1-specificity is plotted. Sensitivity is defined as P (Pred = positive | True = positive) and is approximated by the fraction of true positives that are predicted as positive. (This is the same as a recall.) Specificity is P (Pred = negative | True = negative). It is approximated by the fraction of true negatives predicted as negatives. This metric is used as a summary statistic.

g. **F-measure (FM):** This metric is defined as the weighted harmonic **mean** of the precision and recall of the learning model. It is a measure of the machine learning model's accuracy on a dataset.

Thus, IDS efficiency can be measured by its capacity to accurately classify attacks as normal or malicious, along with other performance parameters. It is mandatory for both industry people and the research community to measure these abilities of IDS. It encourages us to tune the IDS in a superior manner and compare its performance with different IDSs.

7.3 Tools for NIDS Implementation

To detect abnormal or malicious traffic, NIDSs monitor and analyze network traffic. The basic steps required to build NIDS are as follows [41]:

1. Collection of traffic information from the network.
2. Analyzing the gathered information.
3. Identification of malicious events.
4. Detection and reporting of malicious activity.

There are two choices for researchers to perform the above steps: utilization of the current tools to install their own NIDS and developing a novel intrusion detection technique. Among available tools, an individual has a choice to select between a) free benchmark datasets to use in off-line mode, b) freely available open-source network sniffers to record or catch network traffic information, and c) free open-source NIDS that can be deployed for achieving desired outcomes. Further, in this section, we provide research about available tools. We begin with free benchmark datasets for NIDS, then free open-source network sniffers, and finally NIDSs. Further, all these tools for NIDS implementation are correlated. The primary role of network sniffers is to gather network traffic data and store it in datasets. Because input is unlabeled, it becomes mandatory for NIDSs to differentiate attacks or normal behavior instances. Network sniffers are usually smaller than NIDSs. NIDSs utilize network sniffers to capture data that can later differentiate attack instances from normal ones.

7.3.1 Free Data Set Available for IoT Security

For implementation or validation of NIDS, free datasets can be utilized. But, to date, there are no specific datasets made for IoT networks. Two techniques are available: either download the dataset available or deploy sniffer software in the network.

The popular datasets used for NIDS are KDDCUP99 and NSL-KDD, which is an updated version of KDD99. A dataset named UNSW-NB15 [29] is the one that seems suitable for developing NIDS for IoT networks. Public datasets such as KYOTO, ISCX 2012, CAIDA, DEFCON, ADFA IDS, and ICS attack datasets accessible for assessment and testing. The most recent datasets are either prepared out of unlabeled information, are blocked off from certain nations, or are specific domain information. Also, datasets experience the ill effects of i) protection issues; ii) large amounts of redundant data; and iii) the absence of the latest or newer security threats [54].

1. **KDD99** [15,42] is a widely used dataset for recognition of "terrible" associations from the "great" ones at the Third International Knowledge Discovery and Data Mining Tools Competition for building the hearty NIDS. This dataset is considered a separate variant of the DARPA dataset (DARPA is a basic dataset). The records in KDD99 are from a military organization condition with infused assaults, which are classified into: i) user to root; ii) probing; iii) denial of service; and iv) remote to user. KDD99 depends on 41 attributes or features for every association alongside the class name utilizing the Bro-IDS device (introduced of late). KDD99 is well known and is the most utilized by scientists for exploratory investigation. Various works [42–44] were built up to decrease the quantity of highlights by choosing the most applicable ones from the underlying 41 highlights. Nonetheless, numerous explores have announced weaknesses of KDD99.
2. **NSL-KDD** [45] is the redesigned form of KDD99 to beat its constraints. To begin with, copied records from the preparation and testing datasets were eliminated. Second, there was an assortment of records chosen from the first KDD99 to accomplish solid outcomes from the classifier frameworks. Third, the issue of uneven likelihood appropriation was dispensed with. The serious issue that perseveres in the KDD99 dataset is the absence of present-day low-impression assault situations.
3. **UNSW-NB15** [43,45] dataset was made in 2015 by the Cyber Range Lab of the Australian Center for Cyber Security (ACCS) with the IXIA PerfectStorm device. It probably produces half breed genuine present-day typical exercises and designed contemporary assault practices. It has around 2,540,044 records, which are kept in four csv documents. These records are produced by capturing 100GB crude traffic through tcpdump instruments (stored in pcap libraries). The UNSW-NB15 dataset contains nine sorts of assaults, specifically worms, shellcode, reconnaissance, exploits, DDoS, backdoors, generic, fuzzers, and analysis.
4. **CICIDS database** [46] is the popular ongoing intrusion detection/intrusion counteraction information database delivered by the Canadian Institute for

Cyber-security, University of New Brunswick to highlight the most recent dangers. It was based on the theoretical conduct of 25 clients dependent on email, SSH, FTP, HTTPS, and HTTP conventions. The dataset is examined with CIC Flow Meter with marked streams dependent on ports, timestamp, beginning as well as last IP, naming conventions, and assaults. To produce the reasonable traffic, the creators proposed a B-Profile way to deal with the layout and to conduct experiments on conventions. The creators executed DDoS assaults, SSH Heartbleed, and Brute-power FTP while catching the information. The assessment system recognized 11 significant highlights important to construct a solid benchmark dataset, dissimilar to the current customary IDS datasets.

5. **CSE-CIC-IDS2018** [46] is a well-known IDS dataset that has been developed to supplant the current problematic datasets that limit IDS/NIDS test assessments. To defeat the utilization of one-time and static datasets, CSECIC-IDS2018 is kind of an oddity-based progressively created dataset comprising interruption in the network traffic. Creators have included around seven assault situations, including i) local organization invasion assaults; ii) web assaults; iii) DDoS; iv) DoS; v) Botnet; vi) Heartbleed; and vii) Brute-power. The assault framework has 50 hubs, and casualty association consists of 5 offices with almost 30 workers and 420 hosts. The creators have removed 80 highlights from the machine logs and network traffic caught through CIC Flow Meter-V3.
6. **RPL-NIDS17** is a type of IDS used in RPL-based 6LoWPAN organizations. It is created by gathering follows subsequent to recreating diverse directing assaults against an RPL steering convention. The dataset has 21 attributes, for example, app_layer_arrival_time, destination_id, source_id, control_packet_type, etc. The dataset is made with highlight encoding as well as without it. We utilize the encoded dataset for training and perform testing. For double grouping, the preparation dataset contains 116,679 records of typical traffic and 33,337 records of assaults. The dataset for testing contains 59,560 records of ordinary traffic and 16,971 records of assaults. The dataset consists of seven classes of assaults (clone-ID, hi flooding, neighborhood fix, specific sending, sinkhole and blackhole, and Sybil) and typical records for multi-class arrangement. The training dataset is disseminated as follows: ordinary, clone ID assault, hi flooding, neighborhood fix, particular sending, sinkhole, blackhole, and Sybil. Similarly, the testing dataset is circulated as follows: typical, clone ID assault, hello flooding, nearby fix, particular sending, sinkhole, blackhole, and Sybil.
7. **N_BaIoT dataset** was created for recognizing IoT botnet assaults utilizing abnormality identification procedures. The dataset consists of genuine traffic gathered through nine business IoT gadgets, namely Provision PT-737E surveillance camera, Ecobee indoor regulator, and Danmini doorbell. The creators traded off the IoT gadgets utilized in their testbed, utilizing Mirai and BASHLITE botnet assaults. The dataset includes five assaults of BASHLITE (COMBO, TCP, UDP, garbage, and examine) and five assaults of Mirai (UDPplain, UDP, Syn, Ack, and check). To demonstrate that the proposed structure can productively distinguish numerous assaults in a few datasets, particularly in IoT datasets, we assess our proposed approach with five

assaults of Mirai on a Danmini doorbell. The dataset is communicated in the CSV design and contains 115 mathematical highlights. The dataset is dispersed as follows: typical traffic, UDPplain, UDP, sweep, syn, and Ack assaults. The N_BaIoT dataset is utilized for twofold and multi-class order. For paired characterization, we utilize the dataset in two classes (typical versus assault) with ordinary traffic and assaults. For multi-class grouping, we utilize the dataset in six classes (UDPplain, Syn, Ack, UDP, check, ordinary) (Table 7.3).

7.3.2 Free Open-Source Network Sniffers

This section introduces free open-source network sniffers that are widely used. The primary function of sniffing tool is to monitor traffic flowing in the network from source to destination. These tools are very well utilized to record, inspect, analyze, and visualize packets in the network.

1. **Tcpdump** [47] is one of the powerful and popularly utilized packet analyzers. It is just like TCP/IP command-line tools whose function includes capturing, examining, saving, and reviewing packet information. Tcpdump was developed by Van Jacobson, Craig Leres, and Steven McCannelt at the Lawrence Berkeley Laboratory, UC, Berkeley. This tool catches live data packets from a flowing network traffic. One important feature in tcpdump is its ability to store the recorded packets in a pcap file for a research analysis. The captured packets are later on stored in the libpcap library. Libpcap can then be referred by other similar programs. The tcpdump instrument is accessible for the majority of Linux/Unix-based working frameworks. Wireshark is the famous open-source GUI in light of tcpdump, which reads tcpdump pcap files, empowering an easy-to-use interface.
2. **Wireshark** [48] is another well-known, open-source, and free packet analyzer, developed under the GNU permission. It is utilized for network sniffing and network examination. Its function is to capture live packet information from network traffic. Because of trademark issues, wireshark was renamed as Ethereal in May 2006. It runs on Unix-like operating systems, Solaris, and Microsoft Windows. It utilizes libpcap as a library to catch and filter packets and later displays records with its GUI. It has an ability to read outputs of tcpdump. Several protocols are decoded using wireshark. It supports fundamental review of attacks in the network. "tshark" is the command line version of this tool.
3. **Ettercap** is another kind of network sniffer used on multiple platforms. For switched LANs, it acts as "a multipurpose sniffer/interceptor/logger", written by Alberto Ornaghi and Marco Valleri. Also, it is capable of launching several powerful attacks. This tool identifies and separates newer attacks and observation procedures inside its interface. It is capable of sniffing live associations and filters packets just as numerous different features in both a static and dynamic way.
4. **Argus** [48] is another tool to capture network flow data and examine it. It can run on several platforms like Linux and Windows. Network activity audit logs are developed by this tool. Also, audit reports for detected network flows with semantic analysis can be created using live and captured traffic data.

TABLE 7.3
Datasets Used in IDS

Datasets	Pros	Cons
KDD99	• KDD99 is famous and the most utilized. • It has labeled information. • It depends on 41 highlights for every association alongside the class name. • It implements user-to-root, remote-to-user, denial-of-service, and probing assaults. • It gives a network traffic file (PCAP).	• KDD99 experiences lopsided arrangement techniques. • The dataset is outdated. • It is not designed for IoT frameworks.
NSL-KDD	• NSL-KDD is a superior form of dataset KDD99. • It defeats KDD99 impediments. • There are no copied records in preparation and test sets.	• It lacks current low-impression assault situations. • It is not designed for IoT frameworks.
UNSW-NB15	• It gives mixture genuine current ordinary exercises and engineered materials contemporary assault practices. • It gives network traffic flow (PCAP) and CSV records. • It contains around nine sorts of assaults, in particular analysis, fuzzers, backdoors, exploits, DoS, reconnaissance, generic, worms, and shellcode.	• It is more mind boggling than the KDD99 dataset because of the comparable practices of the cutting-edge assault and typical organization traffic.
CICIDS	• It contains labeled information. • It is used for artificial intelligence purpose. • It gives network traffic flow (PCAP) and CSV files. • It implements attacks such as DDoS, DoS, Brute- force FTP, Brute-force SSH, Infiltration, Heartbleed, Web Attack, and Botnet.	• It is not available publicly. • It is not appropriate for IoT frameworks.
CSE-CIC-IDS 2018	• It has labeled information. • It is used for artificial intelligence purpose. • It gives network traffic (PCAP), log, and CSV files. • It implements local network infiltration attacks, DDoS, DoS, Brute-force, Heartbleed, Botnet, and Web attacks. • The dataset is generated dynamically. • It is extensible and can be modified and reproduced as required.	• It is not available publicly. • It is not appropriate for IoT frameworks.

RPL NIDS17	• It is based on 6LoWPAN networks with RPL protocol. • The labeled network flows with binary and multi-class. • It is utilized for machine learning as well as deep learning purposes. • It gives network traffic (PCAP), log, and CSV files with 21 features and 1 label. • The dataset has attacks of seven classes (Sybil, sinkhole and blackhole, sinkhole, selective forwarding, local repair, hello flooding, and clone-ID) and normal records.	• It is not available publicly. • It is restricted to RPL protocol.
N_BaIoT	• It is based on collected data traffic in real conditions using commercial devices in the IoT. • It provides network traffic (PCAP), CSV files with 115 features and 1 label • The labeled network flows with binary and multi-class. • It is for machine learning and deep learning purposes. • It includes five attacks of BASHLITE (COMBO, TCP, UDP, junk, and scan) and five attacks of Mirai (UDPplain, UDP, Syn, Ack, and scan).	• It is restricted to commercial devices, not for wide IoT-based WSN.

It measures libpcap and ERF packet information, which helps the user to get an idea about what is happening on a network. Information related to the parameters, such as load, rate, duration, retransmission, delays, and so forth, can be obtained.

5. **EtherApe** was developed in 2000 by Riccardo Ghetta and Juan Toledo. It is a graphical packet sniffer and network monitoring tool that can run only on the Unix platforms. It has an ability to visually represent data flows and packets connections by coloring hosts and links related to protocols. This tool has an additional functionality of network troubleshooting. Also, via standard formats, network packets can be displayed in real time. Network traffic can be recorded on your own network either port-to-port (TCP) or end-to-end (IP) (Table 7.4)

In the above table various free as well as open-source network sniffers are analyzed. It is observed that the most popular network sniffer is tcpdump. It has a long life which is frequently updated and can be extended with multiple features. Tcpdump is basically developed for data capturing and not for network analysis like other tools. It does not have GUI. Wireshark and etherApe are powerful in terms of graphical features. Both of these tools can display real- time and captured network files. Next is Argus, one of the popular tool for network audit activities. Various protocols are decoded for reports and audits. Regarding ettercap, besides performing sniffing operations, it also can modify the network data and raise various MITM attacks. Also it has ability to gather passwords, break connections, insert fake packets and commands during active connections. Therefore, it is considered more of a hacker tool than a network sniffer. In this part summary of network sniffers is highlighted, hence next section will mention and elaborate about open-source NIDS.

7.3.3 Open-source NIDS

Several free open-source NIDS devices are available which can be used for sniffing, investigating and recognizing malicious activities in network traffic. Examples of NIDS are Snort, Bro-IDS, Suricata, etc. Following are the widely used open source NIDS.

1. **Snort** [45] IDS was first developed in 1998. It is a lightweight and well-known intrusion prevention system used for analyzing traffic in real time. It runs on FreeBSD, Fedora, Centos and Windows operating systems. It is a signature-based NIDS that uses the most popular open source rule list called as Talos. There are three modes in which snort command line operates:
   - Sniffer mode wherein packets are read and captured from the network and later on its output is printed on the console.
   - Packet Logger mode moves the packets to disk.
   - Network Intrusion Detection System (NIDS) mode identifies and investigates network traffic by using a configure file which contains signatures to detect buffer overflows, stealth port scans, etc. It can also detect different application layer attacks like SQL injection attack and cross-site scripting attack.

TABLE 7.4

Comparison of Different Free and Ppen-Source Network Sniffers

Network sniffers	Advantages	Drawbacks
Tcpdump	• It has large number of features and possess long product life with various updates. • Captures live packet flow from a network traffic and saves it. • Has good community support and documentation • With the help of Telnet connection can remotely access the data • Can work with several platforms • Less noisy than Ethereal. • It is Lightweight as far as installation is considered.	• It lacks basic investigation. • Invalid packets are discarded. • GUI or administrative console does not exist.
Wireshark	• Has good community support and documentation. • Operates on multiple platforms • Supports majority of protocols. • Has interactive GUI • Captures live packet flow from a network traffic and saves data files. • Information of protocol is provided in detail.	• There is no notification of abnormal behavior. • It gathers data but cannot control the network. • Requires more resources in terms of installation.
Ettercap	• Compatible with several platforms. • Utilized for LAN hacking procedures. • Several protocols can be decoded. • Gathers passwords for numerous applications • Controls the network by breaking connections, injecting packets and commands into active connection. Thus manipulates it. • It is extensible with additional plug-ins.	• Here Sniffing is considered as an optional feature. • It can be utilized as a hacker tool. • Other network devices can easily detect it.
Argus	• Operates on multiple platforms. • Decodes several protocols. • Audits reports about the network. can be generated. • Native record framework as well as MySQL support. • Works efficiently with huge network traffic.	• Not too clear to even think about mastering.
EtherApe	• Network activity is displayed in terms of graphics with a shading coded protocols mode. • Hosts along with their connections change in size with traffic. • Can filter network packets. • Supports real-time network traffic. • Good performance is observed among the system administrator community.	• Works only with Unix OS. • Command line version does not exist. • Packet headers are only captured.

2. **Suricata** is another signature-based NIDS which was released in December 2009. It can be used mainly for intrusion detection in real time, inline intrusion prevention, network security checking and offline pcap processing. The owner of this tool is Open Information Security Foundation (OISF). It can run on FreeBSD, Linux, OpenBSD, macOS and Windows operating systems.
3. **Bro-IDS tool** [49] is one more popular network analysis tool for review of malicious activities occurring in network traffic. It is a kind of both signature and anomaly-based IDS. Several application layer protocols like, HTTP, FTP, DNS, SMTP, etc are included here. Bro system was created by Vern Paxson of ICSI's Center for Internet Research (ICIR). This IDS runs on Linux, FreeBSD and Mac OS X operating systems.
4. **Kismet** [50] is used as a wireless sniffer, network detector and an intrusion detection system. This tool operates well on multiple platforms such as Android, Windows BSD, Linux, etc. Also it operates with Bluetooth powered devices and Wi-Fi (IEEE 802.11[2]) cards for scanning discoverable BT and BTLE gadgets, the RTL-SDR radio for detecting wireless sensors, thermometers, and switches, and a growing collection of other capturing hardware.
5. **OpenWIPS-ng** was developed by Thomas d'Otreppe de Bouvette, the developer of Aircrack software and basically runs on commodity hardware. It is kind of modular Wireless IDS/IPS which monitors wireless network traffic to detect and identify signature based attacks. It consist of sensors to capture wireless traffic and send it to the server and a server aggregates data from all sensors, detect and analyze intrusions and send responses or alerts and finally a GUI is there to manage the server and display threat information. For more flexibility OpenWIPS-ng has few extension plugins but it does support only Linux systems.
6. **Security Onion** works well on Linux platform for intrusion detection, network security monitoring (NSM) and log management. This IDS has set of specific security tools like Bro, Snort, Suricata, Sguil, Elasticsearch, Logstash, etc. which operates either independently or together to detect malicious activity in virtual LANs and visualized networks. Its primary features includes full packet capture, NIDS and HIDS;and powerful analysis tools.
7. **Sagan** IDS is developed by Quadrant Information Security. This IDS is termed as real-time log analysis and correlation engine which has multi- threaded architecture to identify malicious activities at both log and network levels with a high performance. It is written in C and works on Unix OS. Sagan is host based IDS and was stretched out to be considered as a signature-based NIDS also. It is compatible with data assembled by Snort, Bro, Suricata and other tools.

The merits and demerits of the above mentioned NIDSs are presented in Table 7.5.

Thus, this section summarizes the tools utilized to develop a NIDS, right from free network datasets to free open source network sniffers, to free and open source NIDSs.

TABLE 7.5

Comparison of Various Open-Source NIDS

IDS	Advantages	Disadvantages
Snort	• It is lightweight intrusion detection system. • Has long product life with various updates, newer features and large number of administrative front-ends. • Well documented with a good community support. • Testing done is successful. • Easy for deployment.	• No real GUI or easy to use administrative console. • Packets are lost when the process rates 100-200 megabytes per second before reaching the processing limit of a single CPU.
Suricata	• Faster network traffic analysis due to multithreaded architecture. • Using graphic cards, network traffic inspection can be done. • File downloading can be detected. • LuaJIT scripting allows to detect complex threats easily. • Logs more than packets like TSL/SSL certs, HTTP requests and DNS requests.	• More memory and CPU resources are required.
Bro-IDS	• Supports both signature and anomaly-based intrusion detection methods. • Greater level of abstraction is achieved while analyzing traffic. • Records data about past event and includes them for analyzing new activity. • Can work efficiently with high speed network.	• It is UNIX based. • There is no GUI. It is based on log files. • Experts are needed for set up.
Kismet	• Its function can be extended via plugins. • Channel hopping method allows to detect as many network as possible. • It can passively monitors wireless networks. • Popular and updated open source wireless monitoring tool.	• IP addresses cannot be directly obtained. • Suitable only for wireless networks.

(Continued)

TABLE 7.5 (Continued)

Comparison of Various Open-Source NIDS

IDS	Advantages	Disadvantages
	• Real-time traffic over HTTP can be lively captured.	
OpenWIPng	• It is modular in nature and additional features are included through various plugins. • Software and hardware required can be built by non-professionals. • Because it supports multiple sensors, it is highly accurate.	• Suitable only for wireless networks. • The traffic is not encrypted. • between sensor and server. • Not much popular and yet developed. • Documentation is not provided and don't have community support.
Onion Security	• It is flexible in nature. • Real-time graphical interface is available hence easy Network Security Monitoring and analysis can be done. • Several functions are pre-installed and easily configurable. • Security levels can be improved through regular updates.	• Drawbacks of each constituent tool are inherited. • After installation it works only as IDS, not IPS.
Sagan	• Has fast processing ability and supports real-time log processing. • Several output formats are supported. • IP addresses of the devices that are geographical located can be obtained. • Processing work can be distributed over several machines. • Requires fewer memory resources and are lightweight. • Easy to install.	• Is basically designed for log analysis in spite of intrusion detection.

7.3.4 Case Study

In this case study, the benchmark dataset UNSW-NB15 created in 2015 was implemented using Random Forest Algorithm with Decision Tree Classifier using Anaconda 3 which is free and open source distribution of Python 3. The proposed model will detect anomalies with a better accuracy using a machine learning technique.

1. **Methodology**

There are a total of 45 attributes in this dataset, and strong parameters are used to achieve better accuracy. Also, feature selection plays a significant role in improving accuracy of the intrusion detection system. The UNSW-NB15 dataset is categorized into 45 features, which includes records of both normal traffic as well as attack types [51]. Features are listed in Table 7.6.

This dataset includes 9 types of attacks, namely analysis, fuzzers, exploits, shellcode, reconnaissance, DoS, backdoors, shellcode, and worms. Analysis of these attacks is represented by the graph in Figure 7.11.

Here, the proposed algorithm is RandomForestClassifier, which is a well-known ensemble classification algorithm. A combination of classifiers is called an ensemble classifier. The value of the target is predicted using multiple classifiers rather than using a

TABLE 7.6
UNSW-NB15 Features

Feature Number	Feature Name	Feature Number	Feature Name
1	id	23	dtcpb
2	dur	24	dwin
3	proto	25	tcprtt
4	service	26	synack
5	state	27	ackdat
6	spkts	28	smean
7	dpkts	29	dmean
8	sbytes	30	trans_depth
9	dbytes	31	response_body_len
10	rate	32	ct_srv_src
11	sttl	33	ct_state_ttl
12	dttl	34	ct_dst_ltm
13	sload	35	ct_src_dport_ltm
14	dload	36	ct_dst_sport_ltm
15	sloss	37	ct_dst_src_ltm
16	dloss	38	is_ftp_login
17	sinpkt	39	ct_ftp_cmd
18	dinpkt	40	ct_flw_http_mthd
19	sjit	41	ct_src_ltm
20	djit	42	ct_srv_dst
21	swin	43	is_sm_ips_ports
22	stcpb	44	attack_cat
		45	label

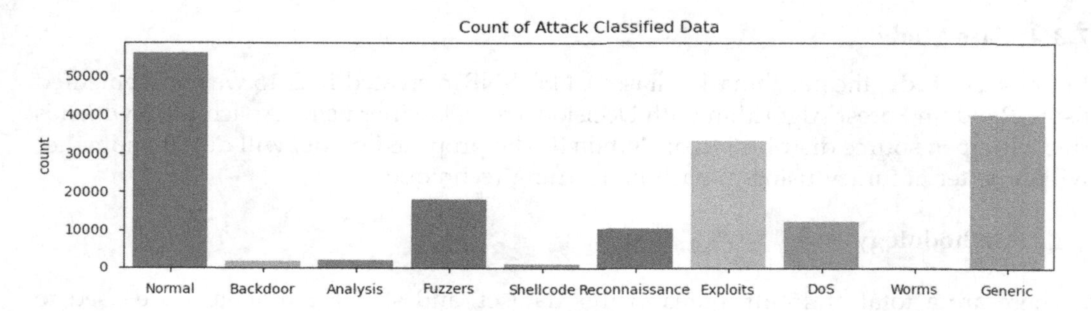

FIGURE 7.11
9 Types of Attacks in UNSW-NB15 Dataset.

single classifier. The randomly created DTs are called classifiers. The target prediction relies on the majority of votes, considering each DT as a single classifier. The maximum number of votes received by the target class is considered to be the final predicted target class. This method gives maximum accuracy with the combination fusion of recursive feature elimination from the Sklearn feature selection method. The top four features are sttl, ct_dst_src_ltm, sbytes, and sload. It provides 98% accuracy.

The correlation matrix of features in the UNSW-NB15 dataset is depicted below, where id feature is not included. This graph in Figure 7.12 clarifies the correlation relationship among the features in the dataset. For example, sttl is closely correlated to feature dttl, and feature sinpkt is closely correlated to dinpkt. Accordingly, the importance of features can be understood using this correlation matrix graph.

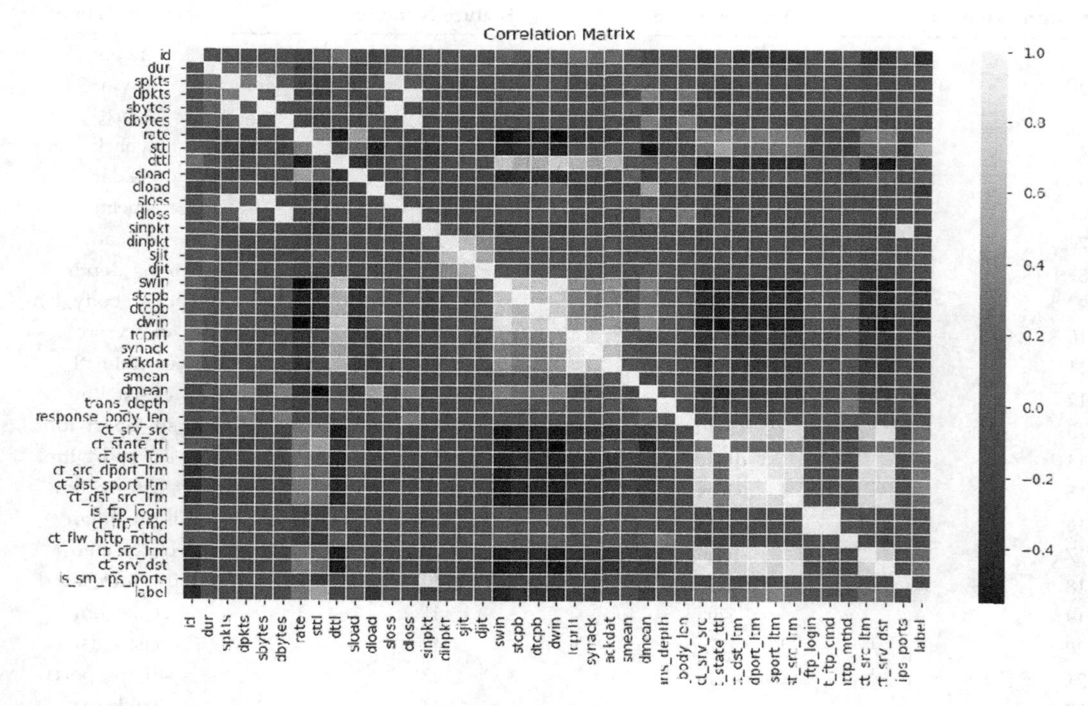

FIGURE 7.12
Correlation Matrix of the UNSW-NB15 Dataset.

	precision	recall	f1-score	support
A	0.66	0.52	0.58	2000
B	1.00	0.20	0.34	1746
D	0.88	1.00	0.94	12264
E	0.99	1.00	0.99	33393
F	0.99	0.99	0.99	18184
G	1.00	0.99	1.00	40000
N	1.00	1.00	1.00	56000
R	1.00	1.00	1.00	10491
S	0.99	0.97	0.98	1133
W	1.00	0.75	0.85	130
accuracy			0.98	175341
macro avg	0.95	0.84	0.87	175341
weighted avg	0.98	0.98	0.98	175341

FIGURE 7.13
Classification Report.

2. **Results and Discussion**
   a. The RandomForestClassifier Classification Report

This tool provides a perspective of the model's performance. It displays the precision, recall, F1, and support scores for the model. The first row **displays** the scores for class A, the second row displays the scores for class B, the third row displays the scores for class D, and so on. A, B, D, E, F, G, R, S, and W represents 9 classes of attacks:
Analysis backdoor DoS exploits fuzzers generic normal reconnaissance shellcode worms (Figure 7.13)

b. Receiver Operational Characteristics Curve (ROC)

ROC curve is utilized to visualize the performance of multi-dimensional classification of a dataset. For evaluating any classification model's accuracy, it is considered one of the distinguished evaluation metrics. ROC curve for Class 1, i.e., Class B, is depicted in Figure 7.14. Similarly ROC curves for other classes can also be obtained.

a. Confusion Matrix

This matrix gives a count of positive and negative predictions and conjointly summarizes the count of normal and malicious attacks; therefore, the below graph is shown with the number of samples being correctly classified for all nine classes. Therefore, the overall confusion matrix outperforms the analysis metrics of this model (Figure 7.15).

Analysis Backdoor DoS Exploits Fuzzers Generic Normal Reconnaissance Shellcode Worms

[172815 173595 161421 141453 156931 135303 119274 164809 174201 175211] True Negative

[1036 354 12261 33287 17967 39741 55999 10442 1101 97] True Positive

[964 1392 3 106 217 259 1 49 32 33] False Negative

[526 0 1656 495 226 38 67 41 7 0] False Positive

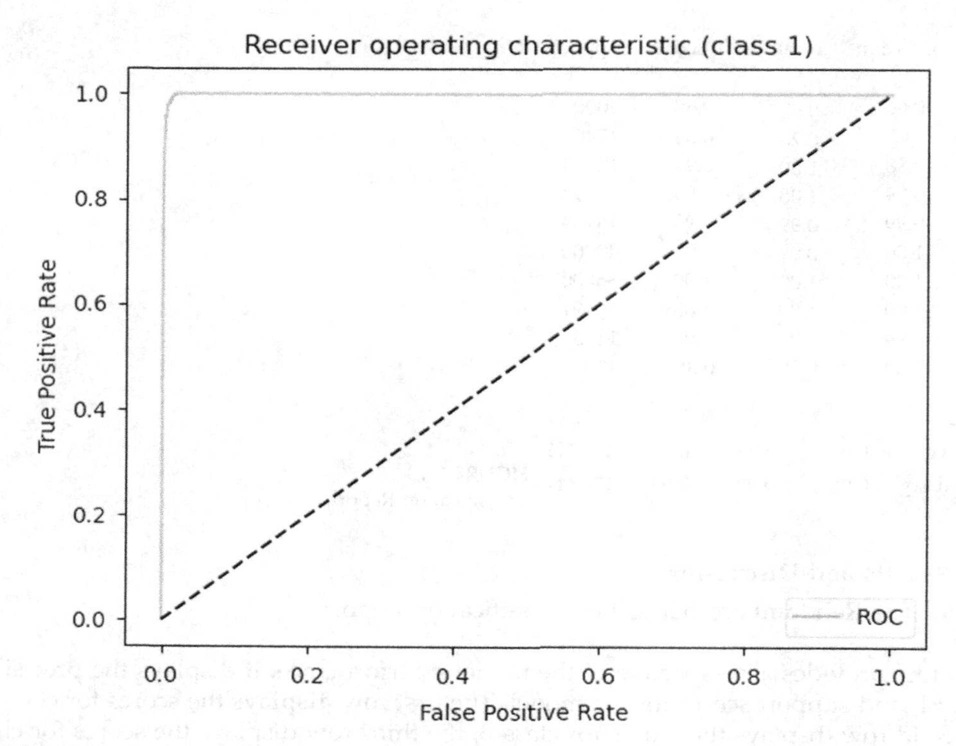

FIGURE 7.14
ROC Curve.

7.4 Challenges and Future Research Directions

Several research works have been carried out related to intrusion detection systems for the IoT. But still, there are numerous open research challenges and issues, especially in terms of implementing machine learning techniques for anomaly-based intrusion detection in the IoT. The most important challenge is that there does not exist any standard mechanism that will ensure that the proposed systems are valid. The majority of research work is carried out on benchmark datasets to evaluate the proposed systems, and it specifically highlights one issue that may not operate on real data as well in the presence of complex problems. Hence, on the basis of literature review done for IDS in the IoT, it is extremely hard to design an IDS that satisfies any one of the most valuable parameters of an effective IDS, is scalable and works effectively on real data, and fulfills all stakeholder requirements. The majority of the published research work presents experimental results tested on benchmark datasets that cover some portion of the system and present results using biased parameters [55].

For any proposed IDS, it is very difficult to define proof of accuracy and completeness. Thus, the conclusion drawn from this chapter is that it is quite difficult to design a complete IDS, which provides good accuracy, robustness, scalability, and security against all types of attacks. A few major difficulties and upcoming challenges that researchers are facing today and will face in the future are mentioned as below. Since the IoT safety measures are yet to be developed, there is tremendous scope for research in this area, especially in anomaly-based intrusion detection using machine learning techniques.

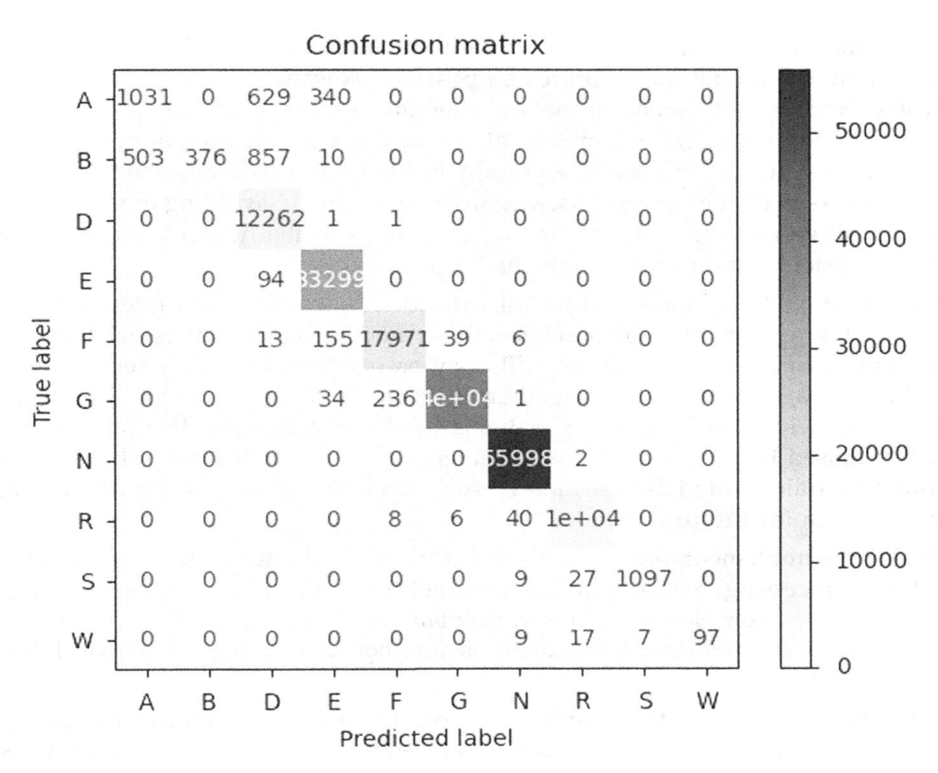

FIGURE 7.15
Confusion Matrix and Its Parameters.

The upcoming challenges identified related to anomaly-based intrusion detection in IoT networks are as follows:

- A decent quality IDS dataset related to IoT is necessary to test and approve proposed NIDS. Such kind of dataset should contain a sensible size of network flow information covering attack as well as normal behavior with the related label. Besides that from every IoT device, normal traffic of data is required to capture, other than the attack data for testing the system. But as discussed in the previous section, several publicly available datasets does not provide the required features such as missing labels, incomplete network features, missing raw pcap files which are hard to understand and additionally have incomplete csv files. Available datasets capture only normal behavior of IoT devices, where training of IDS is restricted on these devices only. Making a dataset which addresses these issues in a real environment is a challenge issue and a upcoming area of research.
- To develop an anomaly-based IDS that operates in real-time IoT networks is a very keen task. It is due to the reason that an IDS would need to become familiar with normal behavior initially to detect malicious events. In the learning stage, it is assumed that attack traffic does not exist during this period, but it cannot be ensured. If these issues are not addressed, such an IDS may raise false alarms.

- As mentioned in the literature, the majority of anomaly-based detection systems aim to build a model that captures all possible patterns of normal traffic in the profile. Such activity seems to be a challenging task as it results in high false-positive rates. Besides, it is impractical to catch all possible normal patterns that may be generated in a network, especially in a heterogeneous environment of IoT frameworks, which increases false-negative rates. Totally avoiding or minimizing false-positive rates as well as false-negative rates in the network intrusion detection system is another research challenge.
- To train models on specific types of IoT devices would be an interesting approach. Using a similar type of device, these models in different associations can be applied to IDSs. This will help different associations to deploy these models and hence saves time needed to gather the information and train the IDSs. Thus, it will be helpful in identifying malignant IoT gadgets, since they are already compromised because they behave different compared to normal behavior generated by trained models. Creation of such models is a major issue and a challenging area for future research.
- During the implementation of machine learning based NIDS, various phases like data-preprocessing, feature reduction, model training, and deployment increase computational complexity. In this way, planning an effective NIDS that means selection of computational resources is another challenging issue for future examination.
- Feature engineering methods utilized for proposed IDSs are appropriate to deal with a particular kind of normal traffic and also to detect a specific type of attack. They may fail to work once the attack scenario sequences change a bit, particularly in the case of the fast-changing IoT environment. Thus, a feature selection method that is non-static as well as computationally efficient and can work for all categories of traffic is a research challenge.
- ML-based techniques are broadly utilized for preparing a learning model on a larger dataset, and they have proved to be effective in handling cyber-attacks. But use of these algorithms for detecting threats in IoT frameworks faces certain challenges that need the attention of researchers; for instance, resource requirement problems with IoT devices restrict the utilization of machine learning algorithms for IoT network security [49]. Another challenge arises in the case of large and distributed networks like that of IoT networks, where use of machine learning techniques faces scalability issues, for example, choice of IDS deployment in various scenarios. Some of the authors suggested a solution to limitations in the use of individual machine learning algorithms is to use an ensemble machine learning classifier that performs better in contrast to a single machine learning classifier. But these methods appears to be expensive in terms of computation, which results in network delay issues, which further cannot be afforded in systems that are critically dangerous to human lives, such as health, autonomous, or Internet of Vehicles (IoVs) systems.
- For designing an IDS for IoT security, semi-supervised learning, reinforcement learning techniques, and transfer learning have not yet been investigated and tested much in order to accomplish significant targets like fast training, real-time, and unified models for detecting anomalies inthe IoT network and therefore are potential areas for future research.

7.5 Conclusion

In the most recent decade, the utilization of IoT devices has expanded dramatically in varying backgrounds because of the IoT's ability to convert objects from several application areas into Internet hosts. Simultaneously, due to IoT security vulnerabilities, privacy and security of users are threatened. More vigorous security solutions are needed for IoT frameworks. Implementing machine learning techniques for IDS is one of the vital methods for securing IoT devices. This chapter presents a review of machine learning based intrusion detection methods utilized in IDS for IoT networks and systems. The detailed architecture of IoT and its threat classification per layers and design challenges are highlighted. Different research work, which summarizes intrusion detection strategies proposed for the IoT, specifically about machine learning algorithms available for IDS in IoT frameworks and its utilization by the researchers, are presented. Likewise, a survey of different datasets accessible for IoT security-related examination is explained. This chapter introduces researchers to detailed but comprehensive and insightful knowledge about the various IoT security challenges presently being faced, along with their possible solutions based on machine learning techniques [52–58].

References

1. Ida, I.B., et al. (2016) "A Survey on Security of IoT in the Context of eHealth and Clouds." 2016 11th International Design & Test Symposium (IDT): 25–30.
2. Babar, S. et al. (2011) "Proposed Embedded Security Framework for Internet of Things (IoT)." 2011 2nd International Conference on Wireless Communication, Vehicular Technology, Information Theory and Aerospace & Electronic Systems Technology (Wireless VITAE): 1–5.
3. Modi, C., et al. (2013) "A Survey of Intrusion Detection Techniques in Cloud." *Journal of Network and Computer Applications* 36: 42–57.
4. Mishra, P., et al. (2017) "Intrusion Detection Techniques in Cloud Environment: A Survey." *Journal of Network and Computer Applications* 77: 18–47.
5. Atamli-Reineh, A. and Martin, A. (2014) "Threat-Based Security Analysis for the Internet of Things." 2014 International Workshop on Secure Internet of Things: 35–43. IEEE.
6. Conti, M., et al. (2016) "A Survey of Man in the Middle Attacks." *IEEE Communications Surveys & Tutorials* 18: 2027–2051.
7. Karlof, C. and Wagner, D. (2003) "Secure Routing in Wireless Sensor Networks: Attacks and Countermeasures." *Proceedings of the First IEEE International Workshop on Sensor Network Protocols and Applications* 2003: 113–127.
8. Wallgren, L., et al. (2013) "Routing Attacks and Countermeasures in the RPL-Based Internet of Things." *International Journal of Distributed Sensor Networks* 9. https://doi.org/10.1155/2013/794326.
9. Bostani, H. and Sheikhan, M. (2017) "Hybrid of Anomaly-Based and Specification-Based IDS for Internet of Things using Unsupervised OPF Based on MapReduce Approach." *Computer Communications* 98: 52–71.
10. Asharf, J., et al. (2020) "A Review of Intrusion Detection Systems Using Machine and Deep Learning in Internet of Things: Challenges, Solutions and Future Directions." *Electronics* 9: 1177.
11. Bou-Harb, E., et al. (2014) "Cyber Scanning: A Comprehensive Survey." *IEEE Communications Surveys & Tutorials* 16: 1496–1519.

12. Zlomislic, V., et al. (2017) "Denial of Service Attacks, Defences and Research Challenges." *Cluster Computing* 20: 661–671.
13. Prabadevi, B. and Jeyanthi, N. (2014) "Distributed Denial of Service Attacks and Its Effects on Cloud Environment – A Survey." The 2014 International Symposium on Networks, Computers and Communications: 1–5. IEEE.
14. Game, S. and Raut, C. (2014) "Protocols for Detection of Node Replication Attack on Wireless Sensor Network." *IOSR Journal of Computer Engineering* 16: 01–11.
15. Araújo, N., et al. (2010) "Identifying Important Characteristics in the KDD99 Intrusion Detection Dataset by Feature Selection using a Hybrid Approach." 2010 17th International Conference on Telecommunications: 552–558.
16. Sujihelen, L., et al. (2015) "Detecting Node Replication Attacks in Wireless Sensor Networks: Survey." *Indian Journal of Science and Technology* 8. DOI:10.1016/j.jnca.2012.01.002.
17. Zhang, Z., et al. (2014) "IoT Security: Ongoing Challenges and Research Opportunities." 2014 IEEE 7th International Conference on Service-Oriented Computing and Applications: 230–234. IEEE.
18. Al-Fuqaha, A., et al. (2015) "Internet of Things: A Survey on Enabling Technologies, Protocols, and Applications." *IEEE Communications Surveys & Tutorials* 17: 2347–2376.
19. Al-Turjman, F. (2018) "QoS - Aware Data Delivery Framework for Safety-Inspired Multimedia in Integrated Vehicular-IoT." *Computer Communications* 121: 33–43.
20. Alabady, S.A., et al. (2018) "A Novel Security Model for Cooperative Virtual Networks in the IoT Era." *International Journal of Parallel Programming* 48: 280–295.
21. Alabady, S.A. and Al-Turjman, F. (2018) "Low Complexity Parity Check Code for Futuristic Wireless Networks Applications." *IEEE Access* 6: 18398–18407.
22. Sicari, S., et al. (2015) "Security, Privacy and Trust in Internet of Things: The Road Ahead." *Computer Networks* 76: 146–164.
23. Yang, Y., et al. (2017) "A Survey on Security and Privacy Issues in Internet-of-Things." *IEEE Internet of Things Journal* 4: 1250–1258.
24. Sikder, A.K. et al. (2017) "6thSense: A Context-Aware Sensor-Based Attack Detector for Smart Devices." ArXiv abs/1706.10220. https://arxiv.org/pdf/1706.10220.pdf.
25. Al-turjman, F., et al. (2017) "Seamless Key Agreement Framework for Mobile-Sink in IoT Based Cloud-Centric Secured Public Safety Sensor Networks." *IEEE Access* 5: 24617–24631.
26. Al-turjman, F. and Alturjman, S. (2018) "Context-Sensitive Access in Industrial Internet of Things (IIoT) Healthcare Applications." *IEEE Transactions on Industrial Informatics* 14: 2736–2744.
27. Ahmed, M., et al. (2016) "A Survey of Network Anomaly Detection Techniques." *Journal of Network and Computer Applications* 60: 19–31.
28. Haider, W. et al. (2017) "Generating Realistic Intrusion Detection System Dataset Based on Fuzzy Qualitative Modeling." *Journal of Network and Computer Applications* 87: 185–192.
29. Casas, P. et al. (2012) "Unsupervised Network Intrusion Detection Systems: Detecting the Unknown without Knowledge." *Computer Communications*35: 772–783.
30. Priya, G. (2017) "Efficient knn classification algorithm for big data." *Neurocomputing* 195: 143–148.
31. Li, L., et al. (2018) "Nearest neighbors based density peaks approach to intrusion detection." *Chaos Solitons & Fractals* 110: 33–40.
32. Liu, Y. and Pi, D. (2017) "A Novel Kernel SVM Algorithm with Game Theory for Network Intrusion Detection." *KSII Transactions on Internet & Information Systems* 11: 4043–4060.
33. Zarpelão, B., et al. (2017) "A Survey of Intrusion Detection in Internet of Things." *Journal of Network and Computer Applications* 84: 25–37.
34. Oh, D., et al. (2014) "A Malicious Pattern Detection Engine for Embedded Security Systems in the Internet of Things." *Sensors (Basel, Switzerland)* 14: 24188–24211.
35. Kasinathan, P., et al. (2013) "Denial-of-Service Detection in 6LoWPAN based Internet of Things." 2013 IEEE 9th International Conference on Wireless and Mobile Computing, Networking and Communications (WiMob): 600–607. IEEE.

36. Ioannou, C. and Vassiliou, V. (2019) . "Classifying Security Attacks in IoT Networks Using Supervised Learning." 2019 15th International Conference on Distributed Computing in Sensor Systems (DCOSS): 652–658. IEEE.
37. Amaral, J.P., et al. (2014) "Policy and Network-Based Intrusion Detection System for IPv6-Enabled Wireless Sensor Networks." 2014 IEEE International Conference on Communications (ICC): 1796–1801. IEEE.
38. Kumar, G. (2014)) "Evaluation Metrics for Intrusion Detection Systems - A Study." *International Journal of Computer Science and Mobile Applications* 2(11): 11–17.
39. Goeschel, K. (2016) "Reducing false positives in intrusion detection systems using data-mining techniques utilizing support vector machines, decision trees, and naive Bayes for off-line analysis." *SoutheastCon* 2016: 1–6.
40. C. I. for Cybersecurity (CIC) (2017) "IDS 2017 Datasets Research Canadian Institute for Cybersecurity UNB" [Online]. Available: https://www.unb.ca/cic/datasets/ids-2017.html
41. Sharma, S. and Dixit, M. (2016) "A Review on Network Intrusion Detection System Using Open Source Snort." *International Journal of Database Theory and Application* 9: 61–70.
42. Sahu, S. et al. (2014) "A Detail Analysis on Intrusion Detection Datasets." 2014 IEEE International Advance Computing Conference (IACC): 1348–1353. IEEE.
43. Chandolikar, N.S. "Selection of Relevant Feature for Intrusion Attack Classification by Analyzing KDD Cup 99."
45. Moustafa, N., and Slay, J. (2015) . "UNSW-NB15: a comprehensive data set for network intrusion detection systems (UNSW-NB15 network data set)."IEEE 2015 Military Communications and Information Systems Conference (MilCIS): 1–6. IEEE. DOI: 10.1109/MilCIS.2015.7348942.
44. Nguyen, H.T., et al. (2010) "Improving Effectiveness of Intrusion Detection by Correlation Feature Selection." IEEE 2010 International Conference on Availability, Reliability and Security: 17–24. IEEE. DOI: 10.1109/ARES.2010.70.
46. Sharafaldin, I., et al. (2018) "Toward Generating a New Intrusion Detection Dataset and Intrusion Traffic Characterization. In Proceedings of the 4th International Conference on Information Systems Security and Privacy (ICISSP 2018), pp. 108–116. ISBN: 978-989-758-282-0.
47. Hoque, N., et al. (2014) "Network Attacks: Taxonomy, Tools and Systems." *Journal of Network and Computer Applications* 40: 307–324.
48. Bhandari, A., et al. (2018) Packet Sniffing and Network Traffic Analysis Using TCP—A New Approach." in A. Kalam et al. (eds.) *Advances in Electronics, Communication and Computing, Lecture Notes in Electrical Engineering,* 443. https://doi.org/10.1007/978-981-10-4765-7_28.
49. Al-Garadi, M. et al. (2020) "A Survey of Machine and Deep Learning Methods for Internet of Things (IoT) Security." *IEEE Communications Surveys & Tutorials* 22: 1646–1685.
50. Thongkanchorn, K., et al. (2013) "Evaluation studies of three intrusion detection systems under various attacks and rule sets." 2013 IEEE International Conference of IEEE Region 10 (TENCON 2013): 1–4. IEEE.
51. Aravind, M. and Kalaiselvi, V. (2017) . "Design of an intrusion detection system based on distance feature using ensemble classifier." IEEE 2017 Fourth International Conference on Signal Processing, Communication and Networking (ICSCN): 1–6. IEEE.
52. Chaabouni, N., et al. (2019) "Network Intrusion Detection for IoT Security Based on Learning Techniques." *IEEE Communications Surveys & Tutorials* 21: 2671–2701.
53. Anwar, S., et al. (2017) "Cross-VM cache-based side channel attacks and proposed prevention mechanisms: A survey." *Journal of Network and Computer Applications* 93: 259–279.
54. Kolias, C., et al. (2016) "Intrusion Detection in 802.11 Networks: Empirical Evaluation of Threats and a Public Dataset."*IEEE Communications Surveys & Tutorial*s 18: 184–208.
55. Granjal, J., et al. (2015) "Security for the Internet of Things: A Survey of Existing Protocols and Open Research Issues." *IEEE Communications Surveys & Tutorials* 17: 1294–1312.

56. Mukherjee, S. and Sharma, N. (2012) . "Intrusion Detection using Naive Bayes Classifier with Feature Reduction." *Procedia Technology* 4: 119–128.
57. Raza, S. et al. (2013) "SVELTE: Real-time intrusion detection in the Internet of Things." *Ad Hoc Networks* 11: 2661–2674.
58. Hande, Kapil, & Shrawankar, Urmila (2021). Role of Machine Learning and Deep Learning Approaches in Designing Network Intrusion Detection System, Algorithms for Intelligent Systems,Proceedings of Integrated Intelligence Enable Networks and Computing (pp. 383–389) 10.1007/978-981-33-6307-6_39.

8

Case Study of Smart Farming Using IoT

Ameya N. Shahu, Chetan R. Wagh, Ritik B. Drona, Rohit A. Suryawanshi, Santosh Kagne, and Shrinit S. Patil
Government College of Engineering, Nagpur

8.1 Introduction

The Internet of Things (IOT) is the greatest, most efficient, and most important technique for answering questions about advancement in farming. IOT evolved from different components that have plenty of sensors, software, network components, and other electronic devices. Monitoring environmental factors is not the entire solution to extend the yield of crops. There are a number of other factors that decrease productivity. Hence, automation must be implemented in agriculture to beat these problems. So, to find solutions to such problems, it is necessary to develop an integrated system that will improve productivity at every stage. But, complete automation in agriculture is not achieved due to various issues. IoT is applicable in various methodologies of agriculture. Applications of IoT are smart cities, smart environment, smart water, smart metering, security and emergency, industrial control, smart agriculture, home automation, e-Health, etc. IoT is predicated on a device that is capable of analyzing the sensed information, then transmitting it to the user. From a survey of the United Nations – Food and Agriculture Organizations, worldwide food production needs to increase by 70% in 2050 for the evolving population [1]. Agriculture is the basis for the human species because it is the main source of food, and it plays an important role in the growth of a country's economy. There are many other factors that affect productivity to a great extent. Factors include attacks from insects and pests, which might be controlled by spraying the right insecticide and pesticide, and also attacks from untamed animals and birds when the crop grows up. The crop yield is declining due to unpredictable monsoon rainfalls, water scarcity, and improper water usage. Though smart farming is implemented at the research level, it is not given to the farmers as a product to induce benefits from the resources. The IOT technology is more efficient due to following reasons:

- Minimum human efforts
- Faster access
- Time efficiency
- Efficient communication

DOI: 10.1201/9781003166702-8

8.2 Literature Review

There are many papers that discuss implementation of computation systems in agriculture for better crop yield, less manual labor, water management, quality of produce, etc. Some of these papers are described below to explain the different concepts of automated farming.

Emerging technologies have made low-power and low-cost wireless sensor networks feasible. Wang, Qiang, & Terzis; Andreas & Szalay; and Alexander designed sensor nodes that are placed completely underground and are used to collect soil temperature and moisture. These sensors use frequency to deliver the measurements to one of multiple relay nodes located above ground. Some of these relay nodes that are capable of long-range communications forward the data collected from the network's sensor nodes to a base node, which is connected to a workstation [2]. This concept does not require an Internet gateway; also, it mainly focuses on soil measurements rather than water management.

Distributed in-field sensor-based irrigation systems offer a potential solution to support site-specific irrigation management that allows producers to maximize their productivity while saving water. Kim, Yunseop, & Evans and Robert & Iversen [3] described details of the design and instrumentation of variable rate irrigation, a wireless sensor network, and software for real-time in-field sensing and control of a site-specific precision linear-move irrigation system. Field conditions were site-specifically monitored by six in-field sensor stations distributed across the field based on a soil property map, and periodically sampled and wirelessly transmitted to a base station. An irrigation machine was converted to be electronically controlled by a programming logic controller that updates georeferenced location of sprinklers from a differential global positioning system (GPS) and wirelessly communicates with a computer at the base station. Communication signals from the sensor network and irrigation controller to the base station were successfully interfaced using low-cost Bluetooth wireless radio communication [3]. This paper describes a generalized methodology to implement a remote controlled irrigation system.

The research study presented by Bhatnagar, Vivek, & Singh; Gulbir & Kumar; Gautam & Gupta; and Rajeev discussed a detailed work of the eminent researchers and designs of computer architecture that can be applied in agriculture for smart farming. This research study also highlights various unfolded challenges of the IoT in agriculture [4].

The IoT refers to the rapidly growing network of connected objects that are able to collect and exchange data using embedded sensors. NB-IOT (narrowband – Internet of Things), LoRa, and Sigfox wireless technologies have been getting a good deal of attention globally as the market for wireless technology matures in light of the prospect of billions of connections. The goal of the LoRa Alliance, LoRaWAN adopters, and SigFox is for mobile network operators to adopt their technology for IoT deployments over both city and nationwide low-power, wide-area networks (LPWANs) [5]. This paper discusses different ways for networking in IoT architecture.

8.3 Problems with Existing Systems

8.3.1 Agricultural Issues

Despite the overwhelming size of the agricultural sector, yields per hectare of crops in India are generally low compared to international standards. Improper water management is another problem affecting India's agriculture. At a time of accelerating water shortages and environmental crises, as an example, the rice crop in India is allocating disproportionately high amounts of water. One result of the inefficient use of water is that water tables in regions of rice cultivation, like Punjab, are increasing, while soil fertility is declining. Aggravating the agricultural situation is an ongoing Asian drought and inclement weather.

Although during 2000–2001 a monsoon with average rainfall was expected, prospects of agricultural production during that period were not considered bright. This was partially because of the relatively unfavorable distribution of rainfall, resulting in floods in certain parts of the country and droughts in others.

Despite the fact that agriculture accounts for as much as 25% of the Indian economy and employs an estimated 60% of the proletariat, it is considered highly inefficient, wasteful, and incapable of solving the hunger and malnutrition problems. Despite progress in this area, these problems have continued to frustrate India for many years. It is estimated that as much as one-fifth of the entire agricultural output is lost because of inefficiencies in harvesting, transport, and storage of government-subsidized crops.[6]

8.3.2 Farmer Issues and Monitoring Issues

1. *Raising agricultural productivity per unit of land:* Raising productivity per unit of land will have to be the biggest area of agricultural growth as virtually all cultivable land is farmed. Water resources are limited, and water for irrigation must manage increasing industrial and concrete needs. All measures to extend productivity will need exploiting, amongst them: increasing yields, diversification to higher value crops, and developing value chains to scale back marketing costs.
2. *Reducing rural poverty through a socially inclusive strategy that comprises both agriculture and non-farm employment:* Rural development must also benefit the poor, landless, women, and scheduled castes and tribes. Moreover, there are strong regional disparities: the bulk of India's poor are in rain-fed areas or within the Eastern Indo-Gangetic plains. Reaching such groups has not been easy. While progress has been made – the agricultural population classified as poor fell from nearly 40% within the early 1990s to below 30% by the mid-2000s (about one-tenth reduction per year) – there is a transparent need for a faster reduction. Hence, poverty alleviation may be a central pillar of the agricultural development efforts of the government as well as the United Nations.
3. *Ensuring that agricultural growth responds to food security needs:* The sharp rise in food-grain production during India's revolution of the 1970s enabled the country to realize self-sufficiency in food-grains and prevented the threat of famine. Agricultural intensification within the 1970s to 1980s saw an increased demand for rural labor that raised rural wages and, along with declining food prices, reduced rural poverty. However, agricultural growth within the 1990s and 2000s caught up, averaging about 3.5% every year, and cereal yields have increased by only 4% every year within the 2000s [6]. The slow-down in agricultural growth has become a significant cause for concern. India's rice yields are one-third of China's and about half those in Vietnam and Indonesia. The same is true for many other agricultural commodities.

Policy makers will thus have to initiate and/or conclude policy actions and public programs to shift the world from the prevailing policy and institutional regime that appears to be not viable and build a solid foundation for a way more productive, internationally competitive, and diversified agricultural sector.

8.4 Proposed Solution

Promoting new technologies and reforming agricultural research and extension: Major reform and strengthening of India's agricultural research and extension systems is the most important need for agricultural growth. These services have declined over time thanks to chronic underfunding of infrastructure and operations, no replacement of aging researchers, and failure to provide broad access to state-of-the-art technologies. Research now has little to produce beyond the time-worn packages of the past. Public extension services are struggling and offer little new knowledge to farmers. There is insufficient connection between research and extension, or between these services and the private sector.

Improving water resources and irrigation/drainage management: Agriculture is India's largest user of water. However, increasing competition for water between industry, domestic use, and agriculture has highlighted the requirement to plan and manage water on a basin and multi-sectoral basis. As urban and other demands multiply, less water will be available for irrigation. Ways to radically enhance the productivity of irrigation ("more crop per drop") must be found. Piped conveyance, better on-farm management of water, and use of more efficient delivery mechanisms like drip irrigation are among the actions that should be taken. There is also a desire to manage as the opposition exploits the utilization of groundwater. Incentives to pump less water, like levying electricity charges or community monitoring of use, have not yet succeeded beyond sporadic initiatives [7].

Other key priorities: Modernizing irrigation and drainage departments to integrate the participation of farmers and other agencies in managing irrigation water, improving cost recovery, and rationalizing public expenditures, with priority to completing schemes with the best returns and allocating sufficient resources for operations and maintenance for the sustainability of investments, are key priorities.

Facilitating agricultural diversification to higher value commodities: Encouraging farmers to diversify to higher value commodities is going to be a major factor for higher agricultural growth, particularly in rain-fed areas where poverty is high. Moreover, considerable potential exists for expanding agro-processing and building competitive value chains from producers to urban centers and export markets. While diversification initiatives should be left to farmers and entrepreneurs, the government can, first and foremost, liberalize constraints to marketing, transport, export, and processing. It may also play a small regulatory role, taking reasonable care that this does not become an impediment.

Promoting high-growth commodities: Some agricultural sub-sectors have particularly high potential for expansion, notably dairy. The livestock sector, primarily thanks to dairy, contributes over 1/4 of agricultural GDP and may be a source of income for 70% of India's rural families, mostly people who are poor and families headed by women.

Growth in milk production, at about 4% per year, has been brisk, but future domestic demand is anticipated to grow by a minimum of 5% per year [6]. Milk production is constrained, however, by the poor genetic quality of cows, inadequate nutrients, inaccessible veterinary care, and other factors. A targeted program to tackle these constraints could boost production and have a decent impact on poverty.

Developing markets, agricultural credit, and public expenditures: India's legacy of intensive government involvement in agricultural marketing has created restrictions on internal and external trade, leading to cumbersome and high-cost marketing and transport options for agricultural commodities. Even so, private sector investment in marketing, value chains, and agro-processing is growing, but much slower than their potential. While some restrictions are being lifted, considerably more has to be done to enable diversification and minimize consumer prices. Improving access to finances for farmers is another need because it remains difficult for farmers to earn credit. Moreover, subsidies on power, fertilizers, and irrigation have progressively come to dominate government expenditures around the world, and they are now fourfold larger than investment expenditures, crowding out top priorities like agricultural research and extension.

8.5 Components and Sensors

8.5.1 DHT11 – Temperature and Humidity Sensor

The DHT11 sensor module is commonly used to measure temperature and relative humidity of atmosphere. This sensor can be easily interfaced with any micro-controller, such as Arduino, Raspberry Pi, etc., to measure humidity and temperature instantaneously. The sensor comes with a dedicated NTC to measure temperature and an 8-bit microcontroller to output the values of temperature and humidity through serial communication. The sensor is also factory calibrated and hence is easy to interface with other microcontrollers. It has an operating voltage range of 3.5v to 5.5v. The sensor can measure temperature from 0°C to 50°C and relative humidity from 20% to 90%, with an accuracy of ±1°C and ±1%. The sampling rate of this sensor is 1Hz, i.e., it gives one reading for every second [7].

8.5.1.1 Working of DHT11

The DHT11 sensor consists of a capacitive humidity-sensing element and a thermistor for sensing temperature. The humidity-sensing capacitor has two electrodes, with a moisture holding substrate as a dielectric between them. When water vapor is absorbed by the substrate, ions are released by the substrate, which increases the conductivity between the electrodes. The change in resistance between the two electrodes is proportional to the relative humidity. The IC processes the resistance values and changes them into digital form [8]. The internal structure of the DHT11 sensor is shown in Figure 8.1.

For measuring temperature, this sensor uses a negative temperature coefficient thermistor, which causes a decrease in its resistance value with increase in temperature.

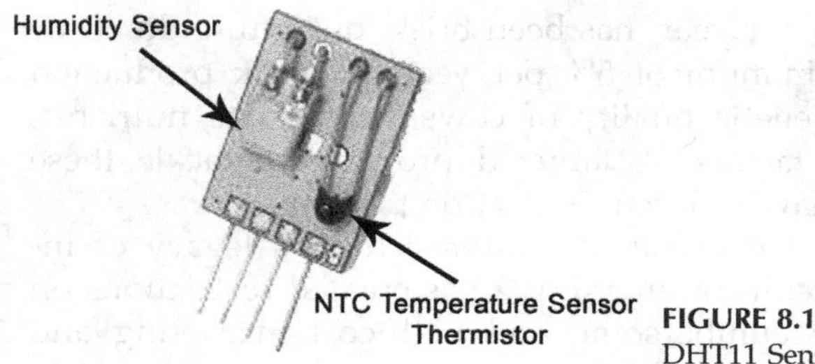

FIGURE 8.1
DHT11 Sensor Internal Structure

To get a value, even for the smallest change in temperature, this sensor is usually made up of semiconductor ceramics or polymers.

8.5.2 Solenoid Valve

In this project, the 12V DC solenoid valve, 2-way type, was used for the simple on/off switch for water flow control. Because this voltage is compatible with many low-voltage systems, the valve is widely used. 12v AC or DC valves may also be found in domestic showers and public display systems, such as water fountains and water curtain display systems. Although ultimately powered by the main AC supply, this type of equipment will use a power supply to provide a 12v supply for the solenoid valves, thereby reducing any possible safety risk by separating potentially hazardous high voltage from the water [9].

In the case of a normally closed (fail-safe closed) solenoid valve, when 12 volts is applied, the solenoid valve opens, allowing flow, as shown in Figure 8.2. When the 12 volts are removed from the solenoid valve, then the valve will automatically close and prevent flow along the pipe [9], as shown in Figure 8.2.

The valve features a solenoid, which is an electric coil with a movable ferromagnetic core (plunger) in its center. In the rest position, the plunger closes off a small orifice. An electric current through the coil creates a magnetic field. The magnetic field exerts an upward force on the plunger, opening the orifice. This is the basic principle that is used to open and close solenoid valves [9].

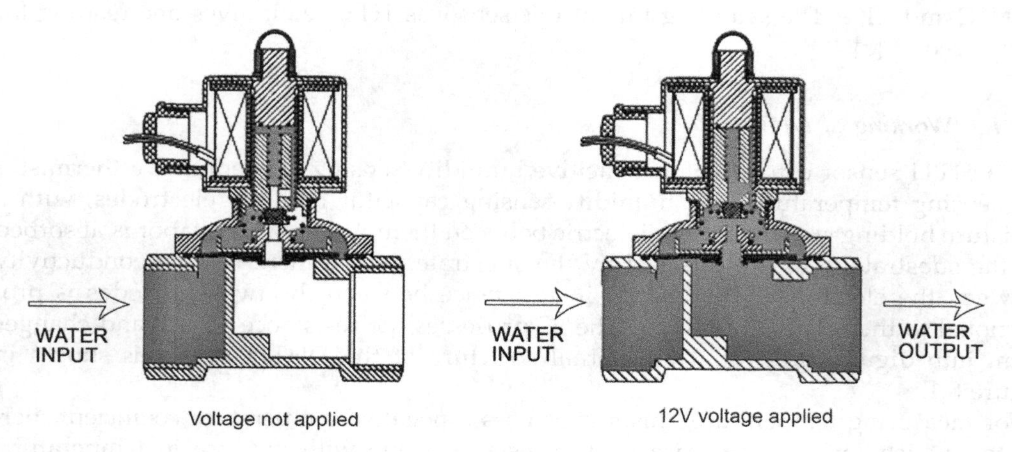

FIGURE 8.2
Normally Closed Solenoid Valve.

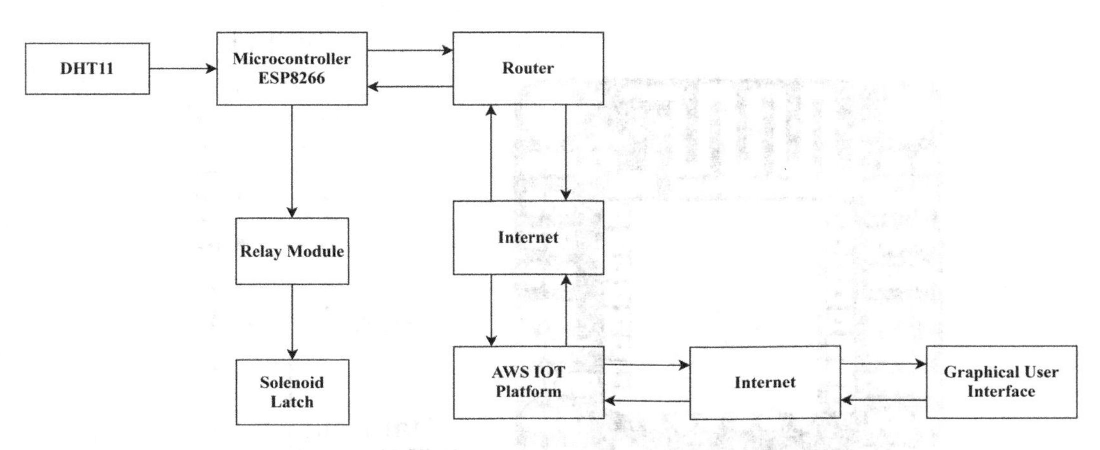

FIGURE 8.3
System Architecture.

8.5.3 NodeMCU (ESP 8266)

The ESP8266 is a low-cost WiFi microchip, with a full TCP/IP stack and microcontroller capability, produced by Espressif Systems in Shanghai, China. ESP8266 has 1 MB of built-in flash, allowing the building of single-chip devices capable of connecting to WiFi. It has 17 GPIO pins with operating voltage of 3.3V. It supports ADC, I2C as well as SPI communication protocol [10].

8.6 IoT System Architecture

Overall architecture, as shown in Figure 8.3, is mainly divided into three components:

1. Communication between DHT11 and microcontroller
2. Communication between solenoid latch and microcontroller
3. Communication between microcontroller and AWS IoT platform (MQTT Broker)

8.6.1 Communication between DHT11 and Microcontroller

ESP8266 is used as a microcontroller that reads temperature and relative humidity from DHT11 through serial communication. DHT11 has three pins: Vcc, ground, and data pins. Vcc is connected with a 3.3v output pin of ESP8266. Ground pin of DHT11 is connected to the GND pin of microcontroller ESP8266 [7]. Data pins of DHT11 can be connected to any of the GPIO pins of ESP8266. In this case, it is connected to GPIO 5 [11].

DHT11 reads data in analog form and internally converts it into digital form. These data are stored in a buffer in DHT11 [12]. Whenever a microcontroller requests data, it transfers to ESP8266 through serial communication. Both temperature and relative humidity are stored in 8-bit form. A circuit diagram of the connection is shown in Figure 8.4.

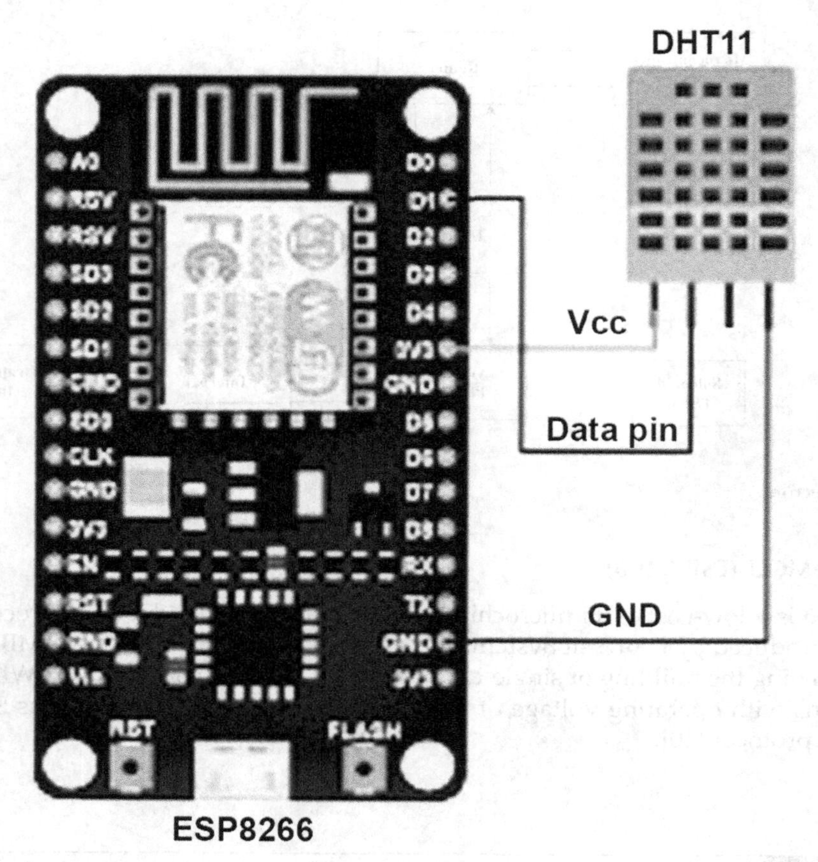

FIGURE 8.4
DHT11 Connection with ESP8266.

8.6.2 Communication between Solenoid Valve and Microcontroller

The solenoid valve operates at 12v, and the microcontroller ESP8266 operates on 3.3v to 5v. The maximum possible output voltage by ESP8266 is 3.3v; hence, the solenoid valve is given an external 12v source, and it is connected with the relay module so that it can be controlled with ESP8266 [13].

A 12v external source is connected to the common relay module. Normally, the closed relay module is connected with a solenoid valve. The ground of solenoid is connected to the ground of the external source. The signal pin of the relay is connected to a GPIO of ESP8266. In this case, it is connected to GPIO 2. GPIO 2 is also connected to the onboard LED of ESP8266. So, whenever the signal pin to relay goes high, the onboard LED also glows. The relay module also has Vcc and ground connected to the battery, which powers ESP8266. The signal pin of relay is flipped to change the state of the solenoid valve [13]. The circuit connection is shown in Figure 8.5.

8.6.3 Communication between Microcontroller and AWS IoT Platform

The microcontroller is connected to the Internet through WiFi. ESP8266 has an onboard WiFi module ESP8266 that uses MQTT protocol to exchange data with the AWS IoT

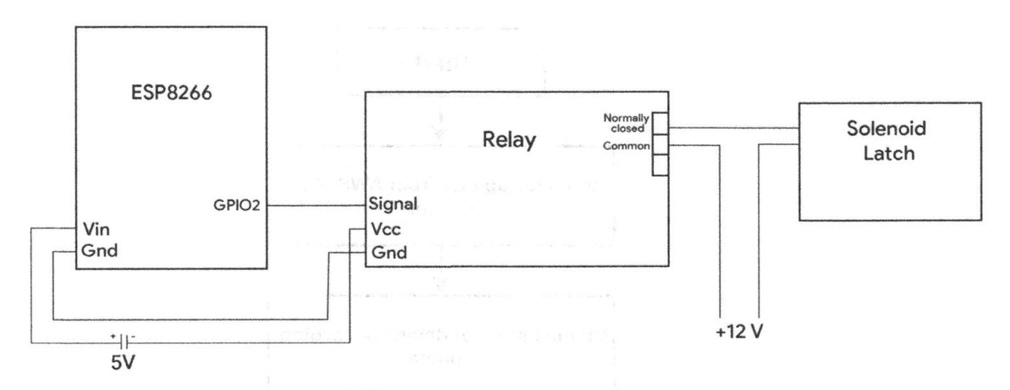

FIGURE 8.5
Connection of Solenoid Latch.

platform. ESP8266 is connected with the AWS IoT platform with a private key, which is uploaded with firmware in ESP8266 [14].

After every 5 sec, ESP8266 reads temperature and relative humidity from DHT11 and publishes a message on the shadow update topic of the AWS IoT broker. Shadow is the document that is maintained at the AWS IoT end. The shadow document is the copy of the device that stores the real-time state of the device. There are different kinds of shadow topics, such as update, accepted, rejected, and delta. When a message is published in the update topic, it updates the shadow document. If update is accepted, the AWS IoT platform publishes the same message on the accepted topic, which is received by all the other clients, also as an acknowledgement for ESP8266. Also, the AWS IoT publishes the difference between received update messages and shadow documents [15]. This message is exchanged in the form of JSON. The message published by ESP8266 has the below format of JSON:

```
{
"ThingID" : {
"Temperature" : value,
"Humidity" :value,
"isValveOn" : true/false
}
}
```

These data published by ESP8266 are represented in the form of a graph on a web portal that is authenticated by AWS IoT. Also, these data are stored in the AWS S3 service. The web portal also has a button through which solenoid valves can be operated manually. Whenever a button is toggled, it publishes a shadow message to change the state of the valve as follows:

```
{
"ThingID" : {
"isValveOn" : Toggled state of button
}
}
```

After the update is received by the AWS IoT, it publishes the accepted or rejected messages on the respective shadow topic, and the difference between the received update and the shadow document on delta topic. The delta topic is subscribed by

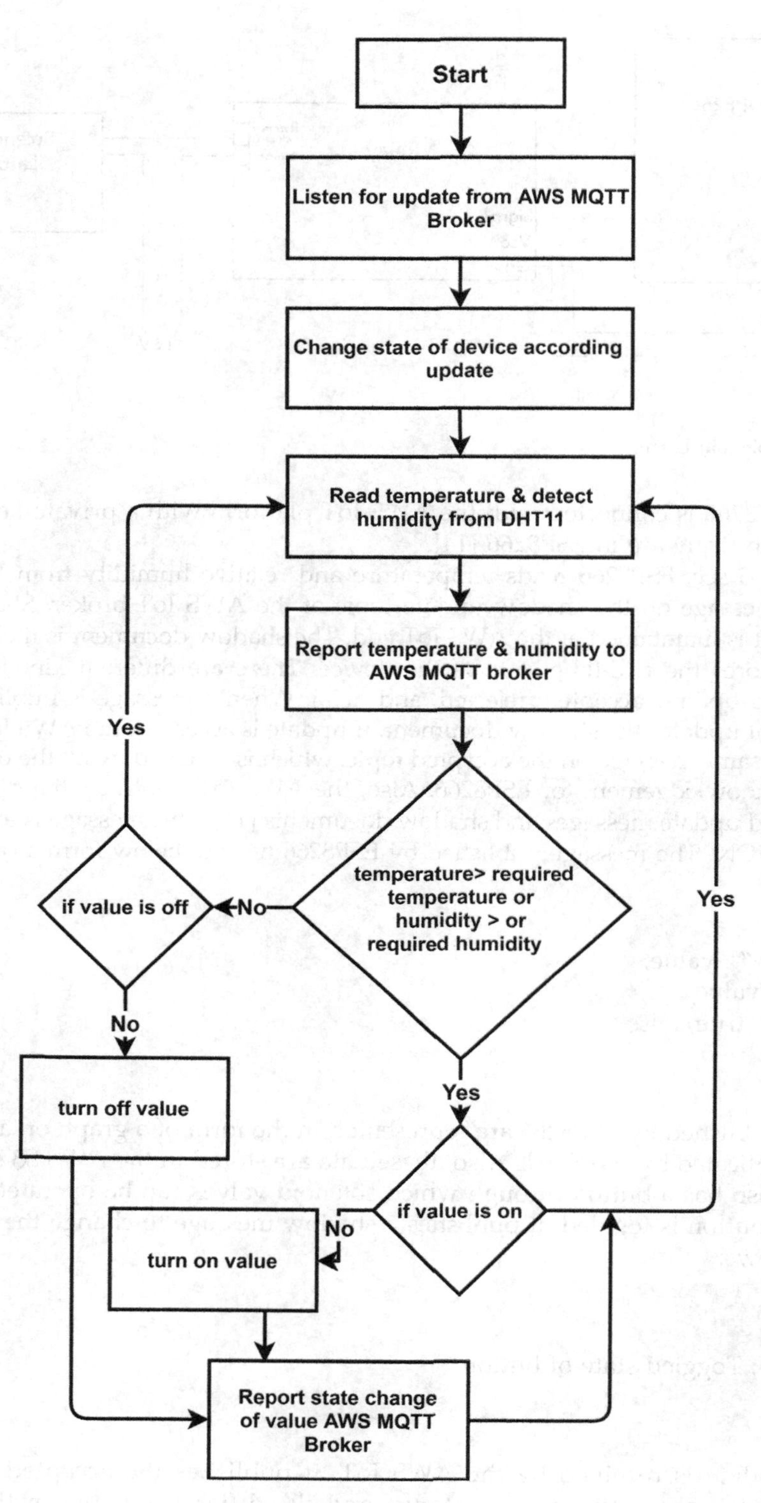

FIGURE 8.6
Flowchart of Firmware in ESP8266.

ESP8266. When ESP8266 receives a message on a delta topic, it changes the signal pin state of relay according to the received message on the delta topic. As soon as the signal pin state changes, it switches on or off the solenoid valve. If the temperature or relative humidity is not in the required range, ESP8266 changes the state of the valve and updates the shadow document [15]. A flow chart for firmware is shown in Figure 8.6.

8.7 Field Implementation

8.7.1 Agricultural Land Preparation for Crop

Land preparation or tillage practice could be an important practice to boost yields from crops. It is among the measures used to prevent crop disease and pest invasion. Tillage practice is the mechanical pulverization or manipulation of the soil to create favorable conditions for the expansion of crops. Proper land preparation reduces weed competition for light and nutrients as well as frost susceptibility during mounding. Also, it improves soil, water, and air conditions and water cultivation. It loosens tight or compacted soil to permit nitrogen fixation [16].

For ploughing, an area of 150 sq.feet was identified to implement the proof of concept. As the plantation field was small in size, it was cultivated with the help of a pickaxe. Then, the cultivated field soil was mixed with cow dung (as manure). Plowing depth was about 15–20 cm. After preparation of the land, the drip pipes were laid down exactly 1.5 feet from each other. The whole field was divided into four rows. The laid-down drip irrigation system is shown in Figure 8.7.

FIGURE 8.7
Actual Farm Irrigation Image.

8.7.2 Planting the Crop

Small pits were dug, and the carrot plants were planted in the pits using manual instruments and sprinkled with water. The plants were planted at a distance of 5–6 cm one by one, making small pits at a distance between two rows at 45 cm to allow for proper growth and equal spreading of the plant's leaves for independent development and to make sure the plants did not hamper each other's growth. Proper distance is important for full-fledged growth and maturing until the full period of crop growth is reached.

8.7.3 Crop Selection According to Climate

Factors to be considered during selection of the crop include environmental conditions, e.g., availability of water and electricity near the fields, and soil conditions, e.g., moisture content, soil content like NPK, and ph level. Considering the above requirements, carrots fulfilled the conditions per our geographical location and weather conditions with all the contents in the soil.

Carrots are sometimes slow to germinate. They can take 2–3 weeks to show any sign of life, so do not panic if your carrots do not appear right away. To keep track of where they were planted, mix carrot seeds with quick-germinating radish seeds or sow radish seeds in rows between carrot rows. The radishes will grow quickly, and by the time the carrots really start to grow, the radishes can be harvested [17].

8.7.3.1 Climate Requirement for Carrot Farming

Carrots are often grown in various forms of soils, but deep, loose, loamy soil is compatible for its cultivation. For optimum yield, soil pH value should be within the range of 5.5 to 7. Together with this, it is also necessary for the soil to be brittle because it makes the carrots long and robust.

To make the soil brittle, level it with the assistance of a leveler after every ploughing. There should be a correct water system as well.

The optimum sowing time for carrot cultivation is October to November. The optimum temperature for growth is between 16 °C and 20 °C. Temperatures below 16 °C affect the color. Temperatures between 15 °C to 20 °C are better for developing attractive roots with excellent red colour and quality. Carrots are mostly eaten as a vegetable in India. They are eaten raw (salad) or cooked but are also utilized in making pickles, jam, juice, etc. Carrots could be a winter season crop cultivated worldwide for their fleshy, edible roots [17,18]. They could be a major vegetable crop and great source of axerophthol. Carrot juice is a rich source of carotene and is typically used for coloring food articles. Carrots have many medicinal properties and also are utilized in Ayurvedic medicine.

The root portion of the carrot is edible to humans, while the surface part is fed to animals. Eating carrots strengthens the immune system and improves eyesight. Karnataka, Punjab, Haryana, Maharashtra, Uttar Pradesh, and Andhra Pradesh are major carrot-growing states in India [17].

8.7.4 IoT System Implementation

The main objective was to develop a model that is easy to install, is sustainable, and will protect all the components inside.

A DHT11 sensor was placed in the soil to measure temperature and humidity. This DHT11 sensor was connected to the microcontroller ESP8266. The ESP8266 microcontroller was kept in the box to avoid direct contact with water. Microcontroller ESP8266 was powered with a battery of 5v and 500mAh [10]. This battery was also kept in the box. The battery was able to power a microcontroller for approximately 15 days.

The microcontroller was connected to the Internet through a WiFi network. The microcontroller read the temperature and humidity data from the DHT11 sensor and sent it to the AWS IoT Platform through the MQTT (message queue telemetry transport) protocol. MQTT is a publish/subscribe, extremely simple, and lightweight messaging protocol designed for constrained devices and low-bandwidth, high-latency, or unreliable networks [19]. The publisher/subscriber model allows MQTT clients to communicate one-to-one, one-to-many, and many-to-one.

The relay module used to control the solenoid valve was connected to the ESP8266 microcontroller. When the server sent a message to the microcontroller, the relay module accordingly turned the solenoid valve on and off. The solenoid valve was an electrically controlled valve in which the on and off valve was controlled by the magnetic field of the solenoid. The relay module was also kept in a box to avoid circuit or component damage due to physical conditions, and it was easy to install. The box had to stay on the ground, connected to WiFi and the solenoid valve. To publish the data on the AWS IoT Platform, first we needed to create things for the microcontrollers so that they could publish the temperature and humidity data from the DHT11 sensor to the cloud. We could then represent the data published in various formats like graph values, etc.

For firmware for ESP8266, Mongoose OS was used. Mongoose OS is a real-time OS tool for programming ESP8266. It provides SDK in C/C++ and Javascript and also provides support for different operating systems such as Windows, MacOS, Ubuntu, etc [20]. For this project, Javascript SDK of mongoose OS was used. For flashing of this code into the microcontroller, Mongoose OS provided a tool called MOS tool. It can be used to flash code into microcontrollers such as ESP8266, ESP32, etc.

As two ESP8266 were involved in the system, we created two different project structures for each of the microcontrollers. For two microcontrollers, two different things on the AWS portal were created and mentioned their thing IDs in their respective projects. Both the microcontrollers were connected to their respective DHT11 sensors, and one of them was connected to the valve or latch that was responsible for allowing or stopping the flow of water into the field. As only one microcontroller was connected to the valve, it had two functions – to send the data (temperature and humidity) sensed by the DHT11 sensor connected to it, and to control the valve based on the data recorded and the conditions required for the crop. Both the sensors were placed diagonally opposite on the field so that variations in temperature and humidity were tracked accurately.

As the temperature required for the farming of carrots should be in the range of 15 °C to 20 °C and 90% to 95% relative humidity [17], if the temperature went above or the relative humidity went lower than the range, then the valve switched on, and water was released into the field in order to decrease the temperature of the atmosphere around the crops [21]. The other thing, i.e., microcontroller did only one job i.e., the work of sending the data to the portal. The data sent by microcontrollers were stored on the portal and could be analyzed using the visualizations provided in the portal. Visualization of the data is shown in Figure 8.8.

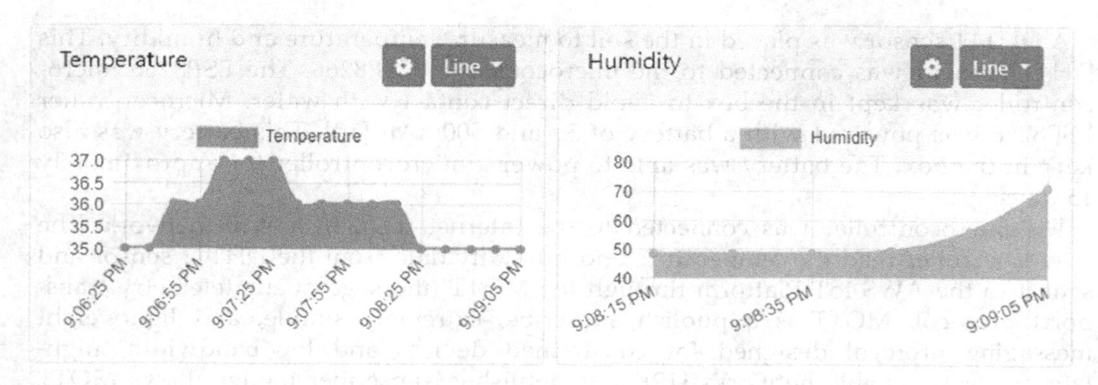

FIGURE 8.8
Visualization of Temperature and Humidity (Graphical).

8.8 Future Scope

Implementation can be made on a larger scale with more field area. Proper messaging and notification systems can be developed to rectify system anomalies and increase the yield and development of crops.

Humidity and temperature are monitored with the help of the IoT. The devices are combined with a lot of smart farming sensors, such as an air quality gas measurement sensor (MQ135), soil moisture sensor, soil pH sensor, and soil nutrient sensor, so they collect various data from the environment and send it to a portal to map the climate condition. With the help of this data, we can solve a lot of problems and make an artificial environment for enhancing the growth of plants. The required amount of sunlight for crops can be produced artificially using halogen bulbs. In addition, the moisture in the air can be maintained using fogger machine systems.

8.9 Conclusion

This proposed system enhances growth and multiplies the field area. The sensor network has enhanced capabilities of smart farming technology versus the conventional method of farming. This implementation system provides proper surveillance of the development of crops and will lead to better nourishment, making a high crop yield with a better profit margin.

References

1. India: Issues and priorities for agriculture. (2012, May 12). Retrieved from http://www.worldbank.org/en/news/feature/2012/05/17/india-agriculture-issues-priorities.

2. Wang, Q., Terzis, A., and Szalay, A. (2010). A Novel Soil Measuring Wireless Sensor Network. *Conference Record – IEEE Instrumentation and Measurement Technology Conference.* 412–415. 10.1109/IMTC.2010.5488224.
3. Kim, Y., Evans, R., and Iversen, W.M. (2008). Remote Sensing and Control of an Irrigation System Using a Distributed Wireless Sensor Network. *Instrumentation and Measurement, IEEE Transactions on* 57: 1379–1387. 10.1109/TIM.2008.917198.
4. Bhatnagar, V. Singh, G.,Kumar, G., and Gupta, R. (2020). Internet of Things in Smart Agriculture: Applications and Open Challenges. *International Journal of Students' Research in Technology & Management*: 1–3. 10.18510/ijsrtm.2020.812.
5. Chaudhary, H. Tank, B., and Patel, H. (2018). Comparative Analysis of Internet of Things (IoT) based Low Power Wireless Technologies. *International Journal of Engineering Research and V7.* 10.17577/IJERTV7IS010001.
6. India – Agriculture. (n.d.). Retrieved December 12, 2020, from https://www.nationsencyclopedia.com/economies/Asia-and-the-Pacific/India-AGRICULTURE.html.
7. Gay, W. W. (2014). DHT11 sensor. Experimenting with Raspberry Pi (pp. 1–13). doi:10.1007/978-1-4842-0769-7_1
8. N. (2019, August 06). Dht11 sensor definition, working and applications. Retrieved December 05, 2020, from https://www.elprocus.com/a-brief-on-dht11-sensor/
9. Solenoid valve - how they work. (n.d.). Retrieved December 16, 2020, from https://tameson.com/solenoid-valve-types.html
10. ESP8266 Technical Reference. (2020). 1.7, 1–104. Retrieved from https://www.espressif.com/sites/default/files/documentation/esp8266-technical_reference_en.pdf.
11. CodeChamp, & Instructables. (2017, July 25). Interface DHT11 (HUMIDITY Sensor) Using nodemcu. Retrieved December 20, 2020, from https://www.instructables.com/Interface-DHT11-Humidity-Sensor-Using-NodeMCU/
12. Agnihotri, N. (2020, August 11). Arduino compatible Coding 15: Reading sensor data from DHT-11 without using a library. Retrieved December 12, 2020, from https://www.engineersgarage.com/microcontroller-projects/articles-arduino-dht11-humidity-temperature-sensor-interfacing/
13. Farmhackunsw, & Instructables. (2019, April 11). Nodemcu relay controlled solenoid valve. Retrieved December 01, 2020, from https://www.instructables.com/NodeMCU-Relay-Controlled-Solenoid-Valve/#:~:text=Introduction%3A%20NodeMCU%20Relay%20Controlled%20Solenoid%20Valve&text=The%20nodeMCU%20uses%20the%20ESP8266,out%20this%20quick%20start%20guide.
14. Sahay, M. R., Sukumaran, M. K., Amarnath, S., and Palani, T. D. (2019). Environmental Monitoring System Using IoT and Cloud Service at Real-Time. 5–6. From https://easychair.org/publications/preprint_open/5Lg1
15. Németi, F., Pauletto, G., Duay, D., and Comtesse, X. (2017). Device Shadow MQTT topics. Retrieved December 07, 2020, from https://docs.aws.amazon.com/iot/latest/developerguide/device-shadow-mqtt.html
16. Reddy, J. (2019, April 22). Land Preparation Types; Methods; Objectives; Advantages. Retrieved January 20, 2020, from https://www.agrifarming.in/land-preparation-types-methods-objectives-advantages
17. Old Farmer's Almanac. (n.d.). Growing Carrots. Retrieved January 06, 2021, from https://www.almanac.com/plant/carrots#:~:text=Carrots%20are%20sometimes%20slow%20to,in%20rows%20between%20carrot%20rows.
18. Gopal Editor. Climate Requirement for Carrot Farming. www.Digitrac.In, 22 Jan. 2021, www.digitrac.in/agricare-blog/climate-requirement-for-carrot-farming/#:%7E:text=The%20temperatures%20between%2015%20to,pickles%2C%20jam%2C%20juice%20etc.
19. Hunkeler, U., Truong, H. L., and Stanford-Clark, A. (2008). MQTT-S — A Publish/Subscribe Protocol for Wireless Sensor Networks. 3rd International Conference on Communication

Systems Software and Middleware and Workshops (COMSWARE '08). doi:10.1109/COMSWA.2008.4554519
20. Introduction to Mongoose OS. (n.d.). Retrieved November 07, 2020, from https://mongoose-os.com/docs/mongoose-os/userguide/intro.md
21. Dunbar, B. (n.d.). Water vapor confirmed as major player in climate change. Retrieved January 05, 2021, from https://www.nasa.gov/topics/earth/features/vapor_warming.html#:~:text=Increasing%20water%20vapor%20leads%20to,increase%20in%20a%20spiraling%20cycle.

Section III

Data Science, Deep Learning, and Machine Learning

9

Stochastic Computing for Deep Neural Networks

Sunny Bodiwala[1] and Nirali Nanavati[2]
[1]*Gujarat Technological University, Ahmedabad, Gujarat, India*
[2]*Sarvajanik College of Engineering and Technology, Surat, Gujarat, India*

9.1 Introduction

Humans have always aspired to create machines that think. People are inclined toward intelligent systems for automating routine labor, speech and image recognition, disease diagnosis, developing self-driving motors, etc. The ability of an artificial intelligence (AI) framework to gain knowledge on its own, by extracting important information from data, is machine learning [1]. Deep learning has arisen as another zone of AI research that permits a machine to consequently gain complex capacities straightforwardly from the information by removing portrayals at numerous degrees of deliberation [2]. Deep neural networks (DNNs) have accomplished remarkable progress in many AI applications, for example, discourse acknowledgment [3] and object detection [4]. Albeit such undertakings are instinctively settled by humans, they initially demonstrated to be a genuine test to computerized reasoning.

In spite of their success, when contrasted with other AI techniques, DNNs require more calculations because of the deep architectural model. Besides, developer's desire for better execution will in general increase the size of the models, prompting longer training and testing time just as more computational resources are required for execution. The overall accuracy of these models depends on the utilization of high-performance infrastructure to implement DNNs. However, high-performance cloud infrastructure incurs huge power utilization and a huge equipment cost, accordingly restricting their deployment for low-cost and low-power applications, for example, implanted and wearable gadgets that need low power and small hardware [5]. Such applications progressively use AI algorithms to perform essential tasks, for example, speech-to-text transcription, natural language processing, and image and video recognition [2,6]. Subsequently, to implement such models in resource constraint frameworks, an alternative option should be found. At times, specific equipment has been designed utilizing Application Specific Integrated Circuits (ASICs) and Field Programmable Gate Arrays (FPGAs) [5,7]. All things considered, an edge of progress exists if the internal structure of models is additionally modified.

DOI: 10.1201/9781003166702-9

Stochastic computing (SC) as an important option in contrast to binary computing is considered in this chapter. SC works on arbitrary bit sequences, in which probabilities are given by the likelihood of a self-assertive bit in the grouping being one. This portrayal is especially alluring as it empowers minimal overhead implementation of key arithmetic units utilizing basic rationale circuits [8]. For instance, addition and multiplication can be performed utilizing a multiplexer (MUX) and an AND gate individually. Stochastic processing offers an extremely low hardware footprint, high level of error resilience, and the capacity to compromise calculation time and exactness with no extra hardware changes [9]. It subsequently can possibly actualize DNNs with fundamentally diminished hardware impression and low cost utilization. SC has a few drawbacks, including accuracy issues because of the inborn fluctuation in assessing the likelihood spoke by stochastic grouping. Besides, sudden increment in the accuracy of a stochastic implementation requires an outstanding expansion of bit stream length [8], consequently expanding the general computation. The more reasonable consideration will be stochastic arithmetic for the application where the precision necessities in the separate calculations are moderately low.

DNNs are described by a characteristic error tolerance, which recognizes them from other AI strategies that require exact calculations and a definite number portrayals. Moreover, Bishop [10] and Murray and Edwards [11] show that the expansion of clamor during the preparation of a neural model improves the model's performance. Designers can naturally support this error tolerance by thinking about the variation of the inclination drop calculation, the stochastic angle drop that is broadly utilized for preparing DNNs. The technique of stochastic inclination drop gives an impartial gauge of the genuine angle dependent on a bunch of tests. By this, the randomization appears to profit the minimization of the target work as it permits a getaway from the local minima.

9.2 Theoretical Background

Literature review is given in this part of the chapter. Related work and deep neural networks are also presented, with subsequent introduction of essential standards of computerized and stochastic math.

9.2.1 *Related Work*

Neural networks have existed for a long time. Notwithstanding, until the beginning of the 21st century, where progresses in innovation of hardware empowered the improvement of competent models. Indeed, DNN training is constrained by the accessible computation even today.

CPUs are all in all incapable of giving enough calculation ability to prepare enormous-scale DNNs. These days, GPUs are the default decision for DNN deployment because of the high computation power and simplicity to utilize advanced frameworks [12]. Facebook AI Group have [13] trained a convolutional neural network (CNN) on multiple GPU. Wen et al. [14] examined the memory effectiveness of different layers of CNN and uncovered exhibition suggestions from data formats and memory access designs. Finally, Cao et al. [15] proposed an execution of a cellular-deep neural network on GPU, a local recurrent neural model (RNN) that is broadly utilized in applications like pattern recognition.

Acceleration based on FPGA of DNN is a state-of-the-art subject. FPGAs can actualize parallel computing and possibly outperform GPU in terms of power and speed efficiency [16]. A principle challenge in acceleration based on FPGA configuration is the absence of advancement frameworks such as keras and Theano. To help the advancement of DNNs on FPGAs, Venieris and Bouganis [17] propose a structure for deploying DNNs on dedicated FPGAs. Moreover, Cho and Kim [18] and Luo et al. [19] propose a hardware-based quickening agent to use the wellsprings of pipelined parallelism to accomplish an efficient execution of a CNN. At last, Zhang et al. [20] present a reconfigurable system for implementing CNNs. While the FPGA and GPU based executions [21] still show a huge margin of progress, they are universally useful computing hardware not explicitly streamlined for DNN execution.

Notwithstanding such methods, DNNs can essentially profit by the SC method, which permits the execution of complex capacities with exceptionally straightforward rationale. Stochastic registering can possibly actualize DNNs with altogether decreased equipment impression when contrasted with a fixed or floating point usage. There have been earlier endeavors to execute DNNs utilizing stochastic representation. Chen et al. [22] use similar representation to execute a DNN based on radial basis function (RBF), essentially decreasing the vital equipment. Nonetheless, the RBF neural model is not, at this point, broadly utilized in deep neural applications as RBF immerses to 0 for a large portion of data, providing inclination-based enhancement testing. In Oh et al. [23], the authors present a neuron design in SC for DNNs and make use of the energy and accuracy arbitration. Reconfigurable enormous deep learning frameworks dependent on SC were planned in Ren et al. [24]. Besides, the authors in Adhikari et al. [25] present stochastic registering hardware plans for the use of CNNs. In Ma et al. [26], weight storage schemes and optimizing strategies to diminish power and area are proposed by the authors. Further, in Ren et al. [5], the authors propose a structure advancement strategy for an overall CNN model to limit area and power utilization while maintaining considerable accuracy.

9.2.2 Deep Neural Networks

This segment presents a brief analysis of DNNs. Beginning from the basic unit of neural models, the perceptron, trailed by an outline of the deep feedforward networks and an investigation of their design. At last, backpropagation, the well-known algorithm for gradient calculation while training, is momentarily introduced.

9.2.2.1 Overview of Feedforward Neural Networks

Multilayer perceptrons, otherwise called DNNs, are the essential networks. As some other AI models, the goal of a deep feedforward network is to an inexact, certain, obscure capacity f^*, regularly alluded to as the objective function. Further, a deep feedforward network characterizes a planning $y = f\,(m,\, \theta)$ and take in the arrangement of boundaries θ that gives the best capacity guess [1]. These models are critical as they structure the premise of some additional deep learning models, for example, recurrent neural network or CNN.

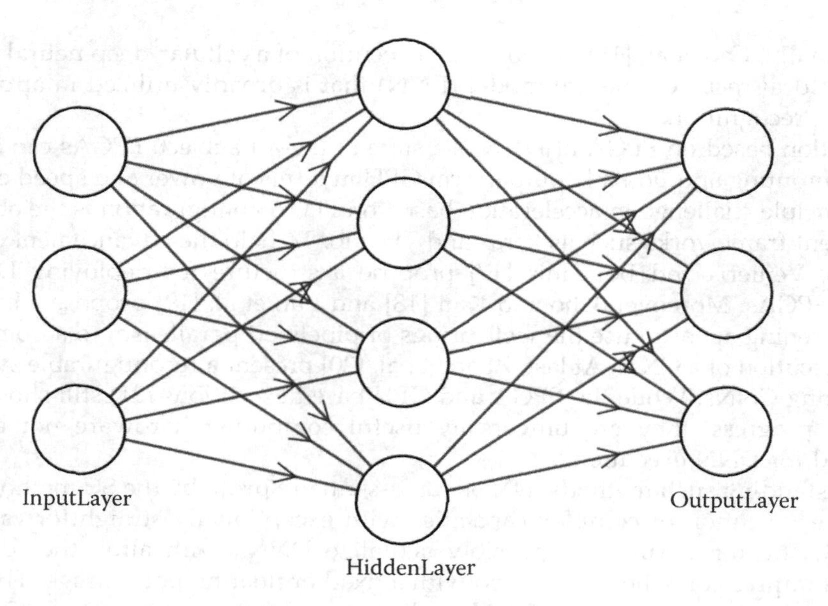

FIGURE 9.1
Deep Feedforward Network.

A deep feedforward neural network is correlated with a coordinated directed acyclic graph (DAG) portraying in what manner the neurons are joined with one another [1]. Figure 9.1 delineates such a chart for a completely associated feedforward model with a solitary hidden layer. There are three input, four hidden, and three output units in the model. In this DNN, the relation $q = f(m) = f^2(f^1(m))$ depicts the connected architecture, where f^1 is the initial layer and f^2 is the next layer of the model. In this, general length of the structure characterizes the network depth. Essentially, the size of each layer (e.g., the quantity of neurons) decides the network's width. While training, the point is to pick network boundaries with the end goal having $f(m) \approx f^*(m)$. The information in training, $I = (m_1, l_1), \ldots, (m_n, l_n)$, basically give uproarious estimations of the objective capacity f^* assessed at various layers [1]. The data trained indicate straightforwardly that the conduct of the yield layer ought to deliver a implementation that is near y_i. Then again, the conduct of the remainder of the layers is not straightforwardly indicated by the trained data. Their conduct is totally indicated by the deep learning techniques, which should conclude how layers can be picked to accomplish an execution that perfectly estimates f^*.

9.2.2.2 Overview of Deep Neural Networks

Summing up, a DNN comprises a series of completely associated layers. Units in each layer compute a weighted amount of the approaching feature, and the outcome is shipped off to an activation function like the tanh or ReLU. At last, the output layer calculates the model output, and the loss is determined showing by how much the computed output goes astray from the actual result. In this part, the main focus is around an SC-based execution of the fundamental activities of a DNN, having activation functions and inner products.

There is no particular rule for picking the hidden layer, albeit amended ReLUs are a superb default decision since they are straightforward and simple to streamline. ReLUs utilize the initiation work $f(x) = max0, x$. Other grounded hidden units utilize the strategic sigmoid actuation work $f(x) = \sigma(x)$ or the tanh function work $f(x) = tanh(x)$. In contrast to ReLU, the sigmoidal units soak across the vast majority of their area, making slope-based learning exceptionally troublesome. This chapter gives the execution of SC for both the ReLU and tanh.

9.2.3 *Principles of Digital Arithmetic*

Number representation is very important when concerned with DNNs because all the operations and implementation are numerical [27]. Basic arithmetic operations are addition, subtraction, division, etc., which are basically performed by arithmetic logic unit (ALU). Mainly, these numbers are classified into two groups, as below [27]:

1. Fixed Value Representation
   - $X = -X_i, \ldots, X_N$ - Integer
   - $N = x/2^n$ - Rational numbers, where $x \in I$ and n is a positive unit
2. Floating Value Representation
   - $a \times b^e$, in which a depicts a rational number, b represents base, and e, is the exponent. Real numbers representation and calculation is done in floating point number with precision.

9.2.3.1 Fixed-Value Representation

The working algorithm of a fixed-point number is generated by a digit vector where each element is represented as a digit and a series of numbers is called *precision of digit* [27]. A positive number Q is given by vector:

$$Q = (Q_{n-1}, Q_{n-2}, \ldots, Q_1, Q_0) \tag{9.1}$$

Several elements are associated with a number system and are given as follows [27]:

1. The digit-vector precision represented as Q.
2. A set of values Xi for the digits Qi. For example, binary number system contains the digit set {0, 1} in binary number representation.
3. An interpretation rule that maps digital vector elements and sets of integers.

Figure 9.2 shows a 5×5 multiplier implementation circuit to show a fixed-point arithmetic operation, containing radix-2 multiplier implementation carried out using carry ripple multiplier. Figure 9.2 contains two modules. The first module uses AND gates for partial product computation, when considering SC multiplier unit. The hardware complexity is much larger and needs much larger hardware space for operations. The fixed-point data occupies a larger hardware area than an AND gate hardware. Finally, as compared to SC, more precision is required on fixed-point data, which requires an additional hardware module to perform overall computation.

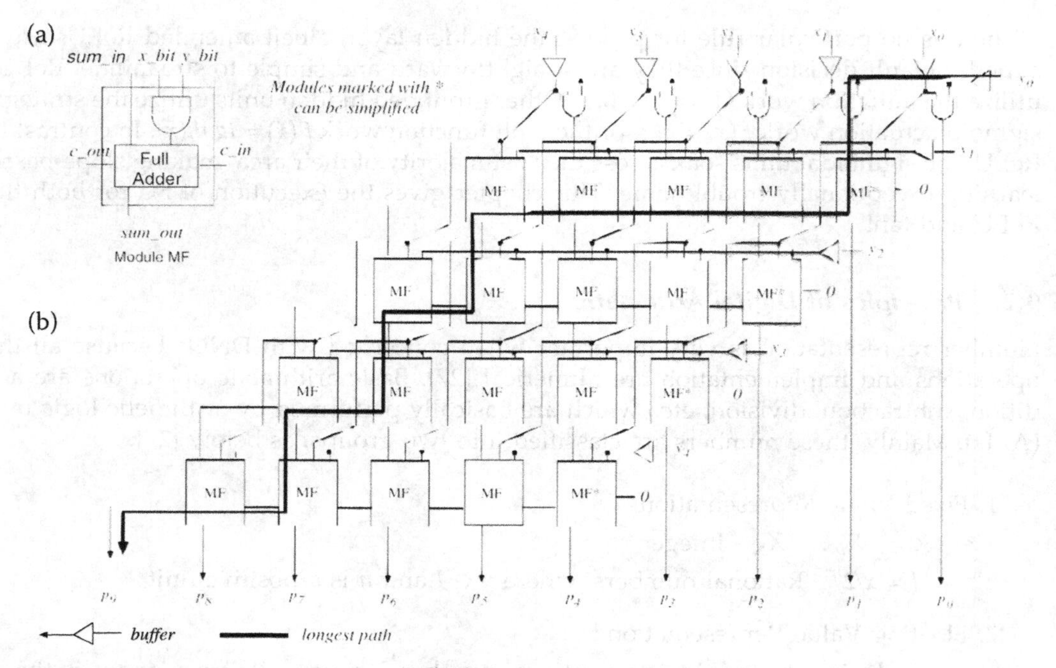

FIGURE 9.2
5×5 Multiplier Operation. (a) Prime Unit (b) Model [28].

9.2.3.2 Floating Value Representation

The general form of floating point number Q has two elements, the mantissa M_x^* and the exponent E, such that

$$q = M_x^* \times b_i^E \tag{9.2}$$

where b is constant [27]. Mantissa can be signed or unsigned, and the exponent is a signed integer. A signed floating number is given by the set of arguments (S_i, E_x, M_x), such that

$$q = (-1)_i^S \times M_i \times b \tag{9.3}$$

where $S_i \in \{0, 1\}$ represents sign bit and M_i depicts magnitude of mantissa [27]. The floating point representation gives more precision in data compared to the fixed-point number. According to Ercegovac and Lang [27], the number can be represented by a dynamic range and the difference between the largest and smallest number. The dynamic range for fixed-point representation is given as,

$$DyRg_f = r^n - 1 \tag{9.4}$$

Contrarily, the dynamic extent of the floating point representation is given as

$$DyRg_{fl} = \frac{M_{max} \times b^{E_{max}}}{M_{min} \times b^{E_{min}}} \tag{9.5}$$

Now, for comparison of two, suppose n digits floating value are apportioned such that m digits are utilized for the important, $n - m$ exponent digits and $b = r$. At that point,

$$DyRg_{fl} = (r^m - 1) \times r^{(r^{n-m}-1)} \tag{9.6}$$

For example, consider a scenario where n = 28, m = 18, r = 2. Then,

$$DyRg_f = 2^{28} - 1 \approx 2.6 \times 10^8 \tag{9.7}$$

$$DyRg_{fl} = 2^{18} \times 2^{28} - 1 \approx 1.5 \times 10^{82} \tag{9.8}$$

We can conclude that floating point numbers gives an effective real number approximation without any major issues. So, we can just ignore the fixed-point number in practical implementation. However, floating point numbers have some demerits, such as rounding errors, precision, and complex hardware requirements.

9.2.4 *Stochastic Computing Overview*

SC depends on the likelihood hypothesis, where a likelihood number is denoted by a bit-sequence of picked length and its value is dictated by the likelihood of a subjective bit in the bit sequence being 1. For instance, a bit sequence having 55% of ones and 45% of zeros gives the value $x = 0.55$, mirroring the way that the likelihood of noticing a one in a discretionary bit area is 0.74. Plainly, when contrasted with a twofold radix portrayal, the SC is not exceptionally conservative. Nonetheless, it prompts exceptionally low-intricacy number juggling units, which was an essential worry previously. For instance, duplication in stochastic number juggling can be carried out by a solitary AND gate. Assume stochastic streams with 2 inputs that are ANDed logically, and expect that the likelihood of detecting one in both bit-streams is p1 and p2 separately. At that point, accepting that the sources of info are reasonably uncorrelated or autonomous, the likelihood of any bit in the AND gate output being a one is $p1 \times p2$. Figure 9.2 delineates this activity.

Another alluring component of SC is its natural error resistance. The likelihood p relies upon the proportion of ones to the size of the stochastic sequence and not on its precise position. For model, (0, 1, 1, 0), (0, 0, 1, 1) just as (0, 1, 0, 1, 0, 1, 0, 1) are, on the whole, conceivable and substantial portrayals of $p = 0.5$. This sort of portrayal offers a great level of error resilience, as a solitary piece flip in a long arrangement will affect the stochastic number less than is addressed. Consider, for example, to some degree flip in the $S2$ of the AND gate, as shown in Figure 9.2. The value would change from 6/8 to 5/8 or 7/8. That is a misstep of 1/8. Of course, a single piece flip in a normal radix-2 processing can cause a tremendous goof, especially if it impacts a high-order bit [8]. In particular, a single flip of bit in fixed point information of $S2$, 0.11, would give a high error rate since 0.25 is given by 0.01. So, that is a 4/8 error. This characteristic of SC is precise because of the way that the entirety of the bits in a bit stream are weighted similarly.

In spite of these attractive highlights, SC has a few drawbacks that have restricted its application. In particular, a stochastic number has a natural fluctuation in calculations [9]. Another drawback of SC is that more exact calculations require an outstanding expansion in the quantity of digits, causing a dramatic expansion in the clock cycles quantity required to achieve a calculation [9]. Casually, one can characterize accuracy of an amount as the number of bits needed to address that value [8]. With precision of m bits, 2 m values can be given. Putting this plan to SC, the arrangement of real numbers inside [0, 1]

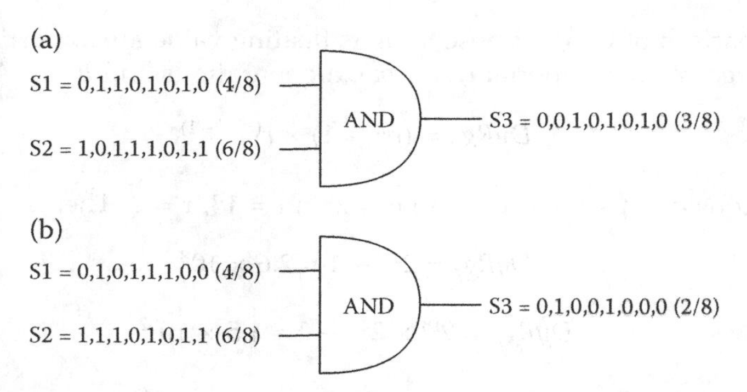

FIGURE 9.3
Multiplication in Stochastic Computing (a) Correct Representation (b) Relative Representation.

addressed with 4-bits if accuracy diminishes to the {1/16, 2/16, 3/16, 4/16…, 16/16} set, and their stochastic portrayal requires bit-surges of 16 length. A stochastic number from 4 to 5 bits requires more stochastic bits for increasing precision, such as a bit sequence of 32. This outstanding increment in the sequence length regarding precision prerequisites is the thing that decreases the speed of calculation in SC as more clock cycles are needed to finish a specific calculation. Note in any case, that an expansion in SC precision does not need extra computational assets. The equivalent AND gate can give more precision in operations like multiplication. The solitary disadvantage is that the dormancy in the calculation will be bigger. Conversely, in fixed value math an increment in precision creates an expansion in the required assets (Figure 9.3).

In rundown, SC has a few benefits over several other techniques [9]: 1) less hardware usage, 2) higher error resilience, 3) simpler hardware executions that can permit high clock rates, and 4) no hardware alterations needed for computation trade-off latency with required accuracy. The issues related to accuracy have restricted the utilization of SC. In any case, as of now, examined DNNs are described by a capacity to bear substantially less precise singular calculations. Consequently, SC can possibly encourage low-cost executions of DNNs on hardware with full parallelism.

9.3 Stochastic Processing Elements

This section examines various stochastic processing utilized in DNNs. As given above, working on arbitrary bit sequence empowers complex tasks to be performed with exceptionally straightforward simple circuits. Earlier, both sequential and combinational circuits were presented for preparing stochastic bits. Earlier work has shown that combinational logic can effectively execute subtraction, addition, squaring, and multiplication in both unipolar and bipolar representation [29]. Then again, linear finite state machines (FSMs) effectively can execute non-linear functions like sigmoid, exponentiation, and saturated linear gain [29]. The stochastic comparator is utilized to productively execute a maximum unit in SC. This is employed to execute a bipolar SC-based ReLU, which is broadly utilized these days as the principle activation function in few DNNs.

The handling blocks are introduced with respect to standard logical components. However, the algorithmic understanding of certain processing units is additionally examined. The investigation of the entirety of the handling components depends on the presumption that the input bit-sequences represent Bernoulli sequences. It suggests that the likelihood of bits in the succession being a 1 is free from all the recently noticed bits. Nonetheless, this does not infer that the stochastic processing unit output is essentially a Bernoulli succession. Finally, it is expected that on account of a unit with different input flags, the inputs are autonomous or uncorrelated.

9.3.1 *Combinational Processing-based Elements*

Fundamental arithmetic units such as scaled addition, multiplication, and subtraction can be executed productively in the stochastic area utilizing basic combinational rationale circuits. The plan and examination of combinational logic computational units is essentially derived from work introduced by BR Gaines [29].

9.3.1.1 Multiplication

As seen earlier, stochastic number multiplication is amazingly straightforward. This is a result of the fact that the multiply operation shows a closed approach for unipolar and bipolar representations on the range [0, 1] or [–1, 1]. In bipolar representation, multiplication is performed by a XNOR gate. The XNOR output is 1 at any point the 2 inputs are either considered 0 or 1. M and N are XNOR gate inputs, and O is the output. By then, the one bipolar signal has

$$\begin{aligned} P_O &= (P_M \cdot P_N) + (P_{\bar{M}} \cdot P_{\bar{N}}) \\ &= (P_M \cdot P_N) + (1 - P_M) \cdot (1 - P_N) \\ &= 2P_M \cdot P_N - P_M - P_N + 1 \end{aligned}$$

Utilizing the way that $P_M = \frac{a+1}{2}$ and $P_N = \frac{b+1}{2}$, *then*

$$Pc = \frac{ab + 1}{2}$$

For bipolar signals, $c = 2P_C - 1$ therefore

$$c = 2\left(\frac{ab + 1}{2}\right) - 1 = a \cdot b \tag{9.9}$$

The stochastic multiplier output gives a fair gauge of the specific outcome, and if M and N are autonomous Bernoulli sequences, at that point output O is likewise a Bernoulli sequence. In case of no error in the estimation of the ideal capacity, the acknowledgment of y is not in general precisely equivalent to the $a\ b$ product. This could be either because of arbitrary vacillations in the sequence of bits or because of quantization error during the adjustment of probabilities a & b to the arrangement of discrete probabilities addressed with SC.

9.3.1.2 Addition and Subtraction

In SC, addition and subtraction are slightly more perplexing activities than multiplication because of the fact that they are not close under range [0, 1] or [–1, 1]. The aftereffect of adding two numbers that exist in [–1,1] does not really exist in [–1,1]. Consequently, a scaled activity is utilized in SC to get the output in range to [–1, 1] from [–2,2]. The amount of two weighted probabilities, $\alpha p1 + (1 - \alpha)p2$, exists in [–1, 1] where $0 \le \alpha \le 1$ and is expressible in the stochastic processing area. Such a calculation can be acknowledged utilizing multiplexer with two-input in which the selection line is guided by the choosing likelihood α [29]. The likelihood of a one showing up at output is equivalent to

$$P_O = P_M \times P_S + P_{N'} P_S$$

Let us take $P_S = 0.5$; bipolar signal one gives

$$y = 2P_O - 1 = \frac{(a + b)}{2} \tag{9.10}$$

The MUX creates an output producing a likelihood with weighted amount of probabilities that is in input. To signify the scaled expansion activity $y = a \oplus b = (a + b)/2$ will be utilized in stochastic processing. Now, bipolar signs $P_S = 0.4$ compares to a stochastic bit-sequence with 60% zeros and 40% ones. Subtraction can be executed with scaling by utilizing the equivalent MUX unit essentially by modifying inputs to be deducted.

9.3.2 *FSM-based Computational Elements*

Combinational logic can productively execute fundamental arithmetic operations in SC, specifically polynomial capacities that map the range [0, 1] to [0, 1] and [–1, 1] to [–1, 1] for the unipolar and bipolar representations. In any case, as examined prior, DNNs utilize non-linear activation functions and other units, for example, the tanH or ReLU. The use of such activation functions with combinational is not possible sometimes and is not easy [30].

9.3.2.1 Stochastic Hyperbolic Function

Before ReLU, the hyperbolic tangent function, was the well-known activation function alongside the sigmoid. Albeit ReLU is a general choice by numerous authors and experts, the hyperbolic tangent is generally utilized as an activation function in numerous applications.

Subsequently, a stochastic estimation of the Stanh is thought of in this segment. As presented in Brown and Card [9], the stochastic guess to the function tanh can be given by s_i, as in,

$$S_i \begin{cases} = 0, & 0 \le n \le \frac{s1}{2} - 1 \\ = 1, & \frac{s1}{2} \le n \le s1 - 1 \end{cases} \tag{9.11}$$

where the bipolar representation is used for input and output values. By this, the tanh function in SC is approximated as shown,

$$Stanh\ (N, x) \approx tanh\left(\frac{N}{2}x\right) \tag{9.12}$$

9.3.2.2 Stochastic Max Pooling

Albeit not discussed above, a regular CNN comprises pooling layers and convolutional layers with connected layers. For feature extraction, the convolutional layer is used through convolution of responsive fields and a bunch of filters [1]. Thereafter, a down examining step is normally performed to total insights of the separated features focusing to lessen the information dimension and alleviate issues of overfitting. This sampling measure is named pooling, and two sorts of pooling are ordinarily utilized: max and average pooling.

The max pooling chooses the maximum values from the probable inputs, and average pooling involves computing the average from the set of values.

9.4 Neural Network Training and Inference in Stochastic Computing

DNN inference comprises of forward propagation of a trained network to recognize, classify, and measure obscure input information. Successfully, the neural network derives data about the new information dependent on the hypothesis work mastered during the training. Subsequently, deduction cannot happen before training. Regardless, this part accepts that a randomly trained neural network exists with known parameters such as weights and biases.

Training will certainly happen utilizing floating point calculations; consequently, the trained DNN will overall have floating point coefficients, operators, and inputs. SC-based model inference comprises a cycle of changing an existent neural network to an SC-based model having N-digit inputs, viable coefficient values, and SC operators. This is comparable to changing over the directed acyclic diagram of a neural network (see Figure 9.1) to an SC-based graph. During the inference, the entirety of the network parameters will be fixed and the values of the coefficient will be given. Subsequently, the above conversion needs to occur only once. After the completion of the conversion, the subsequent model can be utilized to measure, in SC, distinctive data units without reconfiguration of hardware.

As an illustrative model, consider Figure 9.4: the four inputs, $x1....$, $x4$, a hidden layer with two units, and an output layer with single output y. Then, think about an individual neuron $x \in R4$ and expect the weight networks in the remaining layers to be given by $W_i^{(1)} \in R^{4\cdot 2}$ and $W_i^{(2)} \in R^{2\cdot 1}$ individually. Moreover, the predispositions are given by $b_i^{(1)} \in R2$. What's more, $b_i^{(2)} \in R$ for hidden and output layer individually. The enactments in the hidden layer are consequently determined as,

$$h_i^{(1)} = \phi(W_i^{(1)T} x + b_i^{(1)}) \tag{9.13}$$

and network output is given as

$$y_i = \phi(W_i^{(2)T} h_i^{(1)} + b_i^{(2)}) \tag{9.14}$$

where ϕ is an activation function.

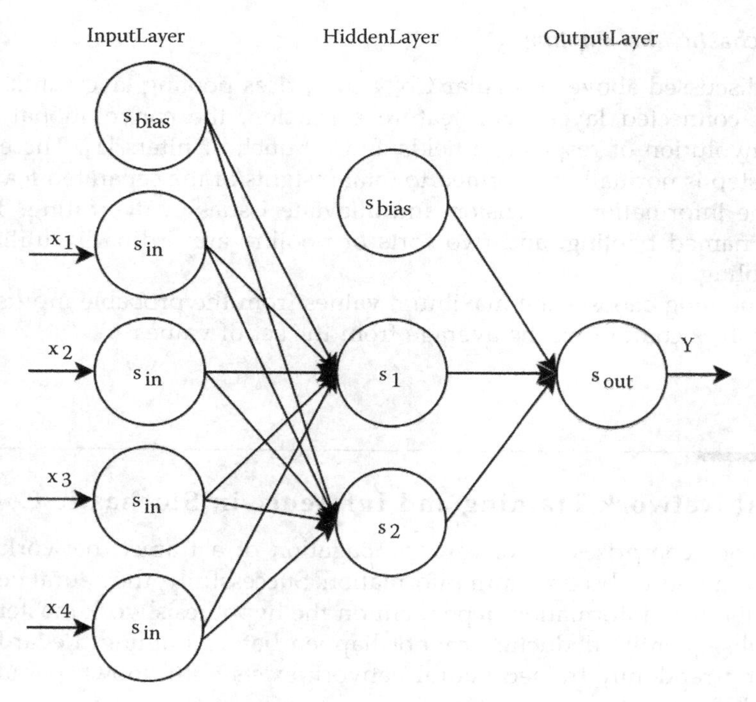

FIGURE 9.4
Deep Neural Data Flow Graph.

In the first place, accepting the input value x_i in some range $[-l, u]$, the info scaling element s_{in} is chosen. In this manner, each info x_i is scaled by s_{in}, and the down-scaled sources of data $x_i = x_i/s_{in}$ are changed over into stochastic streams.

9.5 SC Elements Implementation

NumPy logical units can be used effectively to simulate combinational logic-based processing elements in Python. It may be worth referencing that DNN computational charts are generally determined using matrix operations, for example, matrix addition and multiplication. In particular, the focal calculation in a DNN is given by the relative change $t^{(i)} = W^{(i)T}t^{(i-1)} + b^{(i)}$. In programming constructs, it is not expected to utilize broadcasting to handle different data in a single statement. Along these lines, the past assertion is generally reached out to $T^{(i)} = T^{(i-1)}W^{(i)} + b^{(i)}$, in which each networks column $T^{(i)}$ and $T^{(i-1)}$ relates to different data points. After eliminating misinterpretation, consider that the final statement is good for nothing in numerical expression as addition is just characterized for objects of the equivalent measurement. The assertion infers the bias vector should be added to each column of the network item. Broadcasting is done when a smaller array in dimension comes across the larger one so that they have viable sizes.

The use of FSM-based computational components in Python is likewise generally direct. Note that all the FSMs were initialized at the focal point of the sequence state. In this kind of FSM, initializing ought not to be a significant issue where the bit sequence length L is adequately huge, for example, $L \leq N$. At that point, the likelihood coded by the bit sequence will be surmised over the long run, and the FSM will join to a proper state value.

9.6 Experiments and Results

This section presents test outcomes from the execution of DNN inference in SC and results achieved by the proposed DNN training using SC. The experiments were done on the handwritten digit MNIST dataset with 10 different classes, from digits 0 to 9. MNIST holds 60,000 and 10,000 training and testing data with 28 28 normalized grayscale images. Accentuation is given on DNNs without essentially following a particular network structure. DNNs are trained and developed in Adam and TensorFlow; optimization of stochastic target functions based on first-order gradient is utilized for limiting the loss while model training.

9.6.1 *Stochastic Computing-based Neural Network Inference*

As inference happens after training, a feedforward network is made up and trained to be additionally utilized for the SC-compatible inference purpose. A model dependent on the specifications given in Table 9.1 was prepared on the MNIST dataset after a simple training procedure. Then, the trained DNN was not aware of the subsequent SC-based inference algorithm implementation. The trained network, in particular Network I, accomplished classification accuracy of 86.03% and 86.01% on the training and testing data. As a side comment, the goal of this section, and venture all in all, is not to create models that will accomplish best-in-class accuracy on specific deep learning applications, similar to the recognition of MNIST digits. All things being equal, the goal of most of the models created in this chapter is to give experiment networks to which the strategies introduced in the previous sections can be applied and tried. However, in each circumstance the "best" classification accuracy is attempted to be accomplished.

TABLE 9.1

Network I Characteristics with 784-128-10 Size

Training Specifications	
Optimizing Algorithm	Adam
Cost Function	Logarithmic loss
Learning Rate	0.1
No. of Epochs	600
Batch Length	100
SC Based Training	No
Objective Function	No

9.6.2 *Stochastic Computing-based Neural Network Training*

After DNN inference in SC, the principle objective of this part is to show different tests with training methodologies and examine their impact on the general execution of the neural network on the detection tasks. Analyses are dependent on the MNIST dataset, and as a beginning stage, a model based on the parameters shown in Table 9.2 and Table 9.3 was developed and trained. For comparison, the training procedure continued utilizing the basic penalty function. For this sort of regularizer, the prepared model accomplishes higher accuracy of 95.83% and 94.81% on training and testing sets compared to the work by Alawad and Lin [31]. Here, outcomes vary greatly, like the ones we got in past tests, where the unexpected addition in the gain coefficients show a sudden jump in the loss mitigation. The histogram graphs for the model weights and outcomes from the SC-based training procedure of this DNN are given in Figure 9.5 separately.

TABLE 9.2

SC-Based DNN Inference Parameters

Parameters	Hidden Layer	Output Layer
Scaled Input	1	32
Inner Product Nodes	8	4
Scaled Intermediate Input	128	512
Intermediate Gain Factor of Inner Product	4	4
Intermediate Saturation Level of Inner Product	32	128
Scaled Inner Product Output	256	512
Output Gain factor of Inner Product	8	2
Saturation Level of Inner Product	32	256
Scaled Biased Addition Input	32	256
Scaled Biased Addition Output	64	512
Direct Gain Level	2	2
Saturation Size	32	256

TABLE 9.3

Network II Characteristics with 784-128-10 Size

Optimizer	Adam
Loss Function	Logarithmic loss
Initial Learning Rate	0.1
Epochs	6,000
Batch Size	600
SC-based Training	Yes
Penalty Function	Yes

(a)

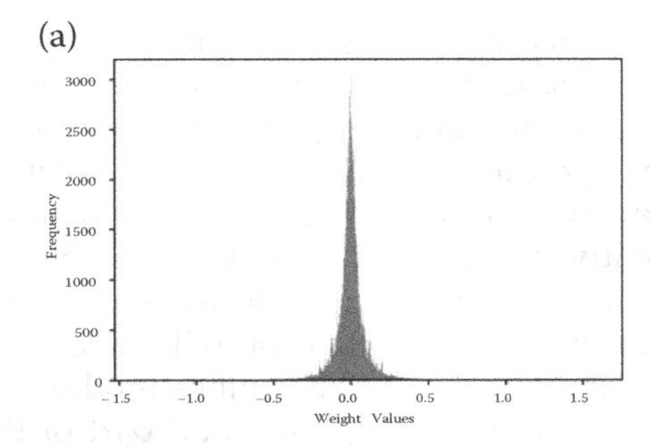

Hidden Layer Weight Distribution

(b)

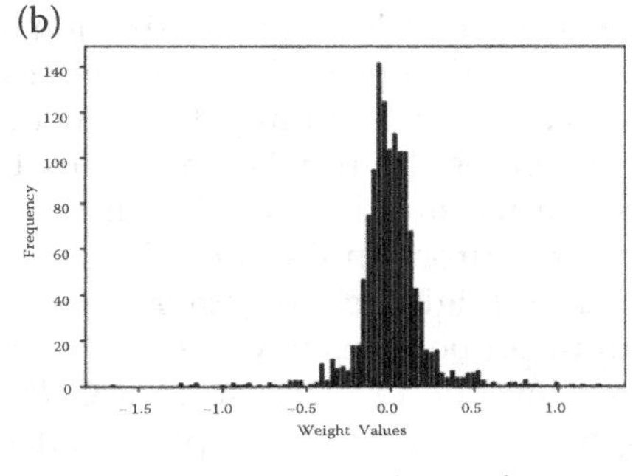

Output Layer Weight Distribution

(c)

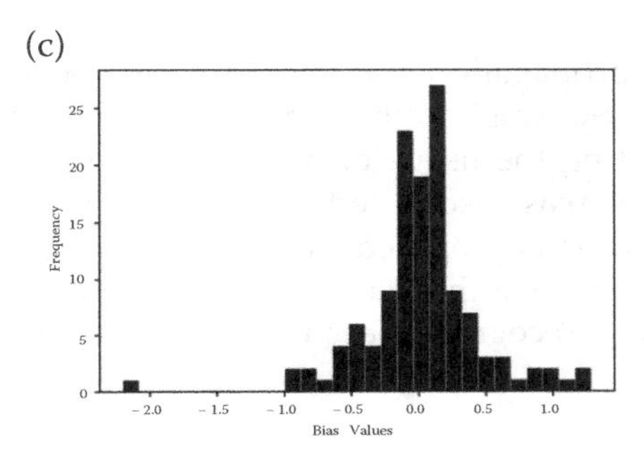

Hidden Layer Bias Distribution

(d)

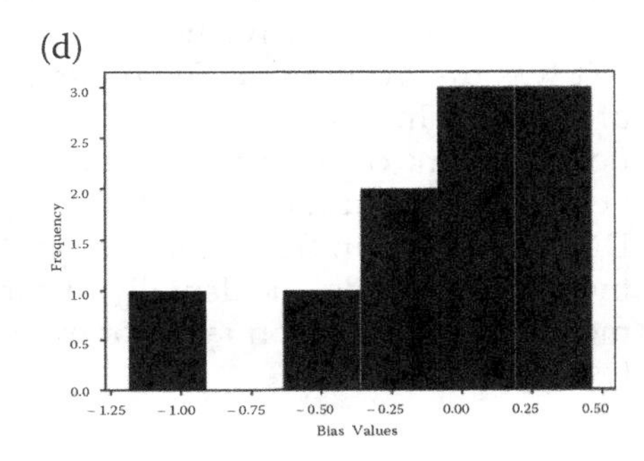

Output Layer Bias Distribution

(e)

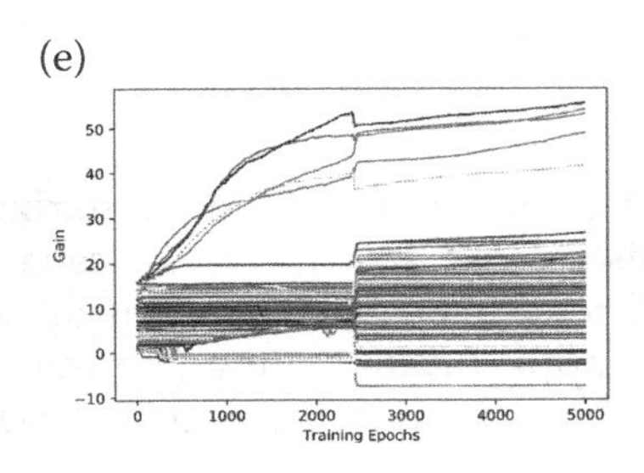

Hidden Layer Weight Distribution

(f)

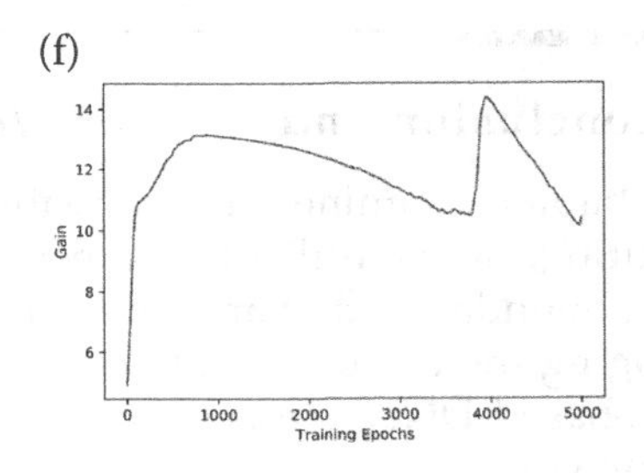

Output Layer Weight Distribution

FIGURE 9.5
Histogram and Results of Network II.

9.7 Case Study

This section presents a basic assessment of the work done in this chapter contrasted with the primary objectives and gives relative merits with additional inadequacies. Beginning from the objectives, the main motive of this study was to look under which conditions SC can be fused inside DNNs.

Indeed, the work introduced in this section presents all the processing elements that are needed for SC-compatible CNN implementation, i.e., activation function and pooling layer. Although not given in detail, it might indeed be "simpler" to implement an SC-based CNN. The motivation to contend for this is given. A significant challenge in the execution of SC inference arises from the huge scalings that are forced all through the chart. Such scalings fundamentally emerge from the internal product unit and increment as the number of inputs to the internal product increments. In a completely connected multi- layered neuron architecture, every neuron will be associated with all other neurons in the preceding layer. Subsequently, the number of inputs to a specific neuron might be altogether huge, prompting a higher scaling level. For instance, if the input proportions have 32 32 3 size, then in the primary layer a neuron will have 1,024 data inputs. In contrast, in a CNN it is not unexpected to utilize a more modest responsive field, for example, to attach every neuron to just a neighborhood part of the input proportions. In these, the scalings presented by all inner products in a CNN might essentially be smaller contrasted with a multilayer perceptron (MLP) relying upon the length of the responsive field.

It was discovered that SC influences forward propagation and backward propagation of a DNN. In spite of the fact that training in SC was viably demonstrated utilizing floating point calculations as opposed to reenacting the usage of model training in SC, very keen conclusions were build. Specifically, it was discovered that the execution of DNN inference in SC can in fact profit if the model is prepared by using the proposed modified methods. Incidentally, it can likewise improve the general performance of the model on the detection task, for example, improve recognition accuracy on training and testing data.

9.8 Conclusion and Further Work

This chapter examined the possibility of whether SC, an alternative to binarized computing, can be utilized to execute DNNs. It was discovered that the worst scenario scaling boundaries that are naturally acquainted by SC tend to be excessively negative, sabotaging the use of neural network SC inference. By appropriate scaling operations, the SC-based DNN can accomplish the same degree of accuracy as the regular binarized network.

Broadening inference of DNNs in stochastic processing, an alternate training procedure was proposed, intending to catch the impediments of the stochastic portrayal inside the model's training phase. Curiously, it was discovered that this permits the organization to build up its own insight, viewing the recognition task as the elective portrayal that we are attempting to force. The model appears to distinguish the limits of stochastic processing and suitably changes its boundaries to address them. As an outcome, an ensuing execution of the inference calculation utilizing SC hardware could profit fundamentally by this strategy. At last, it was discovered that the proposed technique can even improve the predictions.

Moreover, further experiments can be conducted. More deep network designs should be considered in addition to testing with different datasets. Also, the use of CNNs needs to be addressed. These augmentations could be concentrated either by methods for a product or by hardware usage. Another major part of future work is to reenact the preparation of a model utilizing SC handling units as op- posed to demonstrating this process utilizing floating point calculations. As contended in the assessment part, such an augmentation will permit direct analyses with respect to the powerful determination of the bit stream length during training, which might actually offer ascent to exceptionally interesting outcomes. As last, it appears normal to consider a hardware execution of an SC-based model on an FPGA. Expanding the current work, hardware will permit quantitative outcomes with respect to the total area devoured by the SC model.

References

1. Goodfellow, I., Bengio, Y., and Courville, A. (2016) *Deep Learning*. MIT Press. http://www.deeplearningbook.org.
2. Deng, L., and Yu, D. (2014) "Deep Learning: Methods and Applications." *Foundations and Trends® in Signal Processing* 7(3–4): 197–387. 10.1561/2000000039.
3. Abdel-Hamid, O., Mohamed, A., Jiang, H., Deng, L., Penn, G., and Yu, D. (2014) "Convolutional Neural Networks for Speech Recognition." *IEEE/ACM Transactions on Audio, Speech, and Language Processing* 22(10): 1533–1545.
4. Simonyan, K. and Zisserman, A. (2015) "Very Deep Convolutional Networks for Large- Scale Image Recognition." 3rd Int. Conf. Learn. Represent. ICLR 2015 - Conf. Track Proc.
5. Ren, A., Li, Z., Ding, C., Qiu, Q., Wang, Y., Li, J., Qian, X., and Yuan, B. (2017) "SC-DCNN: Highly-Scalable Deep Convolutional Neural Network Using Stochastic Computing." In *Proceedings of the Twenty-Second International Conference on Architectural Support for Programming Languages and Operating Systems*, ASPLOS '17 (pp. 405–418). New York, NY, USA: Association for Computing Machinery, (). 10.1145/3037697.3037746.
6. Lecun, Y., Bengio, Y., and Hinton, G. (2015) "Deep learning." *Nature Cell Biology* 521(7553): 436–444.
7. Dinelli, G., Meoni, G., Rapuano, E., and Fanucci, L. (2020) "Advantages and Limitations of Fully On-Chip CNN FPGA-Based Hardware Accelerator." In *2020 IEEE International Symposium on Circuits and Systems (ISCAS)*, 1–5. IEEE.
8. Alaghi, A., and Hayes, J.P. (2013) "Survey of Stochastic Computing." *ACM Transactions on Embedded Computing Systems* 12(2s). 10.1145/2465787.2465794.
9. Brown, B.D., and Card, H.C. (2001) "Stochastic Neural Computa- tion II: Soft Competitive Learning." *IEEE Transactions on Computers* 50(9): 906–920. 10.1109/12.954506.
10. Bishop, C.M. (1995) "Training with Noise Is Equivalent to Tikhonov Regularization." *Neural Computation* 7(1): 108–116. 10.1162/neco.1995.7.1.108.
11. Murray, A.F., and Edwards, P.J. (1994) "Enhanced MLP Performance and Fault Tolerance Resulting from Synaptic Weight Noise during Training." *IEEE Transactions on Neural Networks* 5(5): 792–802.
12. Vaverka, F., Mrazek, V., Vasicek, Z., and Sekanina, L. (2020) "TFApprox: Towards a Fast Emulation of DNN Approximate Hardware Accelerators on GPU." In *2020 De- sign, Automation & Test in Europe Conference & Exhibition (DATE)* (pp. 294–297). IEEE.
13. Yadan, O., Adams, K., Taigman, Y., and Ranzato, M. (2013) "Multi-GPU Training of ConvNets." Cite arxiv:1312.5853Comment: Machine Learning, Deep Learning, Convolutional Networks, Computer Vision, GPU, CUDA. http://arxiv.org/abs/1312.5853.
14. Wen, N., Guo, R., He, B., Fan, Y., and Ma, D. (2021) "Block-Sparse CNN: towards a Fast and Memory-Efficient Framework for Convolutional Neural Networks." *Applied Intelligence* 51(1): 441–452.
15. Cao, Q., Balasubramanian, N., and Balasubramanian, A. (2017) "MobiRNN: Efficient re- current neural network execution on mobile GPU." In *Proceedings of the 1st International Workshop on Deep Learning for Mobile Systems and Applications*: 1–6.
16. Mittal, S. (2020) "A Survey of FPGA-Based Accelerators for Convolutional Neural Networks." *Neural Computing and Applications* 32(4): 1109–1139.

17. Venieris, S.I. and Bouganis, C.-S. (2016) "fpgaConvNet: A Framework for Mapping Convolutional Neural Networks on FPGAs." In *2016 IEEE 24th Annual International Symposium on Field-Programmable Custom Computing Machines (FCCM)*, May, 40–47. Institute of Electrical and Electronics Engineers (IEEE). 10.1109/FCCM.2016.22.
18. Cho, M. and Kim, Y. (2020) "Implementation of Data-optimized FPGA-based Accelerator for Convolutional Neural Network." In *2020 International Conference on Electronics, Information, and Communication (ICEIC)*, 1–2. IEEE.
19. Luo, C, Sit, M.-K., Fan, H., Liu, S., Luk, W., and Guo, C. (2020) "Towards Efficient Deep Neural Network Training by FPGA-Based Batch-Level Parallelism." *Journal of Semiconductors* 41 (2): 022403.
20. Zhang, S., Cao, J., Zhang, Q., Zhang, Q., Zhang, Y., and Wang, Y. (2020) *2020 IEEE 3rd International Conference on Electronics Technology (ICET)*. "An fpga-based reconfigurable cnn accelerator for yolo." (pp. 74–78). IEEE.
21. Trinh, Q.-K., Duong, Q.-M., Dao, T.-N., Nguyen, V.-T., and Nguyen, H.-P. (2020) "Feasibility and Design Trade-Offs of Neural Network Accelerators Imple- mented on Reconfigurable Hardware." In *International Conference on Industrial Networks and Intelligent Systems* (pp. 105–123). Springer.
22. Chen, W., Han, X., Li, G., Chen, C., Xing, J., Zhao, Y., and Li, H. (2018) "Deep rbfnet: Point Cloud Feature Learning using Radial Basis Functions." *arXiv preprint arXiv:1812.04302.*
23. Oh, J., Neugebauer, F., Polian, I., and Hayes, J. (2020) "Retraining and Regularization to Optimize Neural Networks for Stochastic Computing." In *2020 IEEE Computer Society Annual Symposium on VLSI (ISVLSI)* (pp. 246–251). IEEE.
24. Ren, A., Li, Z., Wang, Y., Qiu, Q., and Yuan, B. (2016) "Designing reconfigurable large-scale deep learning systems using stochastic computing." In *2016 IEEE International Conference on Rebooting Computing, ICRC 2016 - Conference Proceedings*, Nov. Institute of Electrical and Electronics Engineers Inc. 2016 IEEE International Conference on Rebooting Computing, ICRC 2016; Conference date: 17-10-2016 Through 19-10-2016.
25. Adhikari, S.P., Kim, H., Yang, C., and Chua, L.O. (2018) "Building Cellular Neural Network Templates with a Hardware Friendly Learning Algorithm." *Neurocomputing* 312: 276–284.
26. Ma, Y., Cao, Y., Vrudhula, S., and Seo, J.-S. (2018) "Optimizing the Convolution Operation to Accelerate Deep Neural Networks on FPGA." *IEEE Transactions on Very Large Scale Integration (VLSI) Systems* 26(7): 1354–1367.
27. Ercegovac, M.D., and Lang, T. (2004). "CHAPTER 1 - Review of Ba- sic Number Representations and Arithmetic Algorithms." In: Ercegovac, M.D. and Lang, T. (eds.) *Digital Arithmetic* (The Morgan Kaufmann Series in Computer Architecture and Design, pp. 3–49). San Francisco: Morgan Kaufmann. http://www.sciencedirect.com/science/article/pii/B9781558607989500038.
28. de Dinechin, F., Ercegovac, M., Muller, J.-M., and Revol, N. (2009) "Digital Arithmetic." In:Wah, B. (ed.) *Wiley Encyclopedia of Computer Science and Engineering* (pp. 935–948). Wiley. https://hal-ens-lyon.archives-ouvertes.fr/ensl-00542215.
29. Gaines, B.R. (1969) *Stochastic Computing Systems* (pp. 37–172). Boston, MA: Springer US. 10.1007/978-1-4899-5841-9-2.
30. Qian, W., Li, X., Riedel, M.D., Bazargan, K., and Lilja, D.J. (2011) "An Architecture for Fault-Tolerant Computation with Stochastic Logic." *IEEE Transactions on Computers* 60(1): 93–105.
31. Alawad, M., and Lin, M. (2018). "Scalable FPGA Accelerator for Deep Convolutional Neural Networks with Stochastic Streaming." *IEEE Transactions on Multi-Scale Computing Systems* 4(4): 888–899.

10

Convolutional Neural Network and Its Advances: Overview and Applications

Jyoti S. Raghatwan[1,2] and Dr. Sandhya Arora[3]
[1]*Research Scholar, SKN College of Engineering, Pune*
[2]*Assistant Professor, RMD Sinhgad School of Engineering, Pune*
[3]*Professor, Cummins College of Engineering for Women's, Pune*

10.1 Introduction

Nowadays, deep learning is emerging as an area of interest among researchers. Convolutional neural network (CNN), representative of neural networks, shows its excellent performance and has achieved remarkable results in deep learning. The term *convolution* was first used by LeCun et al. [1], who proposed CNN for handwritten zip code recognition, which was the first version of the famous CNN LeNet-5. Subsequently, many improvements were carried out on different components of CNN. It is considered more than a tradition model due to its important characteristics such as reduced number of training parameters using weight sharing and improved generalization. Second, the feature extraction and classification stage both use learning processes. Third, large network implementation is easier in CNN than any other model of artificial neural network. Another important characteristic of CNN is the ability to obtain abstract features when input passes through the deeper layers. Due to all these characteristics, it is widely used in various domains like computer vision, natural language processing, speech processing, and radiology. This chapter focuses on the basic concepts of CNN with their advances and applications in various domains. The aim of this study is to provide concrete details of CNN, which will help researchers to develop applications in the field of CNN.

DOI: 10.1201/9781003166702-10

10.2 Elements of Convolutional Neural Networks

10.2.1 Preliminary Mathematical Concepts

1. Vector and Tensors

Here, we show the different symbols used to represent the vector and tensors. Vector representation is given as $X \in \mathbb{R}^D$ to indicate the column vector having D elements. Capital letter X is used to denotes the matrix. Matrix representation is given as $X \varepsilon \mathbb{R}^{H \times W}$ to indicate the matrix consists of H rows and W columns. Higher order matrices are called *tensors.* For example, the representation given as $X \in \mathbb{R}^{H \times W \times D}$ indicates order three tensor, and each element of the tensor is indexed by triplet (i, j, d) where $0 \leq i \leq H$, $0 \leq j \leq W$, $0 \leq d \leq D$. Tensors are very important in CNN as all inputs, parameters, and intermediate representations are in the form of tensor only.

A tensor can be converted into a vector by using the notation "vec", which is shown in the example below.

$$A = \begin{pmatrix} 3 & 5 \\ 4 & 6 \end{pmatrix} \quad vec(A) = (3, 4, 5, 6)^T$$

A recursive process is applied for vectorization of an order three matrix into an order two and then an order one matrix, i.e., vector. The same process is applied for order four and higher order matrixes.

2. Chain Rule and Vector Calculus

Chain rule and vector calculus is very important in the learning process of CNN. For example, z is scalar, i.e., $z \varepsilon \mathbb{R}$ and given vector is $y \varepsilon \mathbb{R}^H$. Suppose z is the function of y. Then, the partial derivative of z with respect to y is a vector, given as,

$$\left[\frac{dz}{dy}\right]i = \frac{dz}{dyi}$$

which shows the size of vector $\frac{dz}{dy}$ is the same as y and $\frac{dz}{dyi}$ is the ith element of vector. If $x \varepsilon \mathbb{R}^W$ is one more vector and y is a function of x, then the partial derivative of y with respect to x is given as,

$$\left[\frac{dy}{dx^T}\right]i, j = \frac{dyi}{dxj}$$

The obtained result is a matrix of size H × W and $\frac{dyi}{dxj}$ denotes the element of the ith row and jth column. Hence, we can say z is a function of x, and the chain rule can be applied as follows:

$$\frac{dz}{dx^T} = \frac{dz}{dy^T}\frac{dy}{dx^T}$$

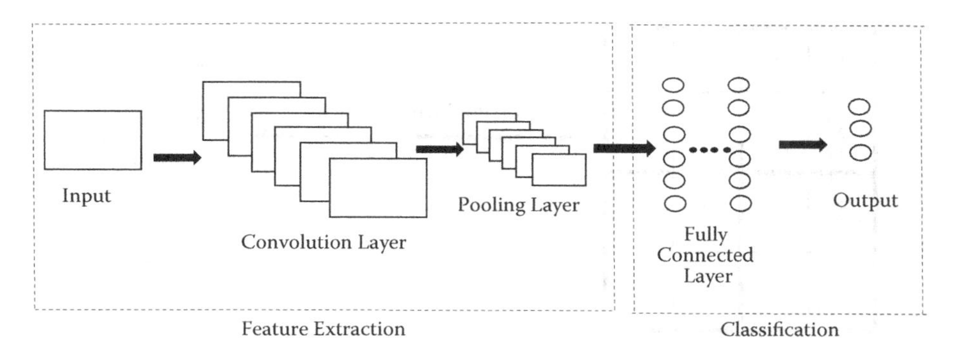

FIGURE 10.1
CNN Architecture.

10.2.2 Basic Components of Convolutional Neural Networks

As shown in the Figure 10.1, CNN architecture consists of three basic building blocks: convolution layers, pooling layers, and fully connected layers. Along with these basic building blocks, other important parameters are the dropout layer and activation function.

1. Convolution Layer

Convolution layer is an important layer of CNN. To minimize the connection overhead [2,3] and to reduce the number weights in a neural network [4,5], the term *convolution* was introduced. Learnable filters or kernels are the parameters used in this layer, which handles the input at full depth. For example, in a classification problem if we provide an image to the input layer, depending upon a feature extracted from the image, output will be a prediction of class label [6]. Some of the neurons of the previous layer are connected to the individual neurons of the next layer. This corelation is known as receptive field [7] and forms a weight vector that is equal for all points of the plane so that similar features are extracted from different locations of input [8]. The output of the convolution layer is stacked feature maps, which are generated by different filters.

Convolution operation is explained in Figure 10.2. Let us assume that convolution kernel size is 2 × 2 and input image is 3 × 4.

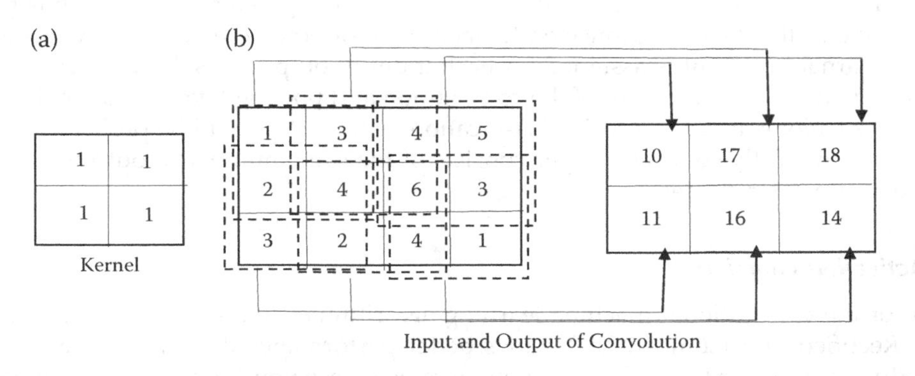

FIGURE 10.2
Convolution Operation.

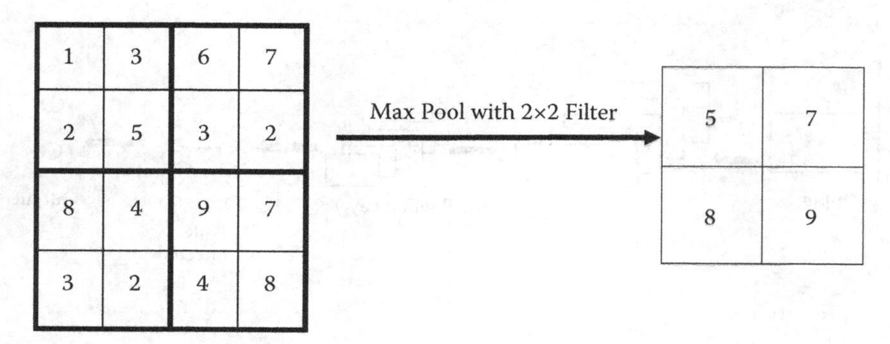

FIGURE 10.3
Max Pool Operation.

Stride is used to control density of convolution. Lower density is achieved by larger size of stride. Stride = 1 means the filter will move one pixel; stride = 2 indicates the pixel will move 2 pixels. Border information of the image may be lost due to the convolution operation because it captures only once. To overcome this limitation, zero padding with input is used to retrieve information as well as to manage output size.

2. Pooling Layer

To reduce the complexity for upcoming layers, the pooling operation is used, which is followed by the convolution layer. After extraction of features, location of feature is less significant, so pooling is applied after convolution [9,10]. The pooling function is applied on selected windows and input elements that reside in the window. The output of the pooling operation is another vector. Max-pooling and average pooling are the most commonly used techniques. In max-pooling, the maximum value of each window is selected in the output vector, whereas average pooling takes the average of each element of the window. The max-pooling operation is illustrated in Figure 10.3; 2 × 2 is a common window size used in this operation.

3. Fully Connected Layer

The fully connected layer is similar to the traditional feedforward neural network. It takes input from the previous layer, i.e., convolution, and the pooling layer, which is flattened. Each node of this layer is connected to each node of both layers, i.e., previous and next layer. Learnable weight is assigned for each input to output. A subset of fully connected layers is mapped to the output of the convolution and pooling layer to the final outputs of CNN. As CNN is mostly used in classification tasks, output will be probabilities of each task. The final fully connected layer consists of the same number of output nodes as the number of output classes.

4. Activation Function

Most of the conventional machine learning algorithms use the sigmoid activation function. Rectified linear unit (ReLU) shows better performance than the former in order to introduce non-linearity. Multi-layer networks use these non-linearities. Training speed is increased by using ReLU. Apart from that, partial derivative calculation of ReLU is easy, and gradients do not disappear with ReLU.

10.3 Advances in CNNs

10.3.1 Convolution Layer

Basic CNNs consist of a generalized linear model as a convolution filter. Here, we summarize work carried out on the convolution layer to improve its representation capacity.

1. Tiled Convolution

The term *convolution* was introduced to limit the parameters of neural networks, but it may limit the learning of other types of invaraiance as well. Ngiam et al. [11] proposed the concept of tiled CNN, in which tiles and multiple feature maps are used to learn scale invariant and rotational features. In the same layer, separate kernels are used for learning and square-foot pooling over neighbouring units, which are used to learn the complex invariances implicitly. Convolution operations are performed on every k unit, i.e., size of tiles. When k = 1, it is the same as traditional CNN because the same weight is shared. k = 2 shows better performance on CIFAR-10 datasets. The best choice of the value of k is equal to the size of the pooling region p, i.e., k = p.

2. Transposed Convolution

Backward pass of a traditional convolution can be seen as transposed convolution. It is also called *fractionally strided convolution* [12] and *deconvolution* [13]. In traditional convolution, multiple input activations connect to the single activation opposite to that; deconvolutions connect to a single input activation with multiple output activations. Deconvolution is widely used in visualization [14], semantic segmentation [14], and recognition [15,16]. The difference between traditional convolution and transposed convolution is shown in Figure 10.4.

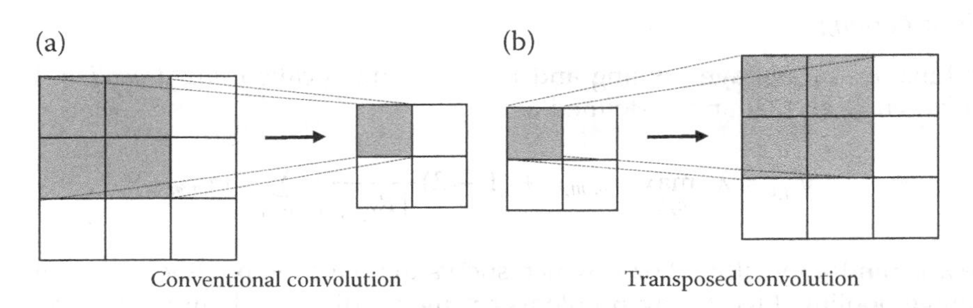

FIGURE 10.4
Conventional Convolution and Transposed Convolution.

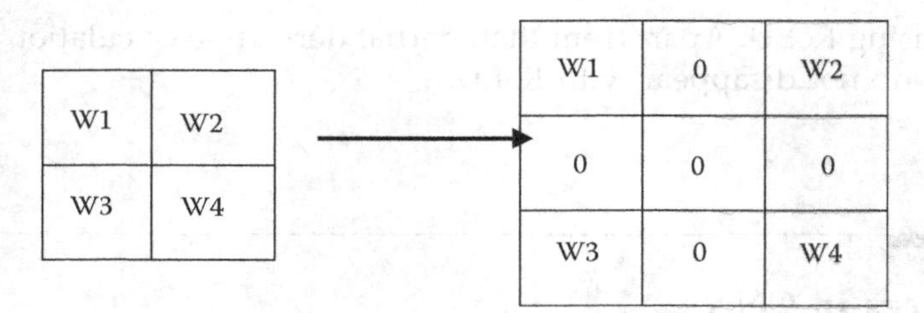

FIGURE 10.5
One Convolution Filter and Its Dilated Filter.

3. *Dilated Convolution*

One more hyper-parameter is introduced in dilated CNN [17]. To cover maximum-related information, zeros are inserted between the filter elements to increase the receptive field size. This technique is widely used if a large receptive field is required to make the prediction. It is mostly used in speech synthesis [18], machine translation [19], scene segmentation [20], and speech recognition [21]. A dilated convolution filter with size 3 × 3 is shown in Figure 10.5.

10.3.2 Pooling

Pooling is another significant operation of CNN. It minimizes the computational complexity by reducing the number of connections among the convolution layers. This section summarizes different pooling operations used in CNN.

1. *Lp Pooling*

Lp pooling is defined as,

$$y_{i,j,k} = \left[\sum\nolimits_{(m,n)\varepsilon R_{i,j}} (a_{m,m,k})^p\right]^{1/p}$$

where $y_{i,j,k}$ is the pooling result of the kth feature map at location (i, j), and $a_{m,n,k}$ is the feature value of the kth feature map at location (m, n) of the pooling region R_{ij}. It was proven in Yu et al. [22] that generalization using Lp pooling is better than max pooling.

2. *Mixed Pooling*

A combination of average pooling and max pooling is called *mixed pooling*. It was proposed by Yu et al. [22] and is defined as,

$$y_{i,j,k} = \lambda \max_{(m,n)\in R_{i,j}} a_{m,m,k} + (1-\lambda)\frac{1}{|R_{i,j}|}\sum_{(m,n)\varepsilon R_{i,j}} (a_{m,m,k})^p$$

where λ is randomly either 0 or 1, which shows the selection of either using max pooling or average pooling. Overfitting problems are more efficiently handled by mixed pooling than by average pooling and max pooling [23].

3. Stochastic Pooling

Stochastic pooling randomly picks the activations according to a multi-nomial distribution so that activation of a non-maximal feature map is possible and utilized. Initially, by normalizing the activations within the region, stochastic pooling calculates the probabilities p for each region Rj, such that $pi = a_i/\Sigma_{K\in R_j}(a_k)$. After getting P (p1,..., p | Rj |), location l within the region is found by sampling on p, and then pooled activation is defined as $y_j = a_l$, where l ~P(p1,..., p | Rj |).

4. Spectral Pooling

Dimensionality reduction is performed in spectral pooling [24]. It uses cropping of input in the frequency domain. Suppose for a given input feature map $X \in \mathbb{R}^{m\times m}$, an expected output feature map is of size h × w. Spectral pooling applies discrete Fourier transform on the input feature map and then only the central h × w submatrix of the frequencies is retained by truncating remaining frequencies. Finally, inverse discrete Fourier transform is performed to get the spatial domain.

10.3.3 Activation Function

Performance of a CNN significantly increases by selecting the proper activation function for a specific task. This section summarizes the activation function used in CNN.

1. ReLU

ReLU is the most commonly used activation function [25]. ReLU activation is represented as,

$$a_{i,j,k} = \max(z_{i,j,k},\ 0)$$

where (i, j) is the location and k is the channel number. ReLU is a linear function where for negative values, it is zero, and for zero and positive values, it is the same or equal to

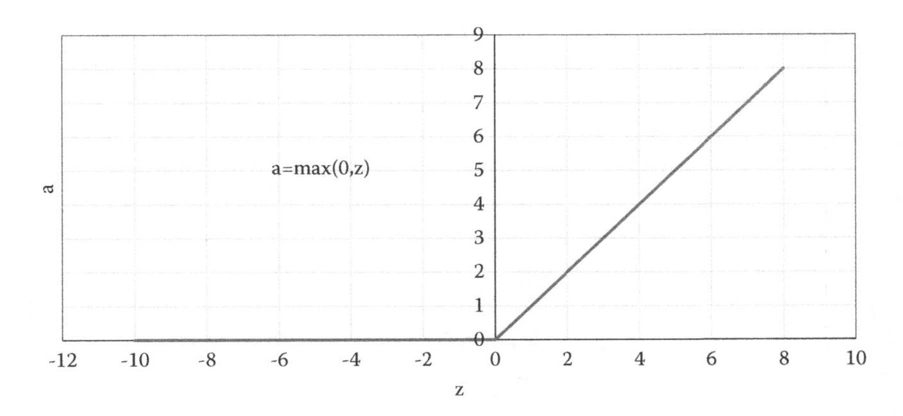

FIGURE 10.6
ReLU Activation Function.

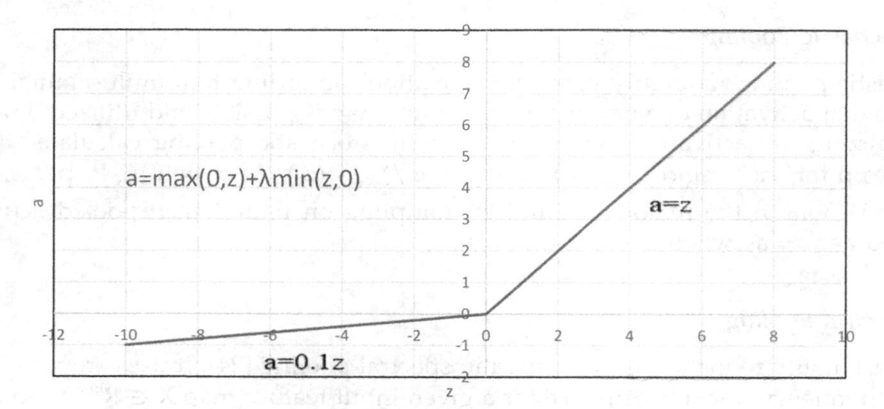

FIGURE 10.7
Leaky ReLu Activation Function.

that value only. It uses the simple max operation; hence, it is faster than the tanh or sigmoid activation functions. Using ReLU deep networks can be trained efficiently [26]. Graphically, it is represented in Figure 10.6.

2. *Leaky ReLU*

One of the disadvantages of ReLU is that it has zero gradient and causes a dying ReLU problem. In a dying ReLU problem, some ReLU neurons basically die for all inputs and remain inactive no matter what input is supplied. Network performance is affected if there is a large number of dead neurons. This problem is overcome by leaky ReLU [27], which is represented as,

$$a_{i,j,k} = \max(z_{i,j,k},\ 0) + \lambda\ \min(z_{i,j,k},\ 0)$$

where the value of λ is predefined and in the range of (0, 1). Instead of mapping the negative part to zero like ReLU, leaky ReLU compresses it. Graphically, it is represented in Figure 10.7.

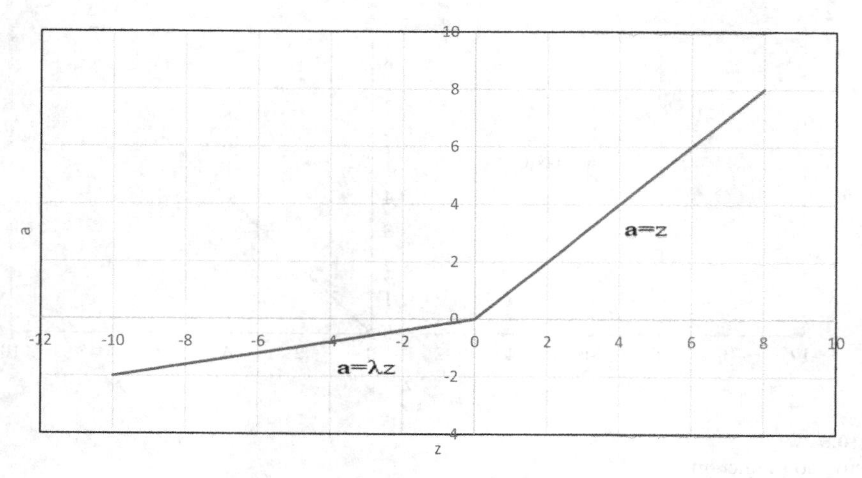

FIGURE 10.8
Parametric ReLU Activation Function.

3. Parametric ReLU

To improve accuracy, He et al. [28] proposed the parametric rectified linear unit (PReLU). Performance of the PReLU is increased by adaptive learning of parameters. PReLU is represented as,

$$a_{i,j,k} = \max(z_{i,j,k}, \ 0) + \lambda_k \min(z_{i,j,k}, \ 0)$$

where λ_k is the kth channel's learning parameter. Though it has an added extra parameter, there is no risk of overfitting, and computational cost is also negligible. Graphically, it is represented in Figure 10.8.

10.3.4 Loss Function

1. Softmax Loss

Softmax loss is a combination of softmax and multi-nomial logistic loss [29]. Suppose a training set is given as $\{(x^{(i)}, y^{(i)}); i \in 1,\ldots,N, y^{(i)} \in 1,\ldots,K\}$, where $y^{(i)}$ is the target class label for $x^{(i)}$, i.e., ith input. So, for the ith input, the prediction of the jth class using the softmax function is given as,

$$P_{(j)}^{(i)} = e^{z_j^{(i)}} / \sum_{l=1}^{K} e^{z_l^{(i)}}$$

where $z_j^{(i)}$ is activation of a densely connected layer and denoted as $z_j^{(i)} = w_j^T a^{(i)} + b_j$. Then, the softmax loss is calculated as,

$$L_{softmax} = -\frac{1}{N}\left[\sum_{j=1}^{N} \sum_{j=1}^{N} 1\{y^{(i)} = j\} log P_j^{(i)}\right]$$

Angular boundary, i.e., θ_j, is presented by large-margin softmax loss. θ_j is the angle between the jth column, i.e., w_j of the weight matrix and the feature vector $a^{(i)}$.

2. Hinge Loss

Large-margin classifiers are trained using hinge loss, e.g., support vector machine. It is calculated as,

$$L_{hinge} = \frac{1}{N}\sum_{i=1}^{N}\sum_{j=1}^{k} [\max(0, 1 - \delta(y^{(i)}, j) w^T x_i)]^P$$

where w is the weight vector and $y^{(i)} \in [1,\ldots, K]$ is a class label:

$$\delta(y^{(i)}, j) = 1 \quad if \quad y^{(i)} = j \ else \ \delta(y^{(i)}, j) = -1$$

10.3.5 Regularization

Overfitting problems of CNN are effectively handled by regularization. The objective function is modified by adding the regularization term, which can affect the model complexity. This section summarizes different regularization techniques used for CNN.

1. Dropout

Dropout was proposed by Hinton et al. [30]. It is applied to fully connected layers. Dropout basically avoids the dependency of the network on a specific neuron. Instead, it powers the network to be more precise, even if certain information is not available. Its output is described as,

$$y = r * a(W^T x),$$

where W is a weight matrix of size n $\times$ d, $x = [x1, x2, \ldots, x_n]^T$ is an input of size n of the fully connected layer, and r is a binary vector having d elements. Bernoulli distribution with parameter p is used to select an element of r, i.e., $r_i \sim \text{Bernoulli}(p)$. Further, several techniques like adaptive dropout [31] and spatial dropout [32] are proposed to improve the performance of dropout.

2. DropConnect

DropConnect [33] is an extension of Dropout. In DropConnect, rather than randomly assigning an output of neuron to zero, it randomly assigns weight matrix (W) elements to zero. The output of DropConnect is given by $y = a((R * W)x)$, where $Rij \sim \text{Bernoulli}(p)$.

10.3.6 Optimization

In this section, we summarize different optimization techniques.

1. Weight Initialization

Parameter initialization is an important task in CNN. Weight initialization should be done properly; otherwise, large output will generate if every layer scales its input by K. In that case, the final output will be K^L times larger than the original input, where L is the number of layers. Russakovsky et al. [34] used zero-mean Gaussian distribution along with standard deviation 0.01 to initialize weight. They set the constant bias term to the fully connected layers as well as the second, fourth, and fifth convolution layers. Glorat and Bengio [35] proposed a random initialization method. Using a number of input and output neurons, the initialization scale is automatically determined. Saxe et al. [36] proved that instead of using Gaussian initialization for linear networks, orthonormal matrix initialization gives a better performance.

2. Stochastic Gradient Descent

Typical training methods use backpropagation algorithms and update the parameter using gradient descent. Standard gradient descent updates the parameter (Θ) of the objective $L(\Theta)$ as follows:

$$\Theta t + 1 = \Theta t - \eta \nabla_\Theta E(L(\Theta t)$$

where η is the learning rate and $E(L(\Theta t)$ is the expectation of $L(\Theta)$.

Stochastic gradient descent (SGD) [37] is defined as follows:

$$\Theta t + 1 = \Theta t - \eta_t \nabla_\Theta E(L(\Theta t; x^{(t)}, y^{(t)})$$

where $(x^{(t)}, y^{(t)})$ is randomly selected for the training set.

For parallel, large-scale machine learning, parallelized SGD methods [38] are used. To improve the performance of a large-scale distributed training process on clusters using many CPUs, downpour SGD [39] is used. Multiple GPUs are also used to calculate gradients asynchronously [40].

10.4 Classic CNN models

In this section, we provide an overview of some representative models.

10.4.1 LeNet-5

In 1998, LeCun et al. [1]. proposed LeNet-5, which consists of two convolution layers, two pooling layers, three fully connected layers, and seven trainable layers. It is the basis of the modern CNN implement weight sharing techniques, and it extracts six different types of feature maps. It is basically designed for handwriting recognition, but apart from this, its other applications are face recognition, machine-printed character recognition, and online handwriting recognition.

10.4.2 AlexNet

AlexNet was proposed in 2012 by Alex et al. [41]. It consists of five convolution layers and three fully connected layers. The key innovations in AlexNet are that it uses dropout to handle the overfitting problem, the ReLU activation function is used, and local response normalization (LRN) was proposed for the first time on CNN. It also makes use of GPUs for speeding up computing. It uses non-saturing neurons for faster training. Its major applications are image classification and object detection.

10.4.3 VGGNets

The Visual Geometry Group (VGG) of Oxford University proposed a series of CNN algorithms that are VGGNets (VGG-11, VGG-11-LRN, VGG 13, VGG-16, and VGG-19) [42]. They proved that the final performance of the network can be improved by increasing the depth of the neural network. Key innovation in VGGNets is instead of 5×5 or 5×5 ones kernel, they used 3×3 convolution kernels, and they removed a layer of LRN.

10.4.4 GoogLeNet

Google researchers proposed a GoogLeNet model in 2014. It is deeper CNN formed by using an inception module. Inception modules are stacked linearly. There are mainly four versions of the inception module, i.e., Inception v1 [43], Inception v2 [44], [45], Inception v3 [45], and Inception v4 [46].

1. Inception v1

Inception v1 consists of 1 × 1, 3 × 3, and 5 × 5 convolution kernels. The key idea behind the different size of kernels is to reduce the number of parameters required to train. A feature map of a different scale can be extracted by convolution kernels of various sizes and then stacked to get more representation.

2. Inception v2

Inception v2 replaced single 5 × 5 convolution layers by two 3 × 3 ones. One n × n convolution layer can be replaced by one 1 × n and one n × 1 convolution layer, and factorization is performed only on the last 3 × 3 convolution of each branch. The internal covariate shift problem is handled by using batch normalization.

3. Inception v3

To increase the depth and non-linearity of CNN and to speed up the training, Inception v3 factorized 5×5 and 3×3 convolution kernels into two one-dimensional ones. It uses RMSProp optimizer.

4. Inception v4 and Inception-ResNet

Inception v4 modules are based on Inception v3. Inception v4 combines the Inception architecture with residual connections. It gives better performance than its predecessors. Additionally, ResNet structure improves performance and training speed.

10.5 Applications of CNN

In the field of deep learning, CNN is one of the important concepts that handles a large amount of data and achieves excellent results in various domains. Hence, it is used in a wide range of applications that handle two-dimensional images as well as one-dimensional and multi-dimensional inputs.

10.5.1 Applications of One-dimensional CNN

One-dimensional data are processed by one-dimensional convolutional kernels of one-dimensional CNN. Typically, one-dimensional CNN is used in an application where it does not matter the feature's location, and features are extracted from a segment of fixed length over all datasets. So, mainly one-dimensional CNN is used in signal identification. Based on the learning feature from training data, CNN distinguishes the input signals. Zhang et al. [47] proposed a method to predict arrhythmia and other diseases using ECG data. They used multi-resolution one-dimensional CNN. In Abdeljaber et al. [48], severity identification and fault detection was performed by one-dimensional CNN. A direct damage identification method was proposed in Abdeljaber et al. [49]. One-dimensional CNN is also used in time series prediction. Urtnasan et al. [50] proposed a method that uses ECG data and atrial fibrillation prediction done automatically by one-dimensional

CNN. Wind direction and speed are predicted in Harbola and Coors [51]. One-dimensional CNN is also used in structural damage detection. Structural damage monitoring and detection is very important in various applications of civil engineering. Most of the research has been done on vibration-based damage detection. In these methods, vibration response is used to locate and detect damage. Abdeljaber et al. proposed first-time vibration-based structural health monitoring, which used one-dimensional CNN [52]. They used a QU grandstand simulator to verify the performance of the proposed system. As one-dimensional CNN automatically learns the extracted features, it removes the necessity of a handcrafted feature's parameter tuning or manual model. This performance of one-dimensional CNN is further enhanced by using multi-core CPU. Other applications of CNN are condition monitoring in rotating mechanical machine parts and fault detection in modular multi-level converters [53]. Ince et al. [54] proposed a new method to detect motor defects caused by bearing faults. Due to this, generated motor current waveform is slightly varied at certain frequencies, which is not easy to detect visually. One-dimensional CNNs detect it with high accuracy in the proposed method.

10.5.2 Applications of Two-dimensional CNN

1. Image Classification

Due to CNN's capabilities, such as the classifier learning joint feature, it achieves more classification accuracy than any other classification method. LeNet-5 [1] was the first CNN, which was used for classification of handwritten digits. In ILSVRC 2012, AlexNet [41] achieved best performance. VGGNets [42] and GoogLeNet [43–46] also showed promising results in the classification task. Tousch et al. [55] built a hierarchy of classifiers to improve the accuracy of an image classification task. To recognize sub-category fine-grained features, learning is important. So, a tree structure of classifiers was used in Wang et al. [56]. Part-based R-CNN [57] was proposed to learn whole objects and sub-categories, i.e., part detectors. For part suggestion, they used selective search [58], and for localizing, they applied non-parametric geometric constraints. Lin et al. [59] proposed a recognition system called deep LAC, which includes part localization, alignment, and classification. In Lin et al. [60], a bilinear model was proposed for fine-grained image classification. They used a two-feature extractor and took the outer product of the output of the two-feature extractor at each location of the image. Finally, the image descriptor was obtained using a pooling operation.

2. Object Detection

Object detection is performed based on image classification. It is done by finding the image category and marking it with a bounding box. There are two approaches to image detection: a one-stage approach, such as SSD [61], YOLO [62–64], or CornerNet [65], which directly predicts class labels using single pipeline detection, and a two-stage approach, in which the region proposals are done first and then classification is done. Examples are R-CNN [66], Fast R-CNN [67], and Faster R-CNN[68].

3. Face Recognition

Face recognition techniques were developed based on the features of the human face. Taigman et al. [69] proposed a model called DeepFace, which consists of detection, alignment, extraction, and classification. DeepID[70], FaceNet [71], and VGGFace [72] are some other examples of face recognition techniques that use different loss functions and network architectures to improve accuracy.

4. Image Segmentation

CNN has shown its excellence in the image segmentation process, which divides the image into multiple segments. Basically, there are three types of segmentation: i) semantic segmentation, ii) instance segmentation, and iii) panoptic segmentation.

Semantic Segmentation: In this type of segmentation, multiple objects of the same class are treated as a single entity. A fully convolutional network for image semantic segmentation was proposed by Long et al. [73], which predicted the input image into label maps. For biomedical image segmentation, U-Net was proposed by Ronneberger et al. [74]. In this architecture, pooling layers in Shelhamer et al. [73] were replaced by up-sampling layers, and hence output resolution was increased. ENet was proposed by Paszke et al. for real-time semantic segmentation where low latency operations are required [75]. It achieved better performance over the existing architectures. Segfast [76] is faster than the method by Shelhamer et al. [73]. It is a combination of depth-wise separable convolution and fire module of SqueezNet.

Instance Segmentation: In this type of segmentation, each instance of an object in the image is identified. Mask-RCNN [77] is a type of instance segmentation that uses a similar approach as used in faster RCNN, in addition to the Region Proposal Network and classification and bounding-box regression, to generate a binary mask for each region of interest. Bolya et al. [78] proposed a method for real-time instance segmentation.

Panoptic Segmentation: This method is the combination of two types of segmentation: semantic segmentation and instance segmentation. Kirillov et al. [79] proposed an architecture for panoptic segmentation that uses an instance segmentation method, Mask-RCNN, with semantic segmentation using the Feature Pyramid Network. OANet [80] is an end-to-end network proposed for panoptic segmentation using a similar approach in Kirillov et al. [79]

5. Object Tracking

Fan et al. [81] proposed shift-variant architecture to be used with CNN as an object tracker. This method considers the images of consecutive frames and extracts the spatial and temporal features. It improves the performance over traditional tracking methods. The limitation of object tracking problems, i.e., shallow classifier structures and handcrafted features, are handled in Chen et al. [82]. To make the decision for updating object appearance models, heuristic schema was used. A visual tracking algorithm was used in Hong et al. [83]. The authors used pretrained CNN with online SVM to learn the appearance of the target with respect to the background and generate a target-specific saliency map.

6. Pose Estimation

DeepPose [84] was first introduced to estimate human poses based on deep neural network. For high-precision pose estimation, the author presented a cascade of deep neural networks. The pose estimation problem is a formulated regression problem toward body joint coordinates. This method takes the whole image as input and pose in a holistic manner, capturing context and reasoning about the human pose. Besides this, different methods for pose estimation and pose refinement were proposed to improve the performance [85–87].

10.5.3 Applications of Multi-dimensional CNN

Theoretically, data with any dimension are handled by CNN, but as multi-dimensional data are hard to understand, CNN is mostly used for up to three-dimensional data. Three-dimensional CNN is basically used in human action recognition, where human action in video is automatically recognized by machine [88,89]. Object recognition/ detection is also done by using three-dimensional CNN [90,91]. In addition to that, three-dimensional CNN is widely used in radiology to handle high-dimensional images, such as CT images and X-rays, in medical research to improve the performance of radiologists and ultimately patient care [92].

10.6 Conclusion

CNNs have given excellent performance across a variety of domains, including image, video, speech, and medical research. This chapter covers a basic understanding of CNNs, along with advances made to CNNs. We have studied several features of CNN, like layer design activation function, loss function, regularization, and optimization. Apart from this, we have also introduced an application of one-dimensional, two-dimensional, and multi-dimensional CNN in various domains. Although CNNs have achieved breakthrough results in various domains, there are still some issues that we need to focus on, like the requirement of a large-scale dataset, large computing power, and complexity reduction. In this chapter, we provided a detailed understanding of CNNs and their applications, which is beneficial for researchers to get future research directions.

References

1. Lecun, Y., Boser, B., Denker, J., Henderson, D., Howard, R., Hubbard, W.E., and Jackel, L. (1989). "Backpropagation Applied to Handwritten Zip Code Recognition. " *Neural Computation* 1(4): 541–551.
2. Guo, Y., Liu, Y., Oerlemans, A., Lao, S.-Y., Wu, S., and Lew, M. (2015). "Deep Learning for Visual Understanding: A Review." *Neurocomputing* 187: 27– 48.
3. J. Wu. (2016). "Introduction to Convolutional Neural Networks."
4. Kwak. (2016). "Introduction to Convolutional Neural Networks (CNNs)."
5. Szegedy, C., Wei, Liu W., Jia, Y., Sermanet, P., Scott, R.S., Anguelov Erhan, D., Vincent, V.A., and Rabinovich, A. (2015). "Going Deeper with Convolutions." The IEEE Conference on Computer Vision and Pattern Recognition (pp 1–9). IEEE.
6. Fang, J., Zhou, Y., Yu, Y., and Du, S. (2017) "Fine-Grained Vehicle Model Recognition using a Coarse-to-Fine Convolutional Neural Network Architecture." *IEEE Transactions on Intelligent Transportation Systems* 18 (7): 1782–1792.

7. Nebauer, C. (1998) "Evaluation of Convolutional Neural Networks for Visual Recognition." IEEE Transactions on Neural Networks 9 (4): 685–696.
8. Palsson, F., Sveinsson, J. R., and Ulfarsson, M.O. (2017) "Multispectral and Hyperspectral Image Fusion Using a 3 -D-Convolutional Neural Network." *IEEE Geoscience and Remote Sensing Letters* 14 (5): 639–643.
9. Lawrence, S., Giles, C., Tsoi, A., and Back, A. (1997). "Face Recognition: A Convolutional Neural Network Approach. Neural Networks." *IEEE Transactions on Neutral Networks* 8: 98–113.
10. Lecun, Y., Bottou, L., Bengio, Y., and Haffner, P. (1998). "Gradient-Based Learning Applied to Document Recognition. Proceedings of the IEEE" (pp. 2278–2324). IEEE.
11. Le, Q., Ngiam, J., Chen, Z., Chia, D., Koh, P., and Ng, A. (2010). "Tiled Convolutional Neural Networks." *National Information Proceeding Systems* 86: 1279–1287.
12. Visin, F., Kastner, K., Courville, A., Bengio, Y., Matteucci, M., and Cho, K. (2015)."ReSeg: A Recurrent Neural Network for Object Segmentation." arXiv abs/1511.07053.
13. Zeiler M.D., Fergus R. (2014). "Visualizing and Understanding Convolutional Networks." In: Fleet D., Pajdla T., Schiele B., Tuytelaars T. (eds.) *Computer Vision – ECCV 2014. ECCV 2014. Lecture Notes in Computer Science*, vol 8689. Springer.
14. Noh, H., Hong, S., and Han, B. (2015). "Learning Deconvolution Network for Semantic Segmentation." ArXiv. 10.1109/ICCV.2015.178.
15. Zhang J., Lin Z., Brandt J., Shen X., and Sclaroff S. (2016). "Top-Down Neural Attention by Excitation Backprop." In: Leibe B., Matas J., Sebe N., Welling M. (eds.) *Computer Vision – ECCV 2016. ECCV 2016. Lecture Notes in Computer Science*, vol 9908. Springer.
16. Yu, F. and Koltun, V. (2015). "Multi-Scale Context Aggregation by Dilated Convolutions." abs/1511.07122
17. Cao, C., Liu, X., Yang, Y., Yu, Y., Wang, J., Wang, Z., Huang, Y., Wang, L., Huang, C., Xu, W., and Ramanan, D. (2015). "Look and Think Twice: Capturing Top-Down Visual Attention with Feedback Convolutional Neural Networks." International Conference on Computer Vision (pp. 2956–2964).
18. Oord, Aa., Dieleman, S., Zen, H., Simonyan, K., Vinyals, O., Graves, A., Kalchbrenner, N., Senior, A., and Kavukcuoglu, K. (2016). "WaveNet: A Generative Model for Raw Audio." 9th ISCA Speech Synthesis Workshop 1, Sunnyvale, USA.
19. Kalchbrenner, N., Espeholt, L., Simonyan, K., Oord, A., Graves, A., and Kavukcuoglu, K. (2016). "Neural Machine Translation in Linear Time." ArXiv, abs/1610.10099
20. Yu F. and Koltun V. (2015). "Multi-Scale Context Aggregation by Dilated Convolutions." abs/1511.07122.
21. Sercu, T., Goel, V. (2016). "Dense Prediction on Sequences with Time-Dilated Convolutions for Speech Recognition." abs/1611.09288.
22. Yu D., Wang H., Chen P., Wei Z. (2014). "Mixed Pooling for Convolutional Neural Networks." In: RSKT 2014. Lecture Notes in Computer Science, vol 8818. Springer.
23. Lee, Chen-Yu, Patrick G., and Zhuowen T.. (2015). "Generalizing Pooling Functions in Convolutional Neural Networks: Mixed, Gated, and Tree." abs/1509.08985.
24. Rippel, O., Snoek, J., and Adams, R. (2015). "Spectral Representations for Convolutional Neural Networks." Proceedings of NIPS, 2 (pp. 2449–2457.)
25. Nair, V. and Hinton, G. (2010). "Rectified Linear Units Improve Restricted Boltzmann Machines." *Proceedings of ICML* 27: 807–814.
26. Russakovsky, O., Deng J., Su, H., Krause, J., Satheesh, S., Ma, S., Huang, Z., Karpathy, A., Khosla, A., Bernstein, M., Ar, B., Li, F.-F. (2014). "ImageNet Large Scale Visual Recognition Challenge". *IJCV*115: 211–252.
27. Maas, A., Hannun, A., and Ng, A. (2013). "Rectifier Nonlinearities Improve Neural Network Acoustic Models."
28. He, K., Zhang, X., Ren, S., and Sun, J. (2015). "Delving Deep into Rectifiers: Surpassing Human-Level Performance on ImageNet Classification, IEEE International Conference on Computer Vision (ICCV)." Santiago (pp. 1026–1034).IERE.

29. Liu, W., Wen, Y., Yu, Z., and Yang, M. (2016). "Large-Margin Softmax Loss for Convolutional Neural Networks." ProC. Int. Conf. Mach. Learn.
30. Hinton G. E., Srivastava N., Krizhevsky, A., Sutskever, I., and Salakhutdinov, R. (2012). "Improving neural networks by preventing co-adaptation of feature detectors". ArXiv, abs/1207.0580.
31. Ba, L.J. and Frey, B. (2013). "Adaptive Dropout for Training Deep Neural Networks." *Advances in Neural Information Processing Systems* 2: 3084–3092.
32. Tompson, J., Goroshin, R., Jain, A., Lecun, Y., and Bregler, C. (2015). "Efficient object localization using Convolutional Networks." IEEE Conference on Computer Vision and Pattern Recognition. pp. 648–656.
33. Wan, L., Zeiler, M., Zhang, S., Lecun, Y., and Fergus, R. (2013). "Regularization of Neural Networks using DropConnect." Proceedings of the 30th International Conference on International Conference on Machine Learning. 28: 1058–1066.
34. Russakovsky, O., Deng, J., Su, H., Krause, J., Satheesh, S., Ma, S., Huang, Z., Karpathy, A., Khosla, A., Bernstein, M.S., Berg, A., and Fei-Fei, L. (2015). "ImageNet Large Scale Visual Recognition Challenge." *International Journal of Computer Vision* 115 :211–252.
35. Glorot, X., and Bengio, Y. (2010). "Understanding the difficulty of training deep feedforward neural networks." 9 (pp. 249–256). Proceedings of Machine Learning Research.
36. Saxe, A., Mcclelland, J., and Ganguli, S. (2013). "Exact Solutions to the Nonlinear Dynamics of Learning in Deep Linear Neural Networks." abs/1312.6120.
37. Wijnhoven, R. and With, P. (2010). "Fast Training of Object Detection Using Stochastic Gradient Descent" International Conference on Pattern Recognition. pp. 424–427.
38. Zinkevich, M., Weimer, M., Smola, A., and Li, L. (2010). "Parallelized Stochastic Gradient Descent". *Advances in Neural Information Processing Systems* 23: 2595–2603.
39. Dean, J., Corrado, G.S., Monga, R., Chen, K., Devin, M., Le, Q., Mao, M., Ranzato, A., Senior, A., Tucker, P., Yang, K., and Ng, A. (2012). "Large Scale Distributed Deep Networks. Advances in Neural Information Processing Systems." *Advances in Neural Information Processing Systems* 23: 2595–2603.
40. Paine, T., Jin, H., Yang, J., Lin, Z., and Huang, T. (2013). "GPU Asynchronous Stochastic Gradient Descent to Speed Up Neural Network Training." abs/1312.6186.
41. Krizhevsky, A., Sutskever, I., and Hinton, G. (2012). "ImageNet Classification with Deep Convolutional Neural Networks." *Neural Information Processing Systems* 25(2): 1097–1105.
42. Simonyan, K. and Zisserman, A. (2014). "Very Deep Convolutional Networks for Large-Scale Image Recognition." arXiv 1409.1556.
43. Szegedy, C., Ioffe, S., Vanhoucke, V., and Alemi, A. (2016). "Inception-v4, Inception-ResNet and the Impact of Residual Connections on Learning." AAAI Conference on Artificial Intelligence. ACM digital Library.
44. Ioffe, S., and Szegedy, C. (2015). "Batch Normalization: Accelerating Deep Network Training by Reducing Internal Covariate Shift." Proceedings of the 32nd International Conference on Machine Learning, 37 (pp. 448–456).
45. Szegedy, C., Liu, W., Jia, Y., Sermanet, P., Reed, S., Anguelov, D., Erhan, D., Vanhoucke, V., Rabinovich, A. (2015). "Going deeper with Convolutions." The IEEE Conference on Computer Vision and Pattern Recognition (CVPR) (pp 1–9). IEEE.
46. Szegedy, C., Vanhoucke, V., Ioffe, S., Shlens, J., Wojna, Z. B. (2016). "Rethinking the Inception Architecture for Computer Vision." *IEEE Conference on Computer Vision and Pattern Recognition* 1: 2818–2826.
47. Zhang, Q., Zhou, D., and Zeng, X. (2017). "HeartID: A Multiresolution Convolutional Neural Network for ECG-Based Biometric Human Identification in Smart Health Applications." *IEEE Access* 5:1.
48. Abdeljaber, O., Sassi, S., Avci, O., Kiranyaz, S., Ibrahim, A., and Gabbouj, M. (2018). "Fault Detection and Severity Identification of Ball Bearings by Online Condition Monitoring." *IEEE Transactions on Industrial Electronics* 66: 8136–8147.

49. Abdeljaber, O., Avci, O., Kiranyaz, S., Gabbouj, M., and Inman, D. (2017). "Real-Time Vibration-Based Structural Damage Detection Using One-Dimensional Convolutional Neural Networks." *Journal of Sound and Vibration* 388: 154–170.
50. Urtnasan, E., Kim, H., Park, J.-U., Kang, D., and Lee, K.-J. (2019). "Automatic Prediction of Atrial Fibrillation Based on Convolutional Neural Network Using a Short-Term Normal Electrocardiogram Signal." *Journal of Korean Medical Science* 34.
51. Harbola, S. and Coors, V. (2019). "One Dimensional Convolutional Neural Network Architectures for Wind Prediction." *Energy Conversion and Management* 195: 70–75.
52. Abdeljaber, O., Avci, O., Kiranyaz, S., Gabbouj, M., and Inman, D. (2017). "Real-Time Vibration-Based Structural Damage Detection Using One-Dimensional Convolutional Neural Networks." *Journal of Sound and Vibration* 388: 154–170.
53. Kiranyaz, S., Avci, O., Abdeljaber, O., Ince, T., Gabbouj, M., and Inman, D. (2019). "1D Convolutional Neural Networks and Applications: A Survey." ArXiv, abs/1905.03554
54. Ince, T., Kiranyaz, S., Eren, L., Askar, M., and Gabbouj, M. (2016). "Real-Time Motor Fault Detection by 1D Convolutional Neural Networks." *IEEE Transactions on Industrial Electronics* 63(11): 7067-7075.
55. Tousch, A.-M., Herbin, S., and Audibert, J.-Y. (2012). "Semantic Hierarchies for Image Annotation: A Survey." *Pattern Recognition* 45:333–345.
56. Wang, Z., Wang, X., and Wang, G. (2018). "Learning Fine-Grained Features via a CNN Tree for Large-scale Classification." *Neurocomputing*: 1231–1240. ArXiv, abs/1511.04534.
57. Zhang, N., Donahue, J., Girshick, R., and Darrell, T. (2014). "Part-Based R-CNNs for Fine-Grained Category Detection." ECCV 2014. Lecture Notes in Computer Science, 8689. Springer.
58. Uijlings, J., Sande, K., Gevers, T., and Smeulders, A.W.M. (2013). "Selective Search for Object Recognition." *International Journal of Computer Vision* 104: 154–171.
59. Lin, D., Shen, X., Lu, C., and Jia, J. (2015). "Deep LAC: Deep Localization, Alignment and Classification for Fine-Grained Recognition." IEEE Conference on Computer Vision and Pattern Recognition (pp. 1666–1674).
60. Lin, T.-Y., Roy Chowdhury, A., and Maji, S. (2015). "Bilinear CNN Models for Fine-Grained Visual Recognition." (pp. 1449–1457).
61. Liu, W., Anguelov, D., Erhan, D., Szegedy, C., Reed, S., Fu, C.-Y., and Berg, A. (2016). "SSD: Single Shot MultiBox Detector." *European Conference on Computer Vision* 9905 (pp. 21–37).
62. Redmon, J., Divvala, S., Girshick, R., and Farhadi, A. (2016). "You Only Look Once: Unified, Real-Time Object Detection." IEEE Conference on Computer Vision and Pattern Recognition (pp. 779–788).
63. Redmon, J. and Farhadi, A. (2017). "YOLO9000: Better, Faster, Stronger." *IEEE Conference on Computer Vision and Pattern Recognition* (pp. 6517–6525).
64. Redmon, J. and Farhadi, A. (2018). "YOLOv3: An Incremental Improvement."
65. Law, H. and Deng, J. (2020). "CornerNet: Detecting Objects as Paired Keypoints." *International Journal of Computer Vision* 128: 734–750.
66. Girshick, R., Donahue, J., Darrell, T., and Malik, J. (2013). "Rich Feature Hierarchies for Accurate Object Detection and Semantic Segmentation. Proceedings of the IEEE Computer Society Conference on Computer Vision and Pattern Recognition." pp. 580–587. IEEE.
67. Girshick, R. (2015). "Fast r-cnn." *IEEE International Conference on Computer Vision* (pp. 1440–1448).
68. Ren, S., He, K., Girshick, R., and Sun, J. (2015). "Faster R-CNN: Towards Real-Time Object Detection with Region Proposal Networks." *IEEE Transactions on Pattern Analysis and Machine Intelligence* 39: 91–99.
69. Taigman, Y., Yang, M., Ranzato, M., and Wolf, L. (2014). "DeepFace: Closing the Gap to Human-Level Performance in Face Verification. Proceedings of the IEEE Computer Society Conference on Computer Vision and Pattern Recognition." IEEE.

70. Sun, Y., Liang, D., Wang, X., and Tang, X. (2015). "DeepID3: Face Recognition with Very Deep Neural Networks."
71. Schroff, F., Kalenichenko, D., Philbin, J. (2015). "FaceNet: A Unified Embedding for Face Recognition and Clustering." *Proc. CVPR* (pp. 815-823).
72. Parkhi, O., Vedaldi, A., and Zisserman, A. (2015). "Deep Face Recognition. 1. 41.1-41.12. 10.5244/C.29.41." *British Machine Vision Conference 2* 1 (pp. 41.1–41.12.
73. Shelhamer, E., Long J., and Darrell T. (2017). "Fully Convolutional Networks for Semantic Segmentation." *IEEE Transactions on Pattern Analysis and Machine Intelligence* 39: 640–651.
74. Ronneberger, O., Fischer, P., and Brox, T. (2015). "U-Net: Convolutional Networks for Biomedical Image Segmentation." *LNCS* 9351: 234–241. 10.1007/978-3-319-24574-4_28.
75. Paszke, A., Chaurasia, A., Kim, S., and Culurciello, E. (2016). "ENet: A Deep Neural Network Architecture for Real-Time Semantic Segmentation."
76. Pal, A., Jaiswal, S., Ghosh, S., Das, N., ans Nasipuri, M. (2018). "SegFast: A Faster SqueezeNet based Semantic Image Segmentation Technique using Depth-wise Separable Convolutions."
77. He K., Georgia, G., Piotr, D., and Ross, G., "Mask R-CNN," *IEEE Transactions on Pattern Analysis & Machine Intelligence*42:1.
78. Bolya, D., Zhou, C., Xiao, F., and Lee, Y. J. "YOLACT: Real-Time Instance Segmentation." *Proceedings of the IEEE/CVF International Conference on Computer Vision*(pp. 9157–9166).
79. Kirillov, A., He, K., Girshick, R., Rother, C., and Dollar, P. (2019). Panoptic Segmentation. 9396–9405. 10.1109/CVPR.2019.00963.
80. Huanyu, L., and Peng, C., Yu, C., Wang, J., Liu, X., Yu, G., and Jiang, W. (2019). "An End-to-End Network for Panoptic Segmentation." IEEE/CVF Conference on Computer Vision and Pattern Recognition (pp. 6165–6174).
81. Fan, J., Xu, W., Wu, Y., and Gong, Y. (2010) "Human Tracking using Convolutional Neural Networks," *IEEE Transactions on Neural Networks (TNN)* 21(10): 1610–1623.
82. Chen, Y., Yang, X., Zhong, B., Pan, S., Chen, D., and Zhang, H. (2016) Cnntracker: Online Discriminative Object Tracking via Deep Convolutional Neural Network, *Applied Soft Computing* 38: 1088–1098.
83. Hong, S., You, T., Kwak, S., and Han, B. (2015) "Online Tracking by Learning Discriminative Saliency Map with Convolutional Neural Network," in: Proceedings of the International Conference on Machine Learning (ICML) (pp. 597–606). ACM.
84. Toshev, A. and Szegedy, C. (2014) Deeppose: Human Pose Estimation via Deep Neural Networks, in: Proceedings of the IEEE Conference on Computer Vision and Pattern Recognition (CVPR) (pp.1653–1660). IEEE.
85. Tompson J. J., Jain A., LeCun Y., and Bregler C. (2014) Joint Training of a Convolutional Network and a Graphical Model for Human Pose Estimation, in: Proceedings of the Advances in Neural Information Processing Systems (NIPS) (pp.1799–1807). ACM.
86. Tompson J., Goroshin R., Jain A., LeCun Y., and Bregler C. (2015) Efficient Object Localization using Convolutional Networks, in: Proceedings of the IEEE Conference on Computer Vision and Pattern Recognition (CVPR) (pp. 648–656). IEEE.
87. Chen X., and Yuille A. (2015) Parsing occluded people by flexible compositions, in: Proceedings of the IEEE Conference on Computer Vision and Pattern Recognition (CVPR) (pp. 3945–3954). IEEE.
88. Stergiou, A., and Poppe, R. (2019). "Spatio-Temporal FAST 3D Convolutions for Human Action Recognition."
89. Huang, Y., Lai, S.-H., and Tai, S.-H. (2019). "Human Action Recognition Based on Temporal Pose CNN and Multi-dimensional Fusion: Subvolume B." IEEE International Conference on Machine Learning And Applications (pp. 183–190).
90. Zhou, Y. and Tuzel, O. (2018). "VoxelNet: End-to-End Learning for Point Cloud Based 3D Object Detection." IEEE/CVF Conference on Computer Vision and Pattern Recognition (pp. 4490–4499).

91. Pastor, F., Gandarias, J., Garcia, A., and Gomez-de-Gabriel, J. (2019). "Using 3D Convolutional Neural Networks for Tactile Object Recognition with Robotic Palpation." *Sensors* 19(24): 5356.
92. Jnawali, K., Arbabshirani, M., Rao, N., and Patel, A. (2018). "Deep 3D Convolution Neural Network for CT Brain Hemorrhage Classification."

11

Convolutional Neural Network: A Systematic Review and Its Application using Keras

Dr. Shilpa Bhalerao[1] and Dr. Maya Ingle[2]
[1]*Professor, CSE, AITR, Indore*
[2]*Professor, SCSIT DAVV, Indore*

11.1 Introduction

Deep learning has gained a lot of attention in the last decade due to abundant data availability on the cloud. Abundant data is in unstructured form that includes images, text, audio, video, and other formats. Deep learning algorithms are used to extract information, making plausible decisions and predictions effectively. Many algorithms have evolved to detect objects from images. Object detection not only involves algorithms to detect the object but also to take care of input received and reduce the input in size so that image processing can be accelerated. A convolutional neural network (CNN) is a deep learning network that inputs images, then assigns weights and biases to the various aspects of the images that in turn help to segment and differentiate objects [1],[2]. In other words, CNN is a multi-layered network with special architecture to detect complex features in data. Figure 11.1 shows an application of CNN, i.e, object detection in image processing and converting text from images.

In this chapter, we discuss the multi-layer architecture of CNN, filters, and pooling strides along with current research trends in CNN.

11.2 Convolutional Neural Network (CNN)

The human brain is constantly observing things through the eyes, analyzing the observed scene, and identifying items based on past experience. CNN is a special type of artificial neural network inspired by the human cortex of the brain. The cortex is a small range of human cells sensitive to the visual field. This idea was coined by Hubel and Wiesel in 1962, who mentioned that individual responses of neuronal cells are only in the presence of edges in the observed view. Also, they pointed out that all of these neurons are organized in a columnar architecture and combined to produce perception [3]. The same idea has been applied in CNN to reduce the image size to make processing of images

DOI: 10.1201/9781003166702-11

FIGURE 11.1
Image Detection with CNN.

easier. In the following sub-sections, we discuss the architecture of CNN followed by the complete process of CNN from training to prediction of objects in images [1].

11.2.1 CNN Architecture

CNN consists of two major components: convolution layers/feature extraction and classification. It works differently than a regular neural network. In CNN, the hidden layer has three dimensions: width, height, and depth, as shown in Figure 11.2. Figure 11.2a is a regular neural network with input layer, output layer, and two hidden layers, whereas Figure 11.2b represents a CNN, which consists of convolutional layers and a pooling layer. These layers reduce the dimensions of the input image, and the reduced dimensions are passed to the fully connected (FC) network. All these layers are discussed in detail in the following sub-sections.

11.2.1.1 Convolutional Layer

Convolution is a linear operation that involves a scalar dot product of weights with the input image weight. A multi-dimensional array used for linear operation is known as a kernel or filter. These specific filters are smaller than the size of the image and are

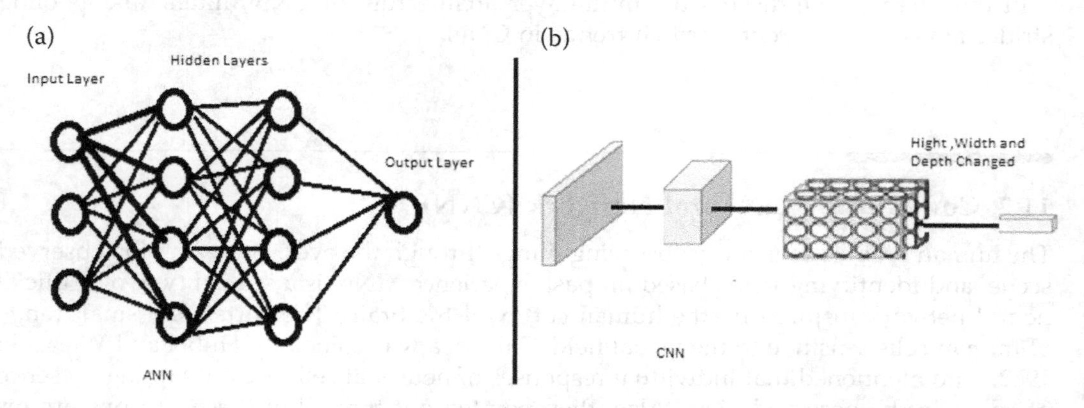

FIGURE 11.2
Regular Neural Network versus Convolutional Neural Network.

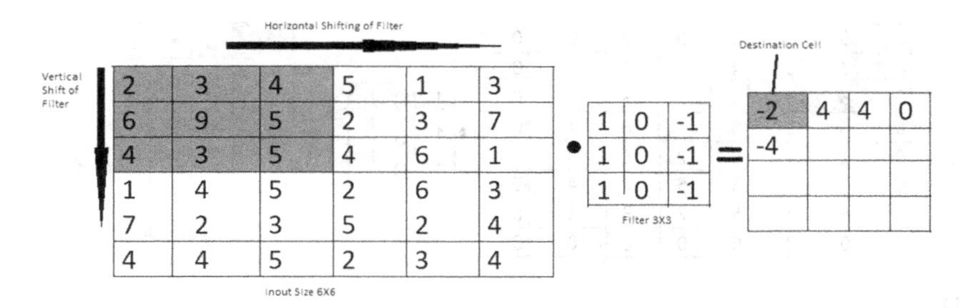

FIGURE 11.3
3X3 Filter Sliding on an Image.

applied to the overall image to extract features of the image such as edges, lines, etc. Multiple filters can be applied to images, which in turn produces multiple feature maps. These multiple feature maps are combined to produce the output of the convolutional layer. Figure 11.3 shows sliding a filter over the entire image, which involves scalar dot products, thereby producing the feature map in 2D. As in Figure 11.3, the dot product is computed by sliding a filter over image.

The distance between two successive positions of a filter is defined as strides. Normally, the filter moves slide by slide or pixel by pixel. In the case of two strides, the filter moves two pixels vertically and horizontally, resulting in downsampling of the feature maps. Pooling is another way of downsampling and is used to reduce the spatial size (i.e., reduces number of parameters), thereby controlling overfitting issues. Generally, the pooling layer is applied independently on every slice of the input image and drastically reduces the size of the image. Sometimes, padding is also introduced with filters to capture the image details minutely at edges of the image. In the absence of padding, the row column of the pixel at the edges is scanned by the filters a fewer number of times compared to the middle pixels. Figure 11.4 shows image size 6X6 with filter size 3X3 applied with single stride. In this case, the filter will move one column. The resultant scalar dot products are also depicted in Figure 11.4. Stride with value two further reduces the size of the feature map. Zero padding has been illustrated in Figure 11.5. It will increase two rows and columns in the image, as shown in Figure 11.5. Padding increases the size of the feature map, but pixels at the corners get equal preference. Pooling involves three classifications: max pooling, sum pooling, and average pooling. Generally, data practitioners use max pooling and average pooling of the feature map as shown in Figure 11.6.

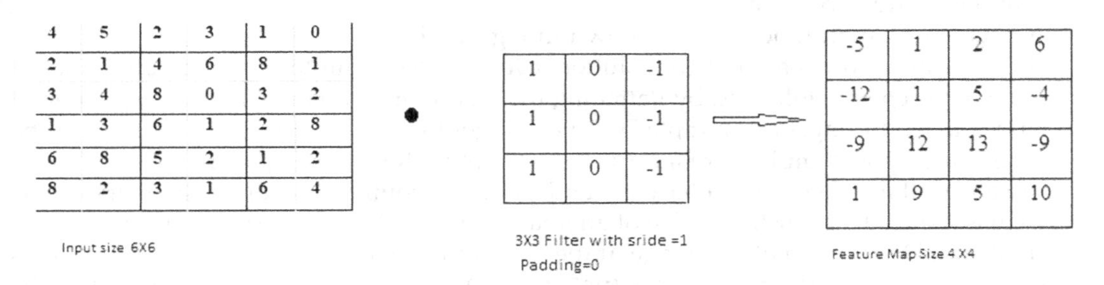

FIGURE 11.4
Filter Shift with Stride = 1 and Padding = 0.

0	0	0	0	0	0	0	0
0	2	3	4	5	1	3	0
0	6	9	5	2	3	7	0
0	4	3	5	4	6	1	0
0	1	4	5	2	6	3	0
0	7	2	3	5	2	4	0
0	4	4	5	2	3	4	0
0	0	0	0	0	0	0	0

•

1	0	-1
1	0	-1
1	0	-1

=

-12	-1	5	5	-7	4

FIGURE 11.5
Padding Illustration in a Feature Map.

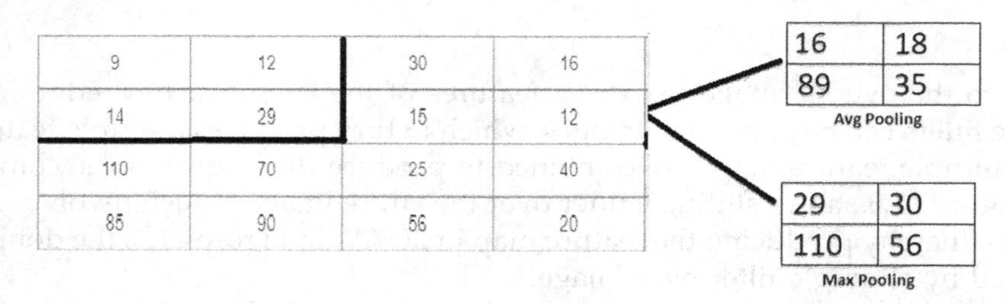

FIGURE 11.6
Max and Average Pooling.

11.2.1.2 Classification

After multiple combinations of the convolution and pooling layers with or without strides, we get the final output as a feature map. The feature map is further flattened into a one-dimensional array for input to one or more FC networks. These FC networks are dense neural networks that classify the object in desired classes based on probabilities. Each FC network uses the activation function for learning and modeling complex data used in videos, images, and audio. Non-linearity in any neural network allows back-propagation as non-linear functions are derivatives. Many non-linear activation functions are evident in literature; commonly used activation functions are rectified linear unit (ReLU), hyperbolic tangent (tanh), sigmoid, and softmax. ReLU selects the maximum value from the function of input x and zero, whereas sigmoid ranges from –1 to +1. Tanh functions are hyperbolic tangent functions. Initially, tanh and sigmoid activation functions are used due to the mathematical representation of human brain neuron behavior. However, the most commonly used activation function is ReLU due to its simplicity in computation functions (Table 11.1).

Various activation functions are shown in Figure 11.7.

Thus, CNN involves a set of convolution layers pooled in multiple combinations based on the requirements followed by flattening the feature maps. The output of the flattened map is passed to single or multiple FC layers to get the desired output in the form of the classification. The complete architecture of a CNN is shown in Figure 11.8.

Based on the image size, filter size, and pooling requirement, dimensionality of the image is reduced. For instance, size of an image is WTXHIXDE, where WT is width, HI is height, and DE is depth of the image. If the size of the filter/kernel is FwXFhXDE, strides used are ST, and padding used is PA (note that PA can have either 0 or 1 value), then the resultant feature map is computed as WT2XHI2XDE2

TABLE 11.1
Activation Function

Name of Activation function	Description
Sigmoid	It has an output between 0 and 1. It is used in binary classification.
Tanh	It is a zero-centered sigmoid.
ReLU	f(x) = max(0,x), where x is the input value.
Leaky ReLU	It resolves dying ReLU issues and has a small negative slope.
ELU	It is an exponential linear unit.
Softmax	It is used in multiclass classification. It ranges from 0 to 1.

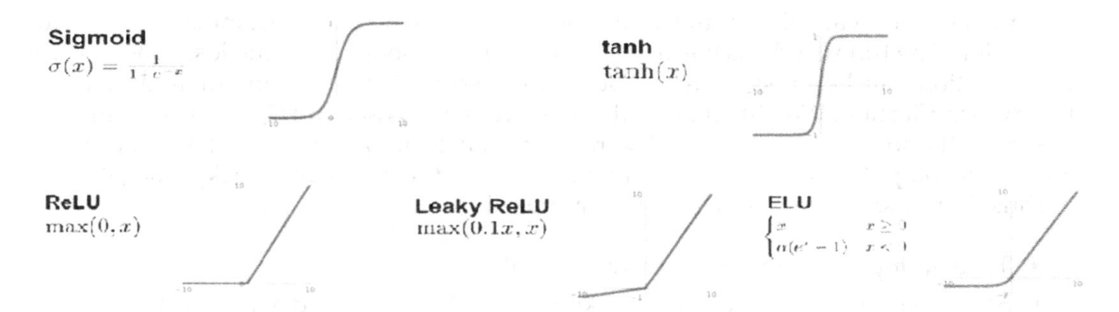

FIGURE 11.7
Activation Functions.

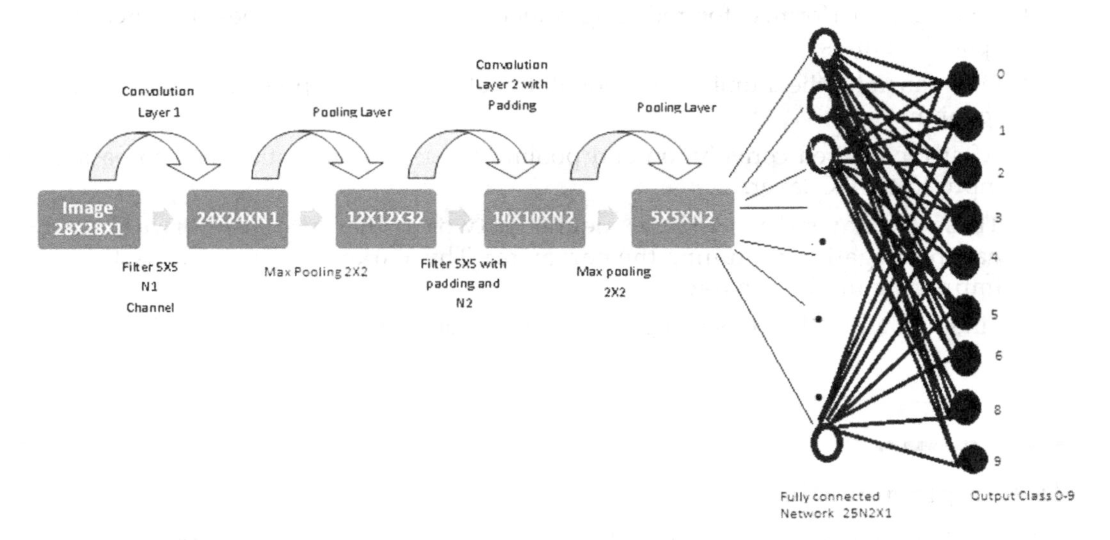

FIGURE 11.8
CNN Architecture.

$$-WT2 = (WT - Fw + 2PA)/S + 1 \quad (11.1)$$

$$-HI2 = (HI - Fh + 2PA)/S + 1 \quad (11.2)$$

(i.e., width and height are computed equally by symmetry)

$$-DE2 = DE \quad (11.3)$$

For example, if image size is 28X28X1, filter size is 4X4X1, and strides used is 2 without padding, then the size of the resultant feature map will be 13X13X1.

11.2.2 Workings of the CNN

Any deep learning application requires sufficient training data to train the model for better prediction. In CNN, sufficient data (images) are required from all classes to train the model. The available dataset is divided into three parts for training, testing, and validation. The training dataset is used for training the model using the loss function. The loss function can be cross entropy for multi-classification or mean squared error for binary classification. Weight and bias (hyper-parameters) used in FC layers are computed based on the loss function. Generally, random values are assigned to these parameters and are recomputed iteratively to minimize the loss function using backpropagation.

Thus, in nut shell, the following steps are used:

- Take an input image for the convolutional layer.
- Select parameters. Apply filters/kernels with strides and padding based on the input image requirements. Convolution is performed on the input image along with the applied ReLU activation function to the matrix for achieving non-linearity.
- Pooling is performed for reducing dimensionality. Pooling does not affect the spatial features.
- One can add the number of convolutional layers and pooling based on the requirements.
- After the desired convolution and pooling layers, output is flattened to pass as input to the FC layer.
- These FC layers are regular neural networks that use feedforward and backpropagation for tuning the parameter. This parameter tuning is useful for improving model accuracy.
- Finally, output the class using an activation function, and classify the images.

11.3 Implementation of CNN

A CNN algorithm is mainly used in object detection, localization, and segmentation, i.e., identifying various objects in a single image. We used the CIFAR-10 dataset for implementation of CNN using keras for classification of images. It is a multiclass classification problem as CIFAR-10 has 10 classes. In this section, we discuss the terminology used in CNN implementation.

11.3.1 Dataset CIFAR-10

CIFAR-10 is a dataset; its characteristics are shown in Table 11.2. The images present in the CIFAR dataset are very small and are used for research purposes only (Figure 11.9).

11.3.2 Loading the Dataset and Importing Required Packages

We have used Google Colaboratory for implementation. First, we have imported required packages such as Numpy, Seaborn, Sklearn, and its sub-packages required for pre-processing and CNN layers. The following code snippet shows packages used in the project:

```
import cv2
import matplotlib.pyplot as plt
import numpy as np
import seaborn as sns; sns.set()
from keras.datasets import cifar10
from sklearn.metrics import confusion_matrix
from sklearn.preprocessing import OneHotEncoder
from keras.layers import Conv2D, MaxPool2D, Flatten, Dense,
Dropout
from keras.models import Sequential, load_model
```

Code for importing packages

11.3.3 Data Pre-processing

Data pre-processing is required to check the integrity of the dataset. Null values and missing values are dropped out from the dataset to get a better result. Higher image size and color images will take more computation, thereby increasing execution time. Therefore, images are converted to greyscale. In Python, we used a cv2 module to convert to greyscale. Data pre-processing involves image processing and label pre-processing also. Further, classes are numbered from 0 to 9 and converted into one hot encoding, as shown in Table 11.3.

TABLE 11.2

CIFAR -10 Dataset

Attributes	Values
Size of dataset	60,000
Number of classes in dataset	10
Size of each image	32X32
Image type	Color
Training dataset under consideration	50,000
Test dataset under consideration	10,000
Size of each batch in training	10,000
Number of batches	5
Mutually exclusive	Yes
Data downloaded	CIFAR-10 Python version

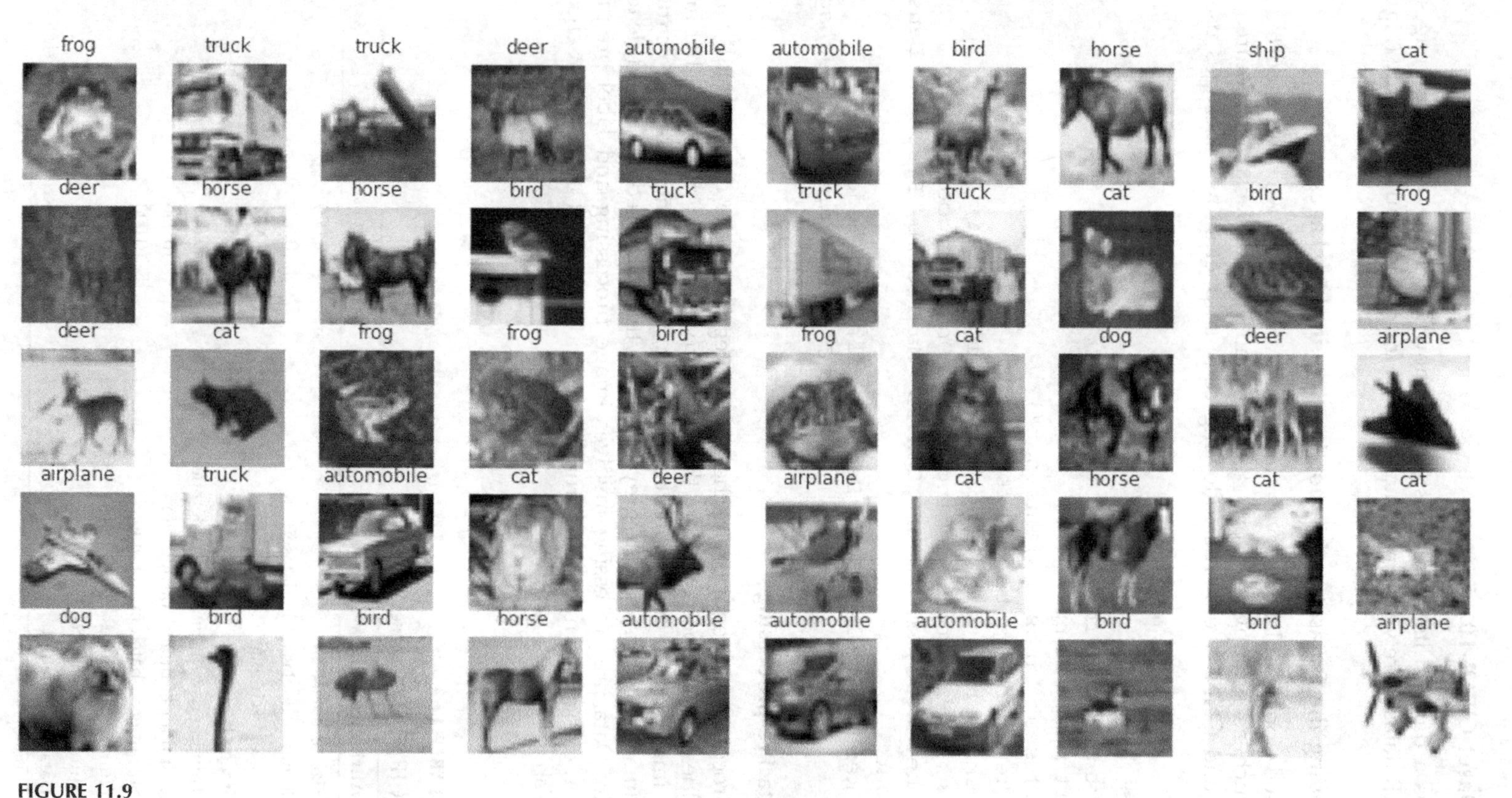

FIGURE 11.9
CIFAR-10 Dataset. (Note: Images are in color in the dataset; we have shown in black and white for printing purposes.)

TABLE 11.3

One hot Encoding

Class	One hot encoding
0	[1,0,0,0,0,0,0,0,0,0]
1	[0,1,0,0,0,0,0,0,0,0]
2	[0,0,1,0,0,0,0,0,0,0]
....	
9	[0,0,0,0,0,0,0,0,0,1]

11.3.4 Construction of a CNN

The dataset is divided into (x_train, y_train) and (x_test, y_test) for training and testing purpose. In the set, x_train and x_test hold the greyscale images, whereas y_train and y_test are labels, as shown in the code snippet. Now, we are ready to construct the CNN. We chose a sequential model. First, we used filter/kernel of size 3X3 with stride value of 1. Filters were increased in subsequent layers before reaching the pooling layers. Two combinations of convolution-max pooling layers were used before the FC network. The ReLU activation function was used in all layers except the last layer to increase accuracy. Finally, in the last layer of 10 neurons, the softmax function was applied to classify the images (Figures 11.10, 11.11, and 11.12). A summary of the model is presented in Table 11.4.

Table 11.5 shows that model accuracy is 84% after 20 Epochs, whereas value accuracy is around 70%. It is evident from the graph of value accuracy and accuracy on the training dataset that the model is converging toward overfitting and accuracy is not changing after 17 Epochs. Experts can introduce the dropouts to avoid the overfitting issues. The confusion metric of the proposed CNN model is depicted in Figure 11.13. The values in the dark square represent the predicted value by the model, which is the same as the actual value. The values in the other squares are misclassified. Figure 11.14 shows the sample output of the implemented model.

11.4 Latest Research Trends in CNN

In this section, we cover CNN's evolution from ImageNet until 2020. We also highlight the dark research areas in CNN. CNN's existing research can be divided into two domains. First is application and implementation of CNN on medical images, videos, and natural language processing (NLP), etc. Second, research revolves around discovering a better activation function, reducing computation with the number of layers, dropouts, etc. We will discuss both aspects of research in this section.

CNN gained its popularity after the ImageNet challenges conducted from 2009 to date for promoting research and development in computer vision. Some of the innovation and image classification algorithms are coined in *ImageNet Large Scale Visual Recognition Challenge (ILSVRC)*. The ILSVRC is an annual computer vision challenge based on a public domain subset of a computer vision dataset called ImageNet. ImagNet is a large collection of annotated images used for research purposes [2]. The ImageNet dataset

```
X_train, y_train), (X_test, y_test) = cifar10.load_data()
Downloading data from https://www.cs.toronto.edu/~kriz/cifar-10-python.tar.gz

170500096/170498071 [==============================] - 2s 0us/step

X_train  = X_train/255
X_test   = X_test/255


one_hot_encoder = OneHotEncoder(sparse=False)
one_hot_encoder.fit(y_train)
X_train = X_train.reshape(X_train.shape[0], X_train.shape[1], X_train.shape[2], 1)
X_test = X_test.reshape(X_test.shape[0], X_test.shape[1], X_test.shape[2], 1)

model = Sequential()
model.add(Conv2D(16, (3, 3), activation='relu', strides=(1, 1),
    padding='same', input_shape=input_shape))
model.add(Conv2D(32, (3, 3), activation='relu', strides=(1, 1),
    padding='same'))
model.add(Conv2D(64, (3, 3), activation='relu', strides=(1, 1),
    padding='same'))
model.add(MaxPool2D((2, 2)))
model.add(Conv2D(16, (3, 3), activation='relu', strides=(1, 1),
    padding='same'))
model.add(Conv2D(32, (3, 3), activation='relu', strides=(1, 1),
    padding='same'))
model.add(Conv2D(64, (3, 3), activation='relu', strides=(1, 1),
    padding='same'))
model.add(MaxPool2D((2, 2)))
model.add(Flatten())
model.add(Dense(256, activation='relu'))
model.add(Dropout(0.5))
model.add(Dense(128, activation='relu'))
model.add(Dense(64, activation='relu'))
model.add(Dense(64, activation='relu'))
model.add(Dense(10, activation='softmax'))
```

FIGURE 11.10
Sample Code for Creating a CNN Model.

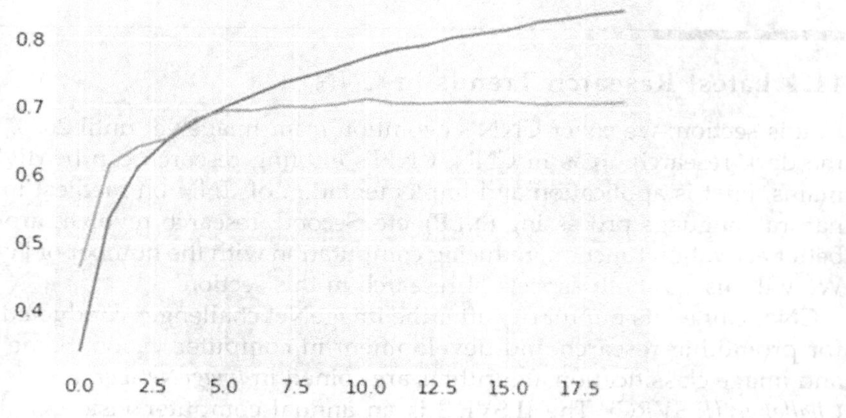

FIGURE 11.11
Accuracy versus Value Accuracy.

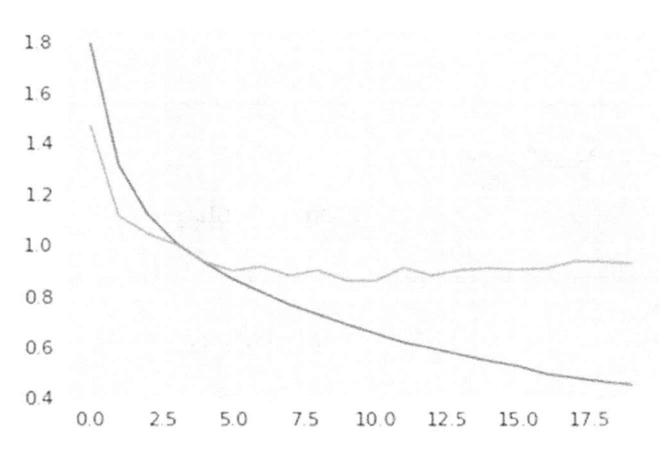

FIGURE 11.12
Loss Function Train and Test Dataset.

contains more than 15 million labeled images with more than 22,000 categories, which created the benchmark in image recognition and classification. Challenges in ImagNet are generally based on image classification, object detection, and single object localization to video labeling. Deep learning experts trained to the deep CNN classify the 1.2 million high-resolution images in the ImageNet LSVRC-2010 contest into 1,000 different classes. Top-1 and top-5 error rates are 37.5% and 17.0%, which is considerably better than the previous state-of-the-art CNNs [4]. Popular ILSVRC winners have created history in deep learning. These are AlexaNet(2012), ZFNet(2013), Inception V1(2015), and VGG(2014). Apart from LSVRC challenges, empirical studies on various domains have been performed. A study of diabetes prediction using retina images has been found in literature, which reveals the importance of number layers for accurate results [5]. Another application of deep neural network found in literature was about skin cancer or lymph node identification in patients from images [6,7]. In 2017, a study explored identification of pulmonary tuberculosis with the use of chest radiography as input to a CNN algorithm. Sensitivity for classifying pulmonary tuberculosis by the proposed algorithm was reported as 97% [8]. In literature, there was a study on medical diagnosis of fatty liver and lesion segmentation using CT scan images [9–11]. A deep convolutional auto-encoder network was trained to identify air, bone, and soft tissue in volumetric head MR image to CT data for training and tested. This deep neural network reduced reconstruction error [12]. Another usage of CNN is observed in field of NLP, where text reports of CT/MR are identified with the use of unsupervised learning. The proposed deep learning model for aforesaid classification was reported as accurate as the traditional NLP model [13].

It has been observed from these deeper networks that performance of the model is improved. By increasing the number of convolutional layers and non-linearity, the model gives accurate results but it also increases the complexity of the network. Increase in complexity in the CNN model makes the network more difficult to optimize and more prone to overfitting. Scientists and researchers are working on various aspects of CNN to improve the performance of CNN. One major aspect is the activation function that plays an important role in the efficiency of the model. Ramachandran and Zoph discovered the activation function, named Swish, which worked better than ReLU on deeper models across a number of challenging datasets [14].

TABLE 11.4

Model Summary

```
model: "sequential_1"

Layer (type)                    Output Shape          Param #
=============================================================
conv2d (Conv2D)                 (None, 32, 32, 16)    160

conv2d_1 (Conv2D)               (None, 32, 32, 32)    4640

conv2d_2 (Conv2D)               (None, 32, 32, 64)    18496

max_pooling2d (MaxPooling2D)    (None, 16, 16, 64)    0

conv2d_3 (Conv2D)               (None, 16, 16, 16)    9232

conv2d_4 (Conv2D)               (None, 16, 16, 32)    4640

conv2d_5 (Conv2D)               (None, 16, 16, 64)    18496

max_pooling2d_1 (MaxPooling2    (None, 8, 8, 64)      0

flatten (Flatten)               (None, 4096)          0

dense (Dense)                   (None, 256)           1048832

dropout (Dropout)               (None, 256)           0

dense_1 (Dense)                 (None, 128)           32896

dense_2 (Dense)                 (None, 64)            8256

dense_3 (Dense)                 (None, 64)            4160

dense_4 (Dense)                 (None, 10)            650

=============================================================

Total params: 1,150,458

Trainable params: 1,150,458

Non-trainable params: 0
```

TABLE 11.5

Model Accuracy in 20 Epochs

```
Epoch 1/20

1563/1563 [==============================] - 350s 223ms/step - loss:
2.0327 - acc: 0.2345 - val_loss: 1.4685 - val_acc: 0.4639

Epoch 2/20

1563/1563 [==============================] - 348s 223ms/step - loss:
1.3797 - acc: 0.5120 - val_loss: 1.1120 - val_acc: 0.6136

Epoch 3/20

1563/1563 [==============================] - 348s 222ms/step - loss:
1.1445 - acc: 0.5985 - val_loss: 1.0432 - val_acc: 0.6397

Epoch 4/20

1563/1563 [==============================] - 349s 223ms/step - loss:
1.0124 - acc: 0.6487 - val_loss: 1.0010 - val_acc: 0.6518

Epoch 5/20

1563/1563 [==============================] - 350s 224ms/step - loss:
0.9203 - acc: 0.6780 - val_loss: 0.9342 - val_acc: 0.6817

Epoch 6/20

1563/1563 [==============================] - 348s 223ms/step - loss:
0.8698 - acc: 0.6963 - val_loss: 0.8979 - val_acc: 0.6945

Epoch 7/20

1563/1563 [==============================] - 348s 223ms/step - loss:
0.8020 - acc: 0.7224 - val_loss: 0.9139 - val_acc: 0.6947

Epoch 8/20

1563/1563 [==============================] - 349s 223ms/step - loss:
0.7553 - acc: 0.7379 - val_loss: 0.8796 - val_acc: 0.7011

Epoch 9/20

1563/1563 [==============================] - 349s 223ms/step - loss:
0.7182 - acc: 0.7500 - val_loss: 0.8996 - val_acc: 0.7005
```

(Continued)

TABLE 11.5 (Continued)

Model Accuracy in 20 Epochs

```
Epoch 10/20

1563/1563 [==============================] - 349s 223ms/step - loss:
0.6809 - acc: 0.7621 - val_loss: 0.8590 - val_acc: 0.7046

Epoch 11/20

1563/1563 [==============================] - 348s 223ms/step - loss:
0.6299 - acc: 0.7816 - val_loss: 0.8588 - val_acc: 0.7124

Epoch 12/20

1563/1563 [==============================] - 350s 224ms/step - loss:
0.6061 - acc: 0.7900 - val_loss: 0.9098 - val_acc: 0.7067

Epoch 13/20

1563/1563 [==============================] - 350s 224ms/step - loss:
0.5734 - acc: 0.7977 - val_loss: 0.8797 - val_acc: 0.7063

Epoch 14/20

1563/1563 [==============================] - 350s 224ms/step - loss:
0.5543 - acc: 0.8088 - val_loss: 0.9015 - val_acc: 0.7079

Epoch 15/20

1563/1563 [==============================] - 347s 222ms/step - loss:
0.5302 - acc: 0.8164 - val_loss: 0.9083 - val_acc: 0.7084

Epoch 16/20

1563/1563 [==============================] - 347s 222ms/step - loss:
0.5111 - acc: 0.8212 - val_loss: 0.9037 - val_acc: 0.7093

Epoch 17/20

1563/1563 [==============================] - 347s 222ms/step - loss:
0.4740 - acc: 0.8337 - val_loss: 0.9081 - val_acc: 0.7065

Epoch 18/20

1563/1563 [==============================] - 348s 223ms/step - loss:
0.4600 - acc: 0.8401 - val_loss: 0.9365 - val_acc: 0.7077
```

TABLE 11.5 (Continued)

Model Accuracy in 20 Epochs

```
Epoch 19/20

1563/1563 [==============================] - 349s 224ms/step - loss:
0.4490 - acc: 0.8450 - val_loss: 0.9340 - val_acc: 0.7072

Epoch 20/20

1563/1563 [==============================] - 347s 222ms/step - loss:
0.4448 - acc: 0.8481 - val_loss: 0.9271 - val_acc: 0.7093
```

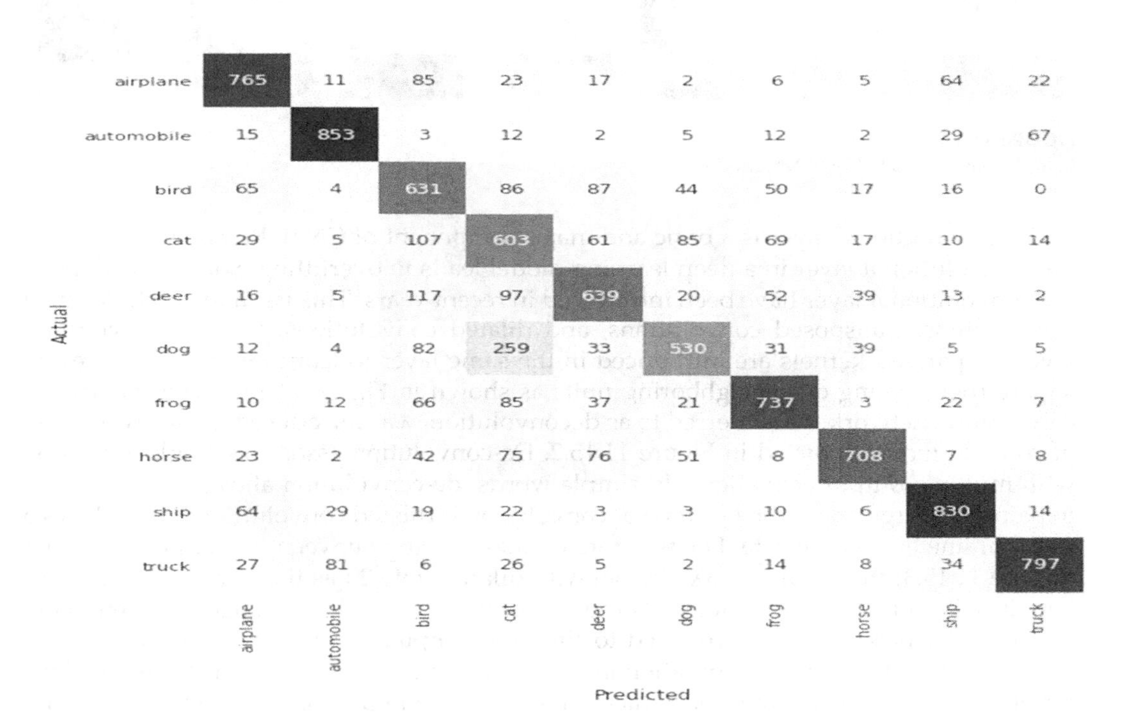

FIGURE 11.13
Confusion Metric CIFAR-10.

$$f(x) = x.\ \text{sigmoid}(\beta x) \tag{11.4}$$

Researchers also proposed that sparsity and neurons operating mostly in a linear regime can be brought together in more biologically plausible deep neural networks. Rectifier units help to bridge the gap between unsupervised pre-training and no pre-training [15]. Network in network was proposed by Lin et al. to enhance model discriminability for local patches within the receptive field [16]. Further research identified improving optimization through the use of Adam, another gradient descent learning algorithm for improving execution time of the model [17–21]. Next, we discuss improvements in the convolutional layer, pooling layer, and loss function.

FIGURE 11.14
Sample Data Predicted by CNN Model.

The convolutional layer is a basic and major component of CNN. Removing or adding one convolutional layer in a deep learning model leads to overfitting. Some variations in the convolutional layer have been introduced in recent years. This variation includes tiled convolutions, transposed convolutions, and dilated convolutions. In tiled convolution layers, separated kernels are introduced in the same layer to learn invariant features by square root pooling over neighboring units as shown in Figure 11.15.1. The transposed convolution network, also referred to as deconvolution, was introduced in backward pass of convolution as depicted in Figure 11.15.2. De-convolution associates single activation with multiple output activations. In simple words, de-convolution allows a small image to paint into large ones. The last type of convolution is dilated convolution. It introduces a new parameter dilation rate. Dilation rate defines spacing between the kernels. A figure (Figure 11.15.3) that depicts a 3X3 kernel with dilation rate 2 has the same effect as a 5X5 kernel without dilation. Dilated kernels are mainly used in image segmentation. Separable kernels are also introduced to simplify computation using dot product.

Another way to reduce computation in CNN is experimenting with activation functions for non-linearity and pooling for reducing the size of computation. In recent years, many activation functions have been introduced by researchers apart from functions discussed in earlier sections. These functions are ReLU, leaky ReLU, randomized ReLU, parmetic ReLU, exponential rectified unit (ELU), and maxout. However, ReLU and leaky ReLU are prominently used by deep learning experts. Experiments are also found in literature on the pooling layer as it reduces the cost of computation drastically. Apart from max, average, and sum pooling, mixed pooling is also used in CNN layers. Mixed pooling is a combination of max and average pooling. Stochastic pooling, inspired by the dropout concept, is also introduced. Stochastic pooling randomly picks the activations according to a multinomial distribution, which ensures that the non-maximal activations of feature maps can also be utilized.

In last five years, the use of CNN has drastically increased in every field of image classification in computer vision. Many researchers worked on medical images received from either X-ray or CT scan to identify the probabilities of various diseases. Some

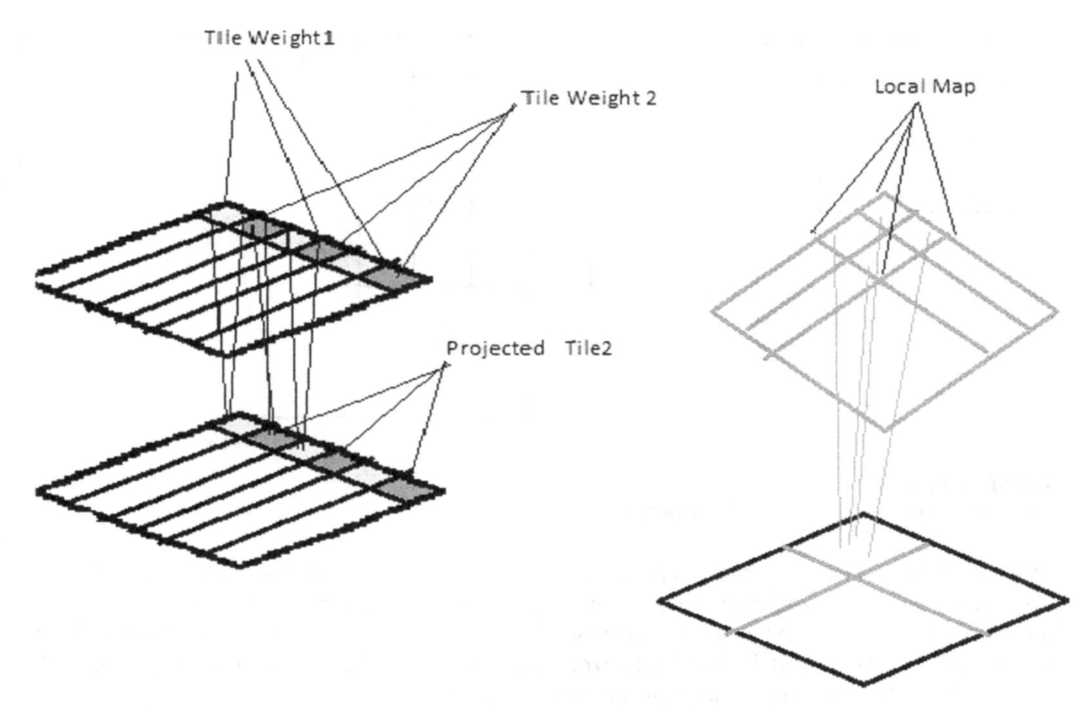

FIGURE 11.15.1
Variation in convolution layer: Tilled Convolution.

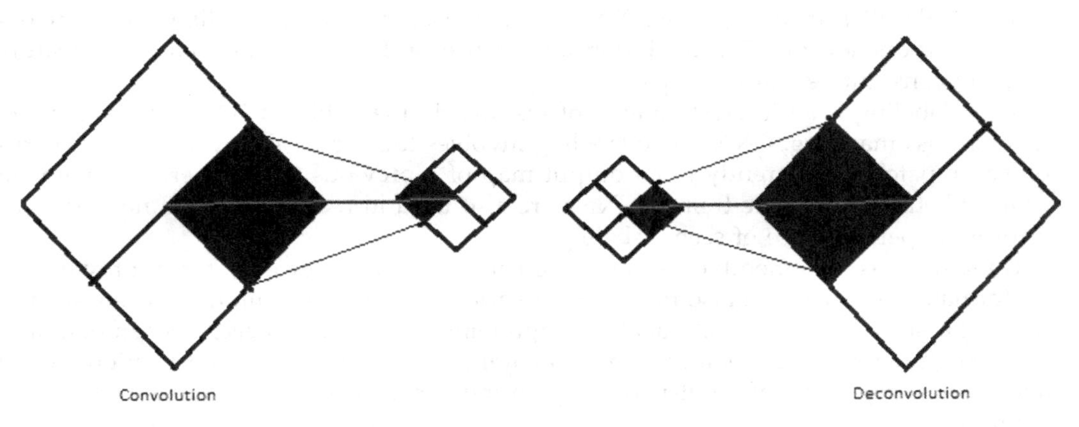

FIGURE 11.15.2
Variation in convolution layer: Deconvolution.

applications were developed to study diabetic retinas and their behavior in diabetic patients. Many datasets are available on the Internet for classification of images using CNN. Sub-category classification is a growing field in image classification based on object parts. Already, success has been recorded in identification of cat, dog, car, birds, and plants. Part localization, part alignment, and classification part were also integrated by Lin et al. However, all these methods are classifications and sub-classification that are based on supervised learning. There is little research found on unsupervised learning for classification.

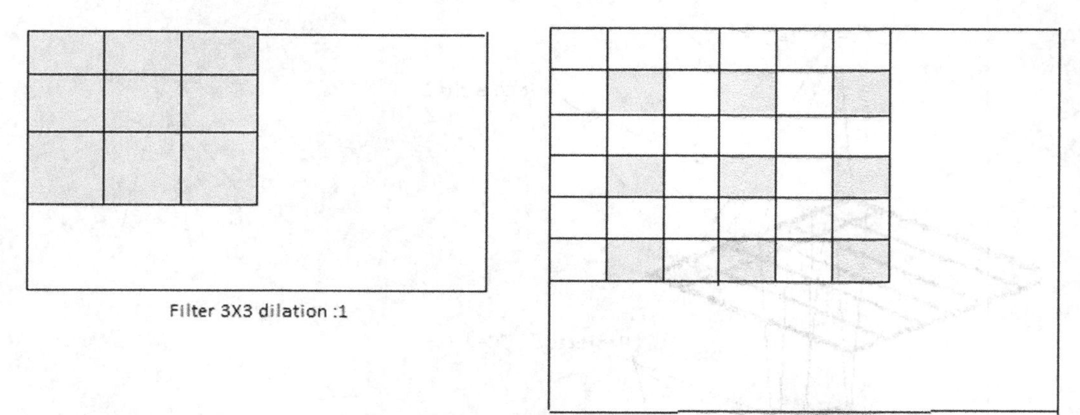

FIGURE 11.15.3
Variation in convolution layer: Dilated Convolution.

Object detection and object tracking in videos are important research areas. They involve not only localization of parts but also movement of objects. YOLO and SSD perform better in object detection with single use of a pipeline network. Pose estimation in still images is already found in the literature but pose estimation in videos requires higher computation, thereby increasing processing time and cost.

Action recognition is another field of research in CNN that can be divided into two categories: action recognition in images and action recognition in videos. Major challenges in the latter one are larger size of videos and new temporal axis in videos. Traditional CNNs are designed for 2D images. Two approaches to handling temporal axis are 1) introduction of 3D convolution and 2) use of 2D convolution and fuse feature maps in consecutive feature maps.

Scene labeling is an interesting area of research in the field of CNN that involves semantic class mapping. CNN scene labeling involves the use of identical CNN networks on scene patches recurrently in an output map of a previous CNN layer. This helps to achieve better results. Pre-trained CNNs are also used in recurrent neural networks for improving performance of scene labeling.

Thus, there is a tremendous scope of research in CNNs not only for improving the performance of CNNs but also for applying various domains of computer vision, such as object classification, detection, labeling, captioning videos and speech recognition, and NLP. High performance requires a deep neural network with high computation facility and high storage. Therefore, it is difficult to implement CNNs in resource-constrained environment.

11.5 Conclusion

CNNs brought one of the major revolutions in the field of object classification and detection. In this chapter, we covered the components of the CNN along with its implementation for one application in Python. We used an existing dataset, CIFAR-10, for classification of an object. In our small project, we observed that the CNN model has issues with overfitting. We also discussed the latest trends and advancements in CNNs due to ImageNet competitions. Popular ImageNets, Alexanet, and other CNNs were also discussed. We highlighted improvements in CNNs on the basis of three aspects; convolutional layer, pooling, and activation functions. We covered recent advancements in the field of object classification, detection, action recognition, and labeling of images and videos. Although CNN has a record of great success in computer vision, it still has major issues in complex computation, huge dataset requirements, and high computation speed. A lot of human effort is required to collect huge labeled datasets. Thus, there is scope of research in unsupervised learning CNNs. The chapter also discussed that computing time is high in existing algorithms so there needs to be more parallel algorithms and high computing GPU clusters. Therefore, it is time to investigate better optimized algorithms without compromising accuracy.

References

1. LeCun, Y., Bengio, Y., Hinton, G. (2015) "Deep Learning." *Nature* 521: 436–444.
2. Russakovsky, O., Deng, J., Su, H. et al. (2015) "ImageNet Large Scale Visual Recognition Challenge." *International Journal of Computer Vision* 115:211–252. Accessed 11 April 2015.
3. Hubel, D. H. and Wiesel, T. N. (1962). "Receptive Fields, Binocular Interaction and Functional Architecture in the Cat's Visual Cortex. *The Journal of Physiology* 160(1): 106–154
4. Krizhevsky, A., Sutskever, I., and Hinton, G. E. (2012). ImageNet classification with deep convolutional neural networks. *Advances in Neural Information Processing Systems* 25. Available online at: https://papers.nips.cc/paper/4824-imagenet-classification-with-deep-convolutional-neural-networks.pdf. Accessed 22 Jan 2018
5. Gulshan, V., Peng, L., Coram, M. et al. (2016). "Development and Validation of a Deep Learning Algorithm for Detection of Diabetic Retinopathy in Retinal Fundus Photographs." *JAMA* 316: 2402–2410.
6. Esteva, A., Kuprel, B., Novoa, R.A., et al. (2017). "Dermatologist-Level Classification of Skin Cancer with Deep Neural Networks." *Nature* 542:115–118.
7. Ehteshami Bejnordi, B., Veta, M., Johannes van Diest, P., et al. (2017). "Diagnostic Assessment of Deep Learning Algorithms for Detection of Lymph Node Metastases in Women with Breast Cancer. *JAMA* 318: 2199–2210.
8. Lakhani, P. andSundaram, B. (2017). "Deep Learning at Chest Radiography: Automated Classification of Pulmonary Tuberculosis by using Convolutional Neural Networks." *Radiology* 284: 574–582.
9. Yasaka, K., Akai, H., Abe, O., and Kiryu, S. (2018). "Deep Learning with Convolutional Neural Network for Differentiation of Liver Masses at Dynamic Contrast-Enhanced CT: A Preliminary Study. *Radiology* 286: 887–896.
10. Christ, P. F., Elshaer, M. E. A., Ettlinger, F., et al. (2016). "Automatic Liver and Lesion Segmentation in CT using Cascaded Fully Convolutional Neural Networks and 3D Conditional Random Fields." In: Ourselin, S., Joskowicz, L., Sabuncu, M., Unal, G., Wells, W. (eds.) Proceedings of Medical image computing and computer-assisted intervention – MICCAI 2016. 10.1007/978-3-319-46723-8_48
11. Kim, K. H., Choi, S. H., and Park, S. H. (2018) "Improving Arterial Spin Labeling by using Deep Learning. *Radiology* 287: 658–666. 10.1148/radiol.2017171154
12. Liu, F., Jang, H., Kijowski, R., Bradshaw, T., and McMillan, A. B. (2018). "Deep Learning MR Imaging-Based Attenuation Correction for PET/MR Imaging. *Radiology* 286: 676–684.

13. Chen, M. C., Ball, R. L., Yang, L., et al. (2018) "Deep Learning to Classify Radiology Free-Text Reports." *Radiology* 286: 845–852.
14. Ramachandran, P., Zoph, B., Le, Q. V. (2017) "Searching for Activation Functions." arXiv. Available online at: https://arxiv.org/pdf/1710.05941.pdf. Accessed 23 Jan 2018.
15. Glorot, X., Bordes, A., and Bengio, Y. (2011). "Deep Sparse Rectifier Neural Networks." In: Proceedings of the 14th International Conference on Artificial Intelligence and Statistics, 15, pp 315–323.
16. Lin, M., Chen, Q., and Yan, S. (2013). "Network in Network." arXiv. Available online at: https://arxiv.org/pdf/1312.4400.pdf. Accessed 22 Jan 2018
17. Qian, N. (1999) "On the Momentum Term in Gradient Descent Learning Algorithms." *Neural Networks* 12(1999): 145–151. Online available http://www.columbia.edu/~nq6/publications/momentum.pdf
18. Kingma, D. P. , and Ba, J. (2014) "Adam: A Method for Stochastic Optimization." arXiv. Available online at: https://arxiv.org/pdf/1412.6980.pdf. Accessed 23 Jan 2018.
19. Simonyan, K., and Zisserman, A. (2015) Very Deep Convolutional Networks for Large-Scale Image Recognition," in: Proceedings of the International Conference on Learning Representations (ICLR), 2015. Cornell University.
20. He, K., Zhang, X., Ren, S., and Sun, J. (2016) "Deep Residual Learning for Image Recognition," in: Proceedings of the IEEE Conference on Computer Vision and Pattern Recognition (CVPR), 2016, pp. 770–778. IEEE.
21. Zhou, B., Khosla, A., Lapedriza, A., Oliva, A., and Torralba, A. (2016) "Learning Deep Features for Discriminative Localization," in: Proceedings of the IEEE Conference on Computer Vision and Pattern Recognition (CVPR), 2016, pp. 2921–2929. IEEE.

12

Big Data Analytics: Applications, Issues and Challenges

Dr. Snehlata Dongre[1], Dr. Kapil Wankhade[2], and Dr. Latesh Malik[3]

[1]*Assistant Professor, Department of Computer Science & Engineering, G H Raisoni College of Engineering, Nagpur*
[2]*Scientist D, Ministry of Electronics & Information Technology, Indian Computer Emergency Response Team(CERT-In), Delhi*
[3]*Associate Professor, Department of Computer Science & Engineering, Govt. College of Engineering, Nagpur*

12.1 Introduction to Big Data

Big data analytics [1,2] benefit businesses as well as individuals. Prior to the 1980s, data creation was a controlled process. Even the rate of data creation was known and manageable. At that time, data creation and processing were at the same location, or co-located. The data were structured and could be stored in tables in relational databases. The data were known a priori to store and process; this was possible because data creation was under control. In the 1990s, data creation was still controlled and still manageable. Relational databases continued to rule the world because data were structured and data processing was still centralized. Then, the Internet came into the picture and changed the whole world. In the Millennium, there was an increase in Internet users. e-Commerce sites became popular. This also increased the data generation rate. Data processing became important to help businesses make important decisions. After that, social networking sites like Facebook, WhatsApp, and Instagram arrived. This brought many challenges, like unstructured data, extremely large dataset generation at a rapid rate, and high data distribution. Figure 12.1 shows an example of structured, semi-structured, and unstructured data. Social media data are unstructured, very large in volume, highly dispersed and distributed, very difficult to move to a centralized location, and difficult to process and generate results for in real time.

Sources of Big Data

Big data come from many sources, such as the following:

DOI: 10.1201/9781003166702-12

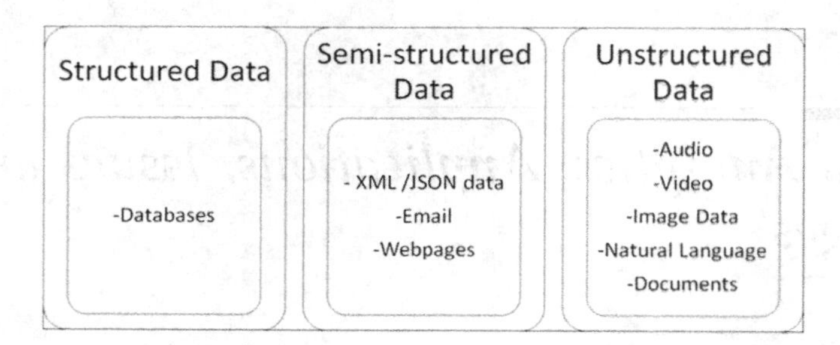

FIGURE 12.1
Structured, Semi-structured, and Unstructured data.

- **Social Networking Sites:** These sites are an important source of big data as they collect huge data for a variety of users. Examples include Twitter, Facebook, Instagram, and LinkedIn. Analyzing the data uncovers the user's personal likes and dislikes.
- **E-commerce Sites:** Even e-commerce sites generate lots of data and are an important source of big data. The user's buying trends can be analyzed via e-commerce data. Examples include Flipkart, Amazon, and Myntra.
- **Weather Stations:** Weather stations and satellites generate continuous data regarding the weather that can be utilized to predict future weather patterns. Forecasting the weather is very useful for farmers.
- **Telecommunications:** Telecommunication records are also an example of big data. Analyzing telecommunication records helps Telecom companies to plan. Examples include Jio, Vodafone, and Airtel.
- **Stock Market:** The Stock Exchange also generates lots of data. Stock market analysis helps to predict the next value of stocks.

Big data analytics [3] has attracted research since the last decade. Big data is a popular field. The importance of big data analytics [4] increases daily due to the increase in online users. Big data have volumes of data of different variety that have velocity as well. That is why big data is said to have volume, variety, and velocity as characteristics, which are generally abbreviated as the 3 V's. A few authors consider two move V's as additional characteristics: veracity and value. These characteristics make analysis of big data difficult. (1) *Volume:* There is a huge volume of data. The size of the data tells us whether the data can be considered to be big data or not. Volume is one of the most important characteristics but not the only criteria. Other characteristics also play a crucial role. (2) *Velocity:* Continuous flowing data with high speed is referred to as velocity of data. Dealing with this kind of data requires proper sampling of the data in small chunks for further processing. (3) *Variety:* This characteristic refers to different types of data. Big data may be structured, unstructured, or semi-structured. As the data come from different sources, their types are different. *Structured data* are well organized. The structure of the data is known a priori and can be organized accordingly. The data can be stored in traditional relational databases. Tables in database management systems are a good example of structured data. *Semi-structured data* are not fully organized or completely unstructured. Log records are good examples of semi-structured data. *Unstructured data* are unorganized. They cannot fit into traditional relational databases. Images, videos, and text are examples of unstructured data. (4) *Veracity:* This characteristic means

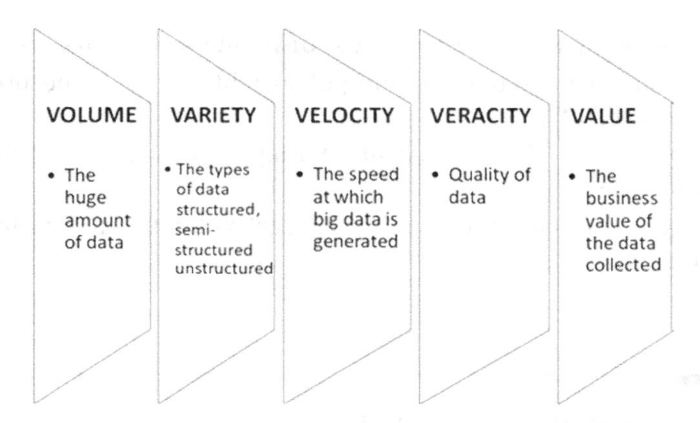

FIGURE 12.2
Characteristics of Big Data 5 V's.

inconsistency and uncertainty in the data, which makes data handling more difficult. (5) *Value:* This characteristic is the last V but the most important one. Raw data are not useful. Dragging out value from a huge amount of data is important.

Understanding all 5 V's is important when dealing with big data (Figure 12.2).

12.2 Big Data Analytics and Its Issues and Challenges

Big data analytics applications often include data from both internal systems and external sources, such as weather data, log data, or demographic data on consumers compiled by third-party providers. Industries are facing issues with handling the data; various life cycle models have been proposed [5]. Streaming analytics applications are becoming common in the big data environment as users look to perform real-time analytics on data fed into Hadoop systems through stream processing engines, such as Flink, Spark and Storm.

There are certain challenges of big data analytics [6]:

1. *Security and privacy of data* [7] [8] [9]: Security and privacy are major issues in big data analytics. In security, data integrity and data confidentiality are important issues. Maintaining privacy is the crucial task. If a company fails to do this, the user's personal information can be misused.
2. *Data diversity* [10]: Data of different variety may come from different sources. Integrating these data is an issue. Even the aggregation of these data is a challenge.
3. *Legal issues* [7]: A few legal issues concerned with big data analytics are data ownership, open data and public sector, database licensing, and copyright infringement.
4. *Processing framework* [7]: Big data are generated from autonomous sources and are highly distributed. To manage this kind of data, the proper framework is required. The processing framework generally needs to handle big data analytics workloads.

5. *Synchronization:* Synchronization across disparate data sources is needed.
6. *Meaningful insights:* Getting meaningful insights through the use of big data analytics is difficult.
7. *Trained professionals:* There is an acute shortage of professionals who understand big data analysis.
8. *Transferring to a platform:* It is difficult to get voluminous data into a big data platform.

12.3 Big Data Analytics using Hadoop

Managing unstructured data is a difficult task. Traditional relational databases are clearly inadequate to store unstructured data. A tool to deal with big data is in high demand. Hadoop is an open-source framework that can perform big data computing. With the increasing need to analyze big data, the Hadoop framework is a blessing for big data analysts. The following characteristics of Hadoop make it popular and suitable for big data analytics.

- Hadoop can process data quickly, even if the volume of data is huge. It can process different types of data also.
- Hadoop uses the distributed computing approach for processing data.
- To ensure fault tolerance, Hadoop uses duplication of data. If a node fails, then data can be assessed because a copy is maintained.
- Hadoop can store data without understating about data.
- Hadoop uses the commodity hardware, so it is low cost. Hadoop is open-source and freely available.
- Scalability is also one feature of Hadoop.

Big Data Analytics Challenges with Hadoop

- MapReduce works well with simple problems. When iterative and interactive tasks are present, Hadoop cannot deal with them properly.
- There is a lack of competent skilled programmers for Hadoop.
- In big data analytics, the main focus is on data. Data security is another challenge in Hadoop.

The Hadoop framework includes many tools to deal with big data. Figure 12.3 shows the ecosystem of Hadoop that includes MapReduce, Hive, Pig, Spark, Flume, Sqoop, etc.

Major Components of Hadoop

HDFS: Hadoop Distributed File System (HDFS) can store a huge amount of data. It is highly scalable. HDFS has two main components: name node and data node. Name node is in the master server. Data nodes appear one per node in the cluster. HDFS provides data storage and replication. It provides robustness.

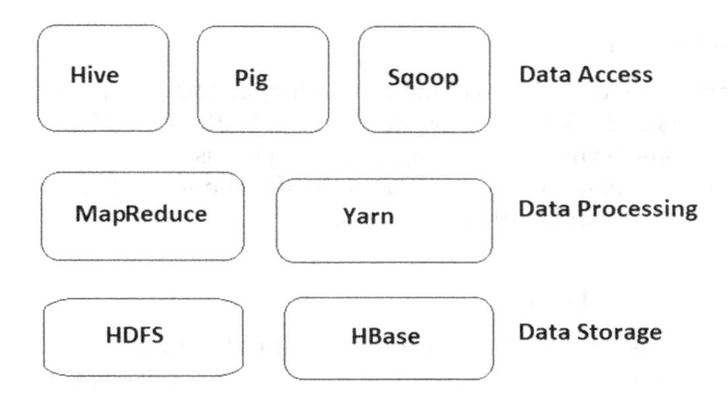

FIGURE 12.3
Big Data Analytics using Hadoop.

MapReduce: MapReduce is a problem-solving technique that uses a divide-and-conquer strategy. Bigger problems are divided into smaller problems, which are solved. These solutions are combined to generate the solution to the bigger problem. In this way, the large-scale problem can be solved using MapReduce. It provides parallel processing and distributed processing and has two main processes: map and reduce. In the map process, the mapper will generate the <key, value> pair. In the reduce process, the combiner will combine the values based on the same keys. It generates the aggregated or combined output.

Hive: Hadoop is difficult to configure and use. Hive lies between the user and Hadoop. It provides a table-like environment for using Hadoop in a more comfortable and familiar way. The user can read and write in the HDFS or local directory using Hive QL, which is similar to SQL-like syntax. Hive is the data warehouse system for Hadoop.

PIG: Pig is the platform to analyze large datasets. It uses the Pig Latin language, which is the data flow language. This provides ease of programming and also has the provision of user-defined functions. It can accept input in many formats.

SQOOP: Sqoop is the framework for transferring the data from Relational Database Management System (RDBMS) to Hadoop. It performs the MapReduce there and even transfers data from Hadoop back into RDBMS. It takes the schema from the databases only, so it performs the import, export, and execution operations.

SPARK: Spark is a fast engine for analysis of a large amount of data. It provides fast processing for batch as well as streaming data. Using Spark is easy because it provides different platforms like Java, Python, R, Scala, and SQL. It can run anywhere. For example, Spark can run on Hadoop, Kubernets, standalone, or in the cloud. Spark is very popular for big data analysis.

12.4 Applications

Big data is used for many applications in numerous fields [11,12,13], such as the following:

I Healthcare [14] [15]

Traditional claims-based analytics alone can no longer address the problems that emerge from too few assets and a large number of patients. With the total amount of clinical data available in electronic health records, big data analytics is vivaciously desired. Healthcare management needs access to these new threads of technology. Big Data Analytics (BDA) plays an incredible role in healthcare system.

II Grid Computing [16,17,18,19]

Power companies also have a tremendous amount of data due to digital measurement devices, such as smart meters. These growing data are an issue for power companies. To increase the efficiency and reliability of grid operations, analytical techniques can be used. A phasor measurement unit (PMU) is a device to measure the frequency of the power grid. Some researchers claim that PMU data analysis can be done using big data analysis. Analyzing the PMU data would be beneficial for the power grid [20,21]. The huge data have been generated by the PMU, and that is why most researchers feel that this is a big data issue [22]. Artificial intelligence techniques are also used to perform big data analytics in the smart grid. This helps to predict energy prices.

III Flight Safety [23]

Flight safety is a crucial issue that cannot be ignored. Passengers' lives and property are at risk due to flight safety problems. Traditional ways of performing flight safety analysis have their own drawbacks. Many flight safety accidents uncover errors made by pilots [24] and well as system errors. Due to conventions, the flight industry has less scope in analyzing flight safety data and supporting decision making.

IV Satellites [25]

Remote sensing data provide information that would be useful for making decisions. They also have various applications like hazard monitoring, urban planning, climate change, etc. These data are huge in volume and require quick processing. Like hazard management, data processing must be very fast to tackle the situation. Big data analytics plays an important role here.

V Quality of Experience (QoE) Monitoring [26]

Quality of Experience (QoE) monitoring is very useful for telecommunications. Mobile Network Operators (MNOs) use data to monitor the quality of services received by their customers as well as their customers' experiences.

VI Government

In the government sector, ball grid array has proved to be highly persuasive. In order to learn how the Indian electorate responds to government action, the Indian Government uses various strategies, just as thoughts for arrangement enlargement.

VII Social Media Analytics

The progress of web-based life has caused a big data upheaval. Different steps have been carried out to analyze various exercises in online life. The big data platform can also recognize the gab. Social networking can provide valuable and constant information that helps companies to change their importance, grow, and combat situations in the same way. Prior to using big data, there needs to be some preprocessing performed. The use of Canny choices from big data is mandatory in this way to consider the buyer's attitude.

VIII Technology

Organizations manage immense amounts of information consistently and put the information to use for business choices, too. Different organizations utilize data warehouses like Hadoop Cluster.

IX Fraud Detection

For businesses who have operations that include transaction processing or any sort of claims, detection of fraud is a mandatory operation. Big data platforms can analyze transactions and claims in real time. Recognizing large-scale patterns across numerous transactions or identifying atypical conduct from a client can change the fraud detection scenario.

X Call Center Analytics

All the questions that a client approaches a call center with get recorded for training purposes. However, not much analysis has been done with this huge stack of unstructured data, which is a significant reason driving the requirement for big data analytics in call centers. With advancement in natural language processing frameworks and intervention of big data analytics, brands can improve client assistance and operational effectiveness for call center operations.

XI Banking

Security concerns are raised by using clients' details. Big data analytics could uncover sensitive personal information by revealing shrouded associations between unrelated pieces of data. Outsourcing of data analysis activities or distribution of customer data across departments often creates security risks for the sake of more extravagant bits of information. Such incidents strengthen concerns about data privacy and discourage customers from sharing personal information in exchange for customized offers.

XII Agriculture

Big data plays a crucial role in activities related to agriculture. To develop crops efficiently, the biotechnology sector makes use of sensor data to develop quality crops. Sensor data collect information related to how plants react to various changes in conditions. The information helps make adjustments to factors, including water levels, temperature, soil composition, growth, and gene sequencing of each plant in the test bed. The information also helps find the best environment for each gene.

XIII Marketing

Advertisers started to utilize facial recognition to figure out how well their promotions succeeded or failed at generating enthusiasm for their items. An ongoing report distributed an architecture that examined outer appearances to uncover what viewers are feeling. The exploration was intended to find out what sorts of advertisements incited watchers to participate in the promotions for their organization, helping advertisers make promotions that are more appealing.

XIV Smartphones

Smartphone users have numerous applications that are readily available that utilize facial recognition. For example, smartphones can recognize event data just from a photograph of an individual, even if the individual has already forgotten the event.

XV Crowd Flow Prediction [27]

Crowd flow prediction is an important application of big data analytics. Mobile data is used for big data analytics based on user detail records.

12.5 Case Study of Big Data Analytics for Vehicle Tracking

Big data analytics has many applications, as we discussed earlier. We present here a case study on big data analytics for vehicle tracking [28]. Vehicle monitoring or tracking is an important task for vehicles that travel long distances. Monitoring vehicles can be done using sensors, such as temperature sensors, pressure sensors, surface sensors, speed sensors, etc. These sensors give information about the vehicle. For example, if the pressure in one of the tires is less than required, then the driver will be notified to maintain sufficient pressure in each tire. Similarly, the surface sensors give information about surface conditions, like dry, wet, snow covered, etc. This is useful for driver safety as well as controlling the breaking system. These sensors produce a huge amount of data and are a good example of big data. This big data is further processed with the help of Hadoop. Analysis of this data has many benefits. If a vehicle is monitored, the driver can be notified regarding needed maintenance. Tracking the vehicle also helps predict expected arrival to the destination. If the arrival is predicted well in advance, then appropriate arrangements can be made. By observing the speed of the vehicle, business owners can tell if the driver is driving safely or over the speed limit. Additional sensors could monitor the health of the driver as well. If they found any abnormalities, they could send an alarm before the driver causes an accident.

12.6 Conclusion

Data Analytics plays an important role in decision making in academics, research, and industry. It is the invisible part of businesses nowadays. In this chapter, we discussed the fundamental concepts of big data. We highlighted the different challenges in big data analytics. We also showed how these issues and challenges can be handled using the Hadoop framework. Also, we explored different applications of big data analytics. Finally, we discussed one case study of big data analytics. This chapter will help beginners to understand the big data domain by demystifying big data computing.

References

1. Singh, S. and Singh N. (2012). Big Data analytics, In *International Conference on Communication, Information & Computing Technology (ICCICT), Mumbai* (pp. 1–4). IEEE.
2. Mittal, S. and Sangwan, O. P. (2019) Big Data Analytics using Machine Learning Techniques In *9th International Conference on Cloud Computing*, Data Science & Engineering (Confluence), *Noida, India* (pp. 203–207). IEEE.
3. Praveena, M. D. A. and Bharathi, B. (2017). A survey paper on big data analytics. In *International Conference on Information Communication and Embedded Systems (ICICES), Chennai* (pp. 1–9). IEEE.
4. Tsai, C., Lai, C., Chao, H. et al. (2015) "Big Data Analytics: A Survey. *Journal of Big Data*: 2–21. https://journalofbigdata.springeropen.com/articles/10.1186/s40537-015-0030-3#Abs1
5. Arass, M. E., Tikito, I., and Souissi, N. (2017) Data lifecycles analysis: Towards intelligent cycle In *Intelligent Systems and Computer Vision (ISCV), Fez* 1–8. IEEE.
6. Komalavalli, C. and Laroiya, C. (2019) Challenges in Big Data Analytics Techniques: A Survey, In *9th International Conference on Cloud Computing, Data Science & Engineering (Confluence), Noida, India* (pp. 223–228).
7. Pandey, K. K. andShukla, D. (2018) Challenges of Big Data to Big Data Mining with their Processing Framework, In *8th International Conference on Communication Systems and Network Technologies* (pp. 89–94). IEEE.
8. Gruschka, N., Mavroeidis, V., Vishi, K., and Jensen, M. (2018) Privacy Issues and Data Protection in Big Data: A Case Study Analysis under GDPR, In *IEEE International Conference on Big Data (Big Data)* (pp. 5027–5033). IEEE.
9. Cuzzocrea, A. (2017) Privacy-Preserving Big Data Stream Mining: Opportunities, Challenges, Directions In *IEEE International Conference on Data Mining Workshops(ICDMW)* (pp. 992–994). IEEE.
10. Jayasingh, B. B., Patra, M. R., and Mahesh, D. B. (2016) Security issues and challenges of big data analytics and visualization, In *2nd International Conference on Contemporary Computing and Informatics (IC3I)* (pp. 204–208). IEEE.
11. Balasupramanian, N., Ephrem, B. G. and Al-Barwani, I. S. (2017) User pattern based online fraud detection and prevention using big data analytics and self organizing maps, In *International Conference on Intelligent Computing, Instrumentation and Control Technologies (ICICICT), Kannur* (pp.691–694). IEEE.
12. Kumar, M. and Nagar, M. (2017) Big data analytics in agriculture and distribution channel In *International Conference on Computing Methodologies and Communication (ICCMC), Erode* (pp. 384–387). IEEE.
13. Mohamed, N. and Al-Jaroodi, J. (2014). Real-time big data analytics: Applications and challenges, In *International Conference on High Performance Computing & Simulation (HPCS), Bologna* (pp. 305–310). IEEE.
14. Reddy, A. R. and Kumar, P. S. (2016) Predictive Big Data Analytics in Healthcare, In *Second International Conference on Computational Intelligence & Communication Technology (CICT), Ghaziabad*, (pp. 623–626). IEEE.

15. Patil, K. H. and Seshadri, R., (2014) Big data security and privacy issues in healthcare, In *IEEE International Congress on Big Data, Anchorage, AK, USA* (pp. 762–765).
16. Chi, M., Plaza, A., Benediktsson, J. A., Sun, Z., Shen, J., Zhu, Y. (2016) "Big Data for Remote Sensing: Challenges and Opportunities." *Proceedings of the IEEE* 104(*11*): 2207–2219.
17. Bhuiyan, S. M. A., Khan, J. F., Murphy, G. V. (2017) Big data analysis of the electric power PMU data from smart grid, *In SoutheastCon, Concord, USA* (pp. 1–5). IEEE.
18. Junaidi, N. andShaaban, M. (2018) Big Data Applications in Electric Energy Systems In *International Conference on Computational Approach in Smart Systems Design and Applications (ICASSDA)*(pp. 1–5).
19. Kadadi, A., Agrawal R., Nyamful, C., and Atiq, R. (2014) Challenges of data integration and interoperability in big data, *In IEEE International Conference on Big Data (Big Data)* (pp. 38–40).
20. Yang, B., Yamazaki, J., Saito, N., Kokai Y., and Xie, D. (2015) Big Data Analytic Empowered Grid Applications- Is PMU a big data issue?, In *12th International Conference on the European Energy Market (EEM), Lisbon, Portugal* (pp. 1–4). IEEE.
21. Roy, V., Noureen, S. S., Bayne, S. B., Bilbao, A., Giesselmann, M. (2018) Event Detection From PMU Generated Big Data using R Programming. In *IEEE Conference on Technologies for Sustainability (SusTech)* (pp. 1–6). IEEE.
22. Kumari, A. and Tanwar, S. (2021) ρReveal: An AI-based Big Data Analytics Scheme for Energy Price Prediction and Load Reduction, In *11th International Conference on Cloud Computing, Data Science & Engineering (Confluence), Noida, India* (pp. 321–326).
23. Li, B., Ming, X., and Li, G. (2017) Big data analytics platform for flight safety monitoring In *IEEE 2nd International Conference on Big Data Analysis (ICBDA)* (pp. 350–353).
24. Dekker, S.W.A. (2002) "Reconstructing the Human Contribution to Accidents: The New View of Human Error and Performance. *Journal of Safety Research* 33(3): 371–385.
25. Shao, J., Xu, D., Feng, C., and Chi, M. (2015) Big data challenges in China centre for resources satellite data and application. In *7th Workshop on Hyperspectral Image and Signal Processing: Evolution in Remote Sensing (WHISPERS)* (pp. 1–4). IEEE.
26. Rueda, D. F., Vergara, D., and Reniz, D. (2018)Big Data Streaming Analytics for QoE Monitoring in Mobile Networks: A Practical Approach, In *IEEE International Conference on Big Data (Big Data)*. 1992–1997
27. Jiang, H., Li, L. Xian, H., Hu, Y., Huang, H., and Wang, J. (2021). Crowd Flow Prediction for Social Internet-of-Things Systems Based on the Mobile Network Big Data. *IEEE Transactions on Computational Social Systems*.
28. Varsha, P. V., Sankar, S. R., and Mathew, R. J. (2016) Predictive Analysis using Big data Analytics for Sensors Used in Fleet Truck Monitoring System. *International Journal of Engineering and Technology (IJET)* 8(2): 714–719.

13

AR-Powered Computer Science Education

Sagar Rane[1], Anant Kaulage[1], and Sanjeev Wagh[2]
[1]*Department of Computer Engineering, Army Institute of Technology, Pune, India*
[2]*Department of Information Technology, Government College of Engineering, Karad, India*

13.1 Introduction

Regardless of the technologies used, field of the computer science and engineering data structures, and algorithms form the most fundamental part of any computer programming-related course. With programming courses now aimed at younger students, it is important to ensure that these critically important concepts are taught clearly. Since data structures and algorithms play a fundamental role in introductory computer science, this only seems to be the first step in creating a better education system for computer science.

Visual aids have proved themselves useful but the usual two-dimensional (2D) visualization proves to be a limitation in understanding complex structures that occur commonly in the field of data structures. To overcome this problem, this chapter looks at augmented reality (AR) as a possible solution.

This chapter aims to make modern computer science education more interesting and digestible for scholars through the use of AR. We feel that visual learning goes a long way in computer science and helps trigger a better response and understanding from the student. If we use AR to make visualizations of various algorithms, it will help both student and teacher at almost every level; AR is interactive and visually appealing, which will help teachers to explain a concept and students to understand it better [1]. Learners can download the tool from (https://argorithm.github.io/), and after updating their settings using the GitHub manual, they can perform the activities. This will enable the learners to visualize computer science concepts.

13.2 Augmented Reality

For quite some time now, AR has been considered to be the future of visualization. It brings an extra level of depth and appeal when compared to existing visualization.

DOI: 10.1201/9781003166702-13

AR is purely about sight but also interacts with other senses as well. The market for AR is expected to grow radically over time based on current trends and research [2]. There have been various studies and experiments over the past decade exploring the benefits and feasibility of AR in the education sector.

Research [3] on AR prototypes has revealed that AR is a powerful tool in the field of education, even more so when it comes to medical training. It used to be extremely difficult for students to learn complex human anatomy from 2D images in textbooks or from limited practice on cadavers. With AR they can freely explore and learn to their heart's content in a boundless, risk-free environment. Additionally, physicians can adopt AR to help patients learn about their ailments through an AR-rendered model and actively partake in medical decision making instead of leaving it to physicians.

A study examined the challenges that the medical education field has faced and estimated that by 2020, the current body of medical knowledge will double every 73 days [4]. This gives the implication that while one goes to medical school for 4–5 years, the learning process is never-ending in the field as there is new information being found every day. In the medical education field, AR has shown immense potential to aid in this learning process: from helping students with their motor skills to see what effects their treatment can have on patients through digitally rendered models, to seeing interactions from the patient's perspective. This is helping students to understand medical concepts. In short, AR has proved to be an engaging resource greatly benefiting the field of medicine. Like with everything, though, there are people both for the AR revolution in the medical industry and against it. Some people argue that it is a distraction that cannot replace actual human interactions and is becoming a hindrance to the field rather than an aid. However, so far it has proves to be an asset.

AR has also been helping medicine by making the mentoring process easier. It allows mentors to experience procedures and treatments they might not be skilled enough to perform themselves. Now, a surgeon with an AR headset can superimpose his hand over a mentor surgeon's hand. Microsoft has launched HoloLens, which uses AR to create holograms of people and project them anywhere, which allows physicians to be in the same space while physically operating in different locations across the globe. Thus, AR can help students to envision themselves performing complex procedures and train themselves with no risks.

AR has proved to have several benefits in education; however, there are a few limitations to it as well. Despite being attractive, AR has a bigger superfluous perceptive load on learners as it forces them to envision physical models in their mind. A 2019 survey showed a significant difference in test scores of students studying neuroanatomy who worked with physical cross-sections of the brain and students who worked with AR; students who worked with AR received lower scores [5]. So, there are some advantages and disadvantages of each type of learning system. This might make it look like the traditional method of learning through textbooks is the better choice. Skeptics have argued that while AR is a beneficial and engaging learning resource, it can never truly compare to the tactical training that one gets from physical practice on cadavers [6]. In short, AR is a new addition and is causing a large revolution in the medical industry; as a result, it is still finding its place among skeptics. Only when the students currently making use of AR-based teaching methods graduate and begin their practice, can we definitively see the effects of AR and whether it should be largely incorporated into the education system.

Invisible concepts [7], such as those on a microscopic or sub-microscopic level, are extremely hard to grasp quickly, which is a situation where AR can prove to be beneficial

[8]. In the model of AR/virtual reality (VR), education is positioned as a knowledge acquisition system. Metal structure material involves numerous ideas such as hard edifices, the seven basic rock crystal arrangements, and metal structures, which is something AR has helped students with. Involving the use of media in the learning process can be more engaging and help students to learn to think critically and be more autonomous in their education [9]. Additionally, AR is commonly used on smartphones, which the current generation of students is well versed with [10]. AR has managed to present information of sub-microscopic elements of metal structure in three-dimensional (3D) form using a marker. It has proved to have several advantages, such as appearance that is offered inductively to stimulate curiosity to develop students' thinking skills.

The procedure for creating AR education mass media on facts about metal structure must refer to drawings to assist in creating AR-based media systems. After the result of concept investigation, a flowchart is formed as an allusion and a storyboard as a design skeleton.

Display learning objectives is something that contains, as the name suggests, learning objectives on metal structure material. They are all arranged based on the students' capabilities on sub-microscopic representation. The main goal of this was to weigh students' reactions to the education methods. The results [11] proved that media are ready to be used in the learning process.

TeachAR is a tool built on AR specifically to help teach children age 4–6 about colors, shapes, and spatial reasoning in English to help them learn the language [12]. It is essentially a desktop-based application (to provide a wider view of the AR scene) to aid young children in learning English vocabulary for certain shapes, colors, and spatial relationships divided into two modules (colors and shapes in module 1 and spatial relationships in module 2). It utilizes the Microsoft Kinect hardware and software to input voice and applies speech recognition, which includes commands such as green, blue, red, pyramid, sphere, square, etc. If it is not possible to utilize speech, this feature can be disabled, allowing children to use an AR marker to initiate the AR scene.

To teach spatial relationships, a jigsaw puzzle metaphor was applied on a playing board, where kids would have to place the color, shape, and preposition markers at the correct place. Learning a new language that is not your native language is always a daunting task; however, through applications like TeachAR, the entire process is more engaging, making it seem less of a chore and more like a hobby.

A study conducted on the effect AR brings to early education [13] has shown that students respond much better to visual aids using AR rather than those without AR and would prefer AR. To fully take advantage of AR applications for younger children, these applications should include games and activities that should focus on attention development, memory, and sensory perception development.

The documented use of AR has only been short term for experimental purposes. There is insufficient data on the feasibility of a curriculum utilizing AR permanently [14]. The benefits of AR are well supported by the above research but very few studies discuss a system meeting the practical requirements on the resources available to education institutes. Nonetheless, with the release of ARCore[1], ARkit[2], and similar software plugins that bring AR to mid-range mobile phones, AR support is now becoming popular among handheld devices[3]. WebAR[4] allows users with handheld devices incompatible with the above plugins to access AR using their web browser. While AR can be better experienced with specialized hardware like Microsoft HoloLens[5], the lighter and easily accessible interface provided by handheld devices is more than enough in most cases. These recent developments, along with the ubiquitous presence of mobile phones, make it possible for all students to access AR with ease. The utilization of AR within the current curriculum should be based on the topics being covered. Fundamental AR learning objects that are easy to design should be used along with traditional methods so that educators can use them as they like. A fundamental learning object is a single digital resource that has no context attached to it [15]. An example is any static 3D object presented in AR. Depending on the context, it can be used to explain various concepts. These are easy to design and implement as well. Gamification of learning situations with the help of AR helps students to understand theoretical paradigms [16].

A study was conducted looking into the application of AR to the preexisting curriculum of bachelor students of civil engineering in Germany [17]. The task given to them focused on submitting AR content to explain infrastructure management. The domain chosen was infrastructure management concerning sustainable development because very few students had practical experience in the subject. This study not only observed the students' engagement with the task but also the quality of their responses and their reviews of the activity as a whole. The quality of students' submissions, the type of submission, the sources utilized, the students' reviews of the activity, and their opinion of AR were taken into account. The study concluded that even the conceptual design of an AR application is a valuable learning activity. All the students in the group believed that AR will play a pivotal role in engineering education. The observations show that the procurement of AR assets for learning activities can be self-sustained in engineering courses, which strengthens the argument for the long-term involvement of AR in programming education.

As per the new education policy in India, 6th standard students have to learn to code computer science concepts. AR will be tremendously beneficial to these students to help them understand the concepts in an easier way. A prerequisite for the incorporation of AR in programming education will be allowing students to bring their own devices. As discussed above, AR has become accessible due to its increased support in handheld devices. Not many educational institutes would support students bringing their own devices to class, especially in schools, as it would seem like a distraction for students. There are also concerns regarding unequal learning opportunities if students are asked to bring their devices as some students might not have devices. A study in Sweden observed school student behavior after they were asked to share mobile phones for an AR game [18]. Students age 10–12 years were engaged in the location-based game Pokémon Go[6] during learning sessions of mathematics and social science. The students were informed that the activity was a learning activity, and the teachers drew parallels to various issues using the game as the medium. The study collected data on the attention given to the task as well as social factors and teachability. Contrary to concerns raised, students were able to identify and understand the benefits of collaboration. The teacher was able to utilize the game by drawing parallels, which reaped more responses than normal. It should be noted that the game became a learning experience because of the teachers. AR should not be perceived as a replacement for teachers but as a powerful asset for teachers, encouraging more engagement in class and the idea of collaborative learning. Allowing students to bring their own devices is a crucial step toward incorporating AR, but one that must be executed properly.

Authoring AR content is another challenge for AR in programming education. For AR to be a feasible option for educators to use, it is not practical to expect them to reuse preexisting static visualizations. Educators should be able to create their AR content comfortably. The next section looks at current trends in data structure visualization and how we can utilize them to design AR content. While gamification of concepts and reusing fundamental learning objects are good options, the existence of machine-readable code gives us the power to generate more advanced customizable learning objects without much manual work. The next section focuses on prevalent data structure visualization techniques that can be used as inspiration for translating code into AR learning objects.

13.3 Data Structure Visualization

Data structures and algorithms largely involve logical sequences using abstract objects, and creating visual content is not an easy task, even without considering the complexities of AR. Even before the recent spike in interest toward computer science education, there was plenty of research conducted on computer science education using visual elements [19,20], which has shown how powerful visual elements can be in this field.

There are different methods for creating visualizations for data structures and algorithms. The first is creating visualizations using design tools and animators as static examples. Creating such assets for specific examples would be tedious manual labor, and creating more fundamental objects would require context and effort to be utilized. Taking an example of a sorting algorithm, a teacher can create a 3D object representing an array and animate it being sorted. Here, the visualization is static to the input given when the content is created. Corner cases are extremely important when explaining logical processes, and not being able to customize the input array is a considerable drawback. The alternative would be to create 3D objects denoting elements of an array and move them in AR. This is a better alternative, but such an implementation would heavily rely on the teacher and limit the interaction of the students with the objects. These kinds of applications make sense when used to represent the usual 2D visualizations but fail to completely utilize AR.

In Figure 13.1, Visualgo [21] is a library of immutable 2D data structure visualizations. These are preferred by educational tech companies or large organizations with manpower and resources. The application benefits from the large library of data structures that cover most use cases. Since the visualizations are immutable, the application relies on having a vast library of visualizations, without which it would prove insufficient. Developing such a library of AR content will be a challenge for smaller institutes.

Another visualization method is writing code in an environment where written code is parsed by the environment. The environment then provides visualizations or a graphic interface to interact with the code. An example of such a tool is BlueJ [22]. BlueJ, in Figure 13.2, is an integrated development environment (IDE) for Java specifically designed for use in early introduction to programming. It is a popular tool that has shown significant performance improvements in students. BlueJ is better at teaching object-oriented concepts and program execution. It is not designed to explain data structures. A drawback in extending IDE-based visualization tools like BlueJ to AR is the challenge of experiencing AR on computers/laptops, as reiterated above.

In Figure 13.3, Jelliot is an example of a visualization tool that takes source code as input and generates visualizations of the objects and their data members and function calls [23]. The program does not require any manual effort to create the visualizations. Jelliot can be easily used by teachers to visualize their programs and also used by

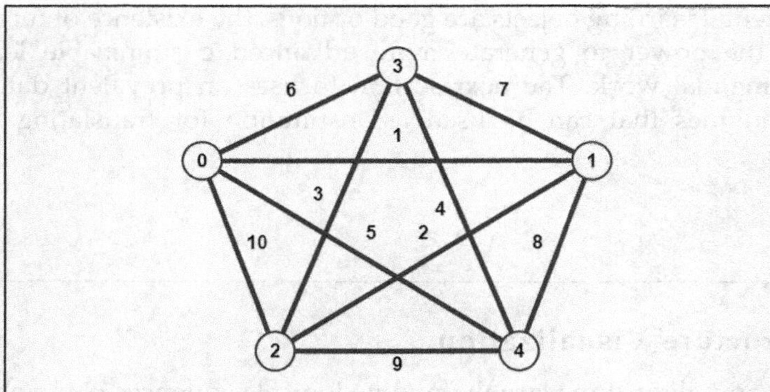

Adjacency Matrix

	0	1	2	3	4
0		1	10	6	5
1	1		2	7	8
2	10	2		3	9
3	6	7	3		4
4	5	8	9	4	

Adjacency List

0:	(1, 1)	(2, 10)	(3, 6)	(4, 5)
1:	(0, 1)	(2, 2)	(3, 7)	(4, 8)
2:	(0, 10)	(1, 2)	(3, 3)	(4, 9)
3:	(0, 6)	(1, 7)	(2, 3)	(4, 4)
4:	(0, 5)	(1, 8)	(2, 9)	(3, 4)

Edge List

0:	0	1	1
1:	0	2	10
2:	0	3	6
3:	0	4	5
4:	1	2	2
5:	1	3	7
6:	1	4	8
7:	2	3	3
8:	2	4	9
9:	3	4	4

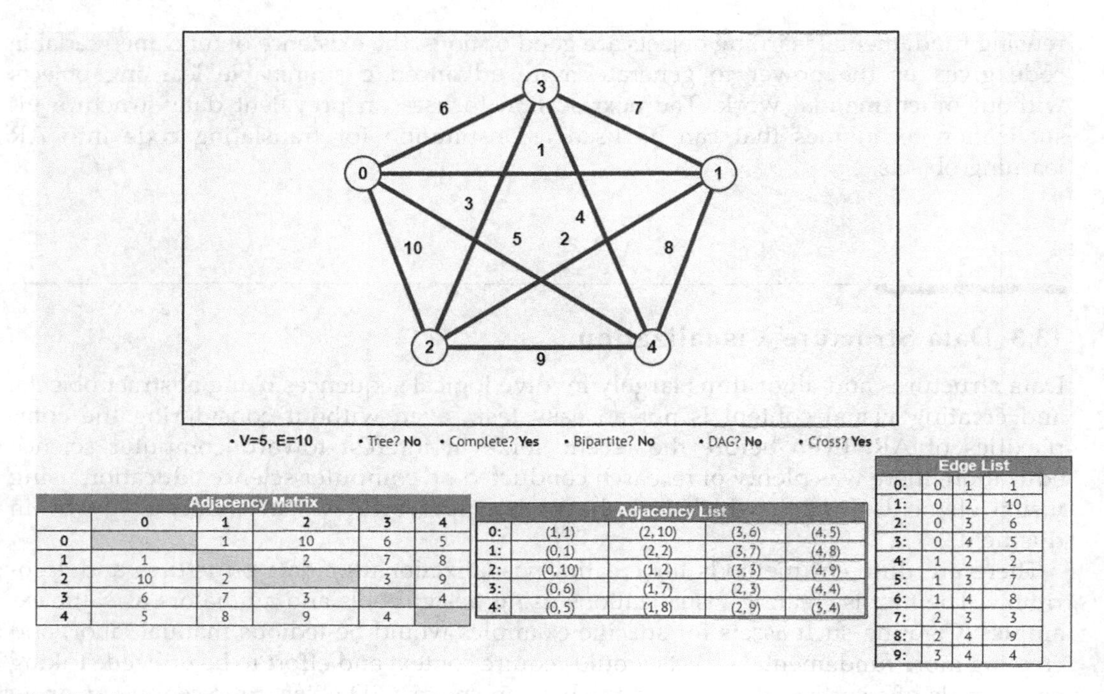

FIGURE 13.1
Visualgo Data Structure Visualization.

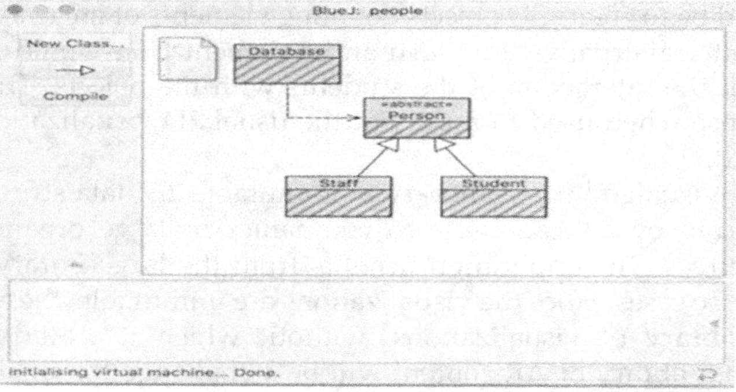

FIGURE 13.2
BlueJ IDE.

students to visualize their code. Jelliot visualizes content in a step-by-step sequence, allowing students to understand data flow better. Like BlueJ, visual representation of different classes is not distinguishable by appearance. That means Jelliot is also better equipped for object-oriented programming.

Both Jelliot and BlueJ have the same drawback, i.e., the inability to visualize data structures in a distinguishable way. This was overcome in DVIS [24], a visualization tool that tries to make object-oriented programming and data structures easier to understand in an active environment with the help of media and animations. The tool captures concepts, behavior, and actions of various elements in a program that learners can

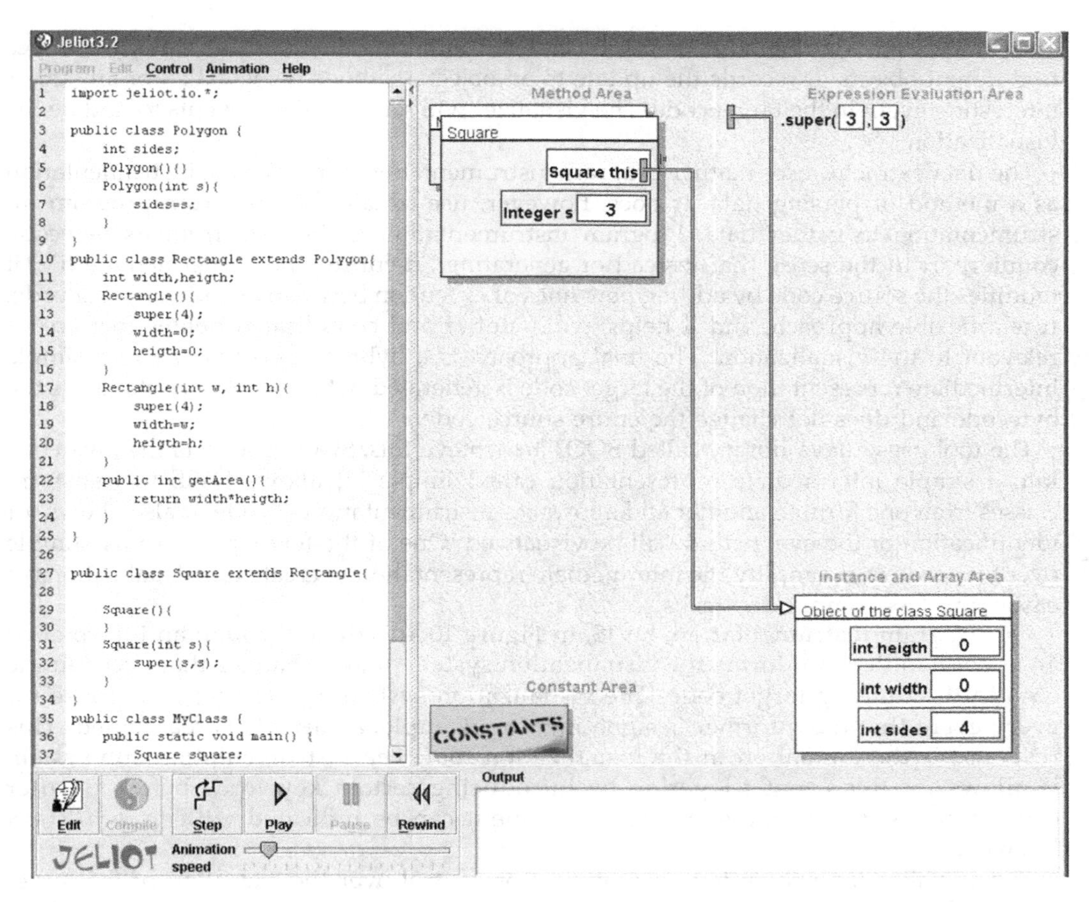

FIGURE 13.3
Jelliot Visualization Tool.

actively control and interact with. The tool is accommodating of logical errors and can help identify and pinpoint them through visualization of the algorithm. DVIS aims to avoid the limitations of the previously discussed methods by making the visualizations straightforward, avoiding extra programming, trying to include more data structures and algorithms, and targeting advanced programming. All this allows for a tool that can help in learning, teaching, and debugging algorithms and is easy to use.

The program to be visualized, called the *target code*, does not need to be logically correct, which makes it easier to identify mistakes. DVIS will not change the source of the target code to avoid confusion that the code being visualized is different from the target code. The tool also displays the content of the stack frame, the status of the heap, and the current active objects.

To fulfill the above promises, the developers of this tool had some interesting and clever design ideas. The basic design divides the overall system into two separate modules:

1. Data extractor
2. Visualizer

The data extractor as a module is responsible for gathering the data and making changes to different objects. It records the update to an object: methods being called or any other interesting event in the target code. The visualizer will then use these events to update the visualization.

The data extractor uses neither program instrumentation nor bytecode instrumentation as a method of parsing data. It does, however, use certain features from program instrumentation to gather data. Program instrumentation is different from its bytecode counterpart in the sense that instead of generating an entirely new intermediate file, it modifies the source code by adding new lines of execution between already existing code. It is a flexible approach, and it helps avoid clutter and complicated details that are irrelevant to the visualization. The final approach is a hybrid approach where a simple intermediate representation of the target code is generated, which is not as detailed as the bytecode and does not change the entire source code.

The tool uses a Java library called SOOT to achieve this. SOOT converts the Java code into a simple intermediate representation called Jimple[7]. It allows DVIS to transform classes from one form to another and allows for instrumentation of code. It also allows for identification of the events that will be visualized. One of the few reasons to use Jimple over bytecode is to simplify the intermediate representation to create the visualizations of easier and less complicated tasks.

For program instrumentation, DVIS, in Figure 13.4, is done through an informer or an event handler. It informs the visualization system of any changes that occur in the data in the running target code. The visualization system will trigger a response for every event that the informer is signaling, for example, a variable is reassigned. This response is the animation in the visualization tool. The instrumentation sub-system is allows for this event triggering by introducing certain keywords before or after code snippets, for example, a return statement can be indicated with *methodReturn* keyword.

The visualization sub-system also allows for control over the execution of the target code and step through each statement. For rendering the graphical interface, yFiles, a graph library, is used. The data and all other details are converted into strings.

This system is quite useful for scholars trying to understand the basic code structure and object interaction, which in turn helps in understanding basic algorithms. A similar system

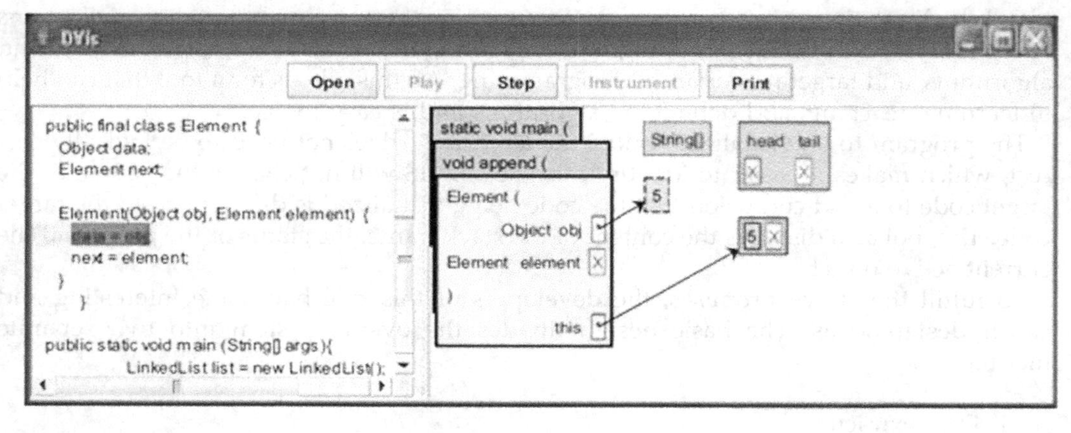

FIGURE 13.4
DVIS Visualization Tool.

can be implemented using fundamental AR objects as basic components and the transformer rendering animations step by step through the code.

13.4 Software Design and Development

Our project considers the perspective of two parties. The teachers and educators that will use the application to create source code for the visualizations will be referred to as *programmers*. The students who will interact with these visualizations will be referred to as *users*. People can belong to both of these categories.

Previously mentioned projects and papers dealt with 2D visualizations. Our project worked with 3D visualizations in AR, which posed a challenge. AR cannot be experienced on the computer properly. For the AR experience, the output device has to be a mobile device. We required an intermediate form that would be transferred from the PC, where the program works, to the mobile phone that would visualize it in AR. This intermediate form would have to monitor how objects are created, modified, and deleted as the program gets executed. Each change in the data structure objects must trigger a message that informs the visualizer that some event has occurred and how that event has affected the objects. We have termed this change a *state*, and the set of states that are triggered along the execution makes up the intermediate form, which is altogether termed *states*.

For the generation of states, in each program, we created an object that records states and, within our tool, provided a set of classes that act as a wrapper around traditionally used data structures to monitor them and create states. At the end of execution, the states are sent to get rendered in AR in the mobile device. When the states arrive at the mobile device, the device will read the metadata in the states and use that information to render objects and animate them in AR.

Users should not have to rely on programmers to adjust the input at execution and run the program for them. Users should have independent control over what input is given to generate states and should be able to execute programs at their convenience. For this purpose, we required another module that stores program code so that it can be executed per user request and generate states for visualization. This stored program code is what we refer to as *argorithm*. We have given the name *ARgorithm* to our developed tool.

Thus, adhering to the requirements of the process, we propose a three-module system:

- Toolkit
- Server
- Mobile application

The toolkit is a library providing classes and functions for data structures that can create states. The toolkit is used by the programmer to make argorithms that are stored in the server. The server can be on the same machine where the argorithm is created or can be in a cloud-based system that can act as a repository of argorithms for an organization. The user would connect to the server using the mobile application and render any argorithm of their choice (Figure 13.5).

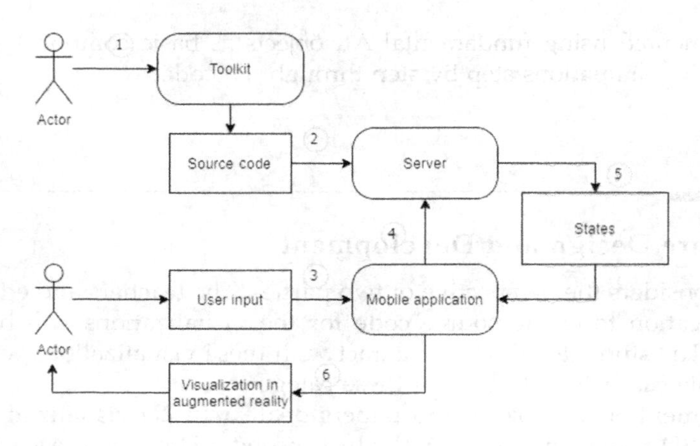

FIGURE 13.5
Logical Representation of End-to-End Process.

Due to its immense popularity, we selected Python as the programming language for the toolkit. Python is not only popular [22] but its easy-to-write style is ideal as the medium for argorithm creation as it makes the process of making argorithms fairly simple for the programmer.

The toolkit will work by creating an object container that stores states. This object container will be termed *stateset*. It contains the final collection of states used as an intermediate state. The toolkit will provide classes for all data structures. When an instance of this class is created, it stores a reference of a stateset object, and whenever there is a change in its value or a method of the class is invoked, the data structure class object will append a state to the stateset for creating more detailed visualizations. We will provide a feature for adding comments to the state, which will act as a helpful textual description when it is all rendered in AR.

13.5 Practical Implementation and Results

The above section provides sufficient evidence that the best approach for authoring AR content is having an intermediate state while converting code to visualizations using a transformer of sorts. An important thing to keep in mind while implementing this is that AR cannot be experienced on computers/laptops without additional hardware, which makes it difficult for everyone to access. As discussed before, AR is more easily experienced on handheld devices. Thus, we propose a server-based architecture in which a server is used to send code files to mobile devices so that teachers can share AR content with their students. In this chapter, we have shown sample results with one sorting algorithm. It is possible to create different visualizations for various computer science concepts with ease. We are continuously adding more concepts into the ARgorithm tool. The details of implementation of each step of an ARgorithm tool are available on https://argorithm.github.io/.

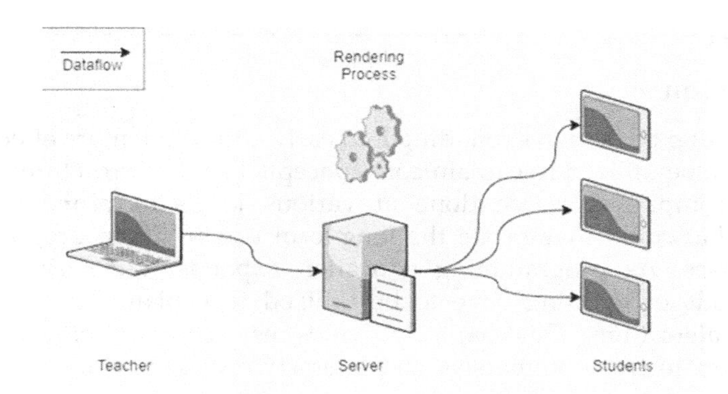

FIGURE 13.6
Proposed Infrastructure for AR-powered Laboratory.

This server can act as a library where teachers can store and share their codes and students can visualize the codes on their handheld devices. The server is used to render the processes given by teachers to students. Right now, we work with Google Collaboratory, but in the future, such an infrastructure can be set up in laboratories at education institutions, which would be the first step in establishing AR as a learning tool in programming education. Figure 13.6 shows the proposed infrastructure of an AR-powered laboratory.

In Figure 13.7, we ran the states of a bubble sort on an array of integers. In this figure, the first picture shows the array initialization. There are five numbers (i.e., 2, 1, 4, 5, and 3). As per the bubble sort logic, the index of [1] is bigger than the index [0], so we swapped them. In the third picture, the index of [2] is smaller than index of [3], so we did not swap them. The last picture shows the final sorted array after all the swapping. With these visualizations, students can easily understand how the bubble sort works as well as many computer science concepts like it.

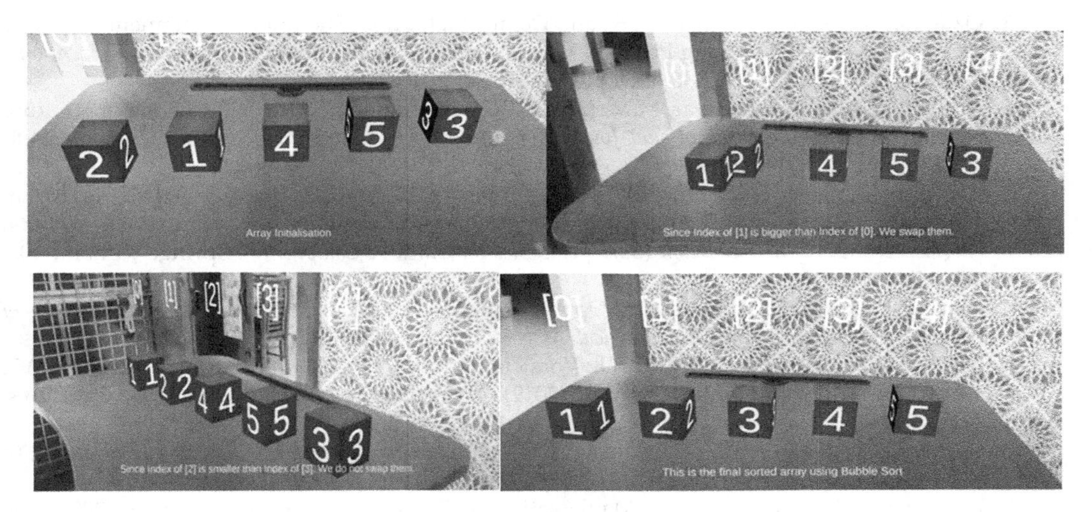

FIGURE 13.7
Visualization of Bubble Sort on Array of Integers in Augmented Reality

13.6 Conclusion

With programming education expanding into early education, more effective methods of teaching have to be utilized to explain core concepts like data structures and algorithms. This chapter compiles research done at various levels to propose a possible implementation that could inaugurate the long-term use of AR in programming and engineering courses. As programming education expands into early education, more effective methods of teaching have to be utilized to explain core concepts like data structures and algorithms. This chapter provides research, requirements, and a tool that can help teachers to create immersive and interactive visualizations using AR.

Notes

1 For more information, see https://developers.google.com/ar
2 For more information, see https://developer.apple.com/arkit
3 For full list of devices, check out https://developers.google.com/ar/discover/supported-devices
4 Create WebXR. Augmented reality on the web. https://www.createwebxr.com/webAR.html
5 Microsoft HoloLens. https://www.microsoft.com/en-us/hololens/
6 Pokemon Go. https://pokemongolive.com/
7 Jimple. https://www.sable.mcgill.ca/soot/doc/soot/jimple

References

1. Özdemir, M., Şahin, Ç., Arcagok, S., and Demir, M. (2018). "The Effect of Augmented Reality Applications in the Learning Process: A Meta-Analysis Study." *Eurasian Journal of Educational Research (EJER)*. n74: 165–186. DOI: 10.14689/ejer.2018.74.9.
2. Alkhamisi, A., and Monowar, M.M. (2013). "Rise of Augmented Reality: Current and Future Application Areas." *International Journal of Internet and Distributed Systems* 1: 25–34. DOI: 10.4236/ijids.2013.14005
3. Campisi, C.A., Li, E.H., Jimenez, D.E., and Milanaik, R.L. (2020) Augmented Reality in Medical Education and Training: From Physicians to Patients. In: Geroimenko V. (eds) *Augmented Reality in Education (Springer Series on Cultural Computing)*. Springer, Cham. 10.1007/978-3-030-42156-4_7
4. Densen, P. (2010). "Challenges and Opportunities Facing Medical Education." *Transactions of the American Clinical and Climatological Association* 122: 48–58.
5. Henssen, D., Heuvel, L., De Jong, G., Vorstenbosch, M., van Cappellen van Walsum, A.-M., Hurk, M., Kooloos, J., and Bartels, R. (2019). *Neuroanatomy Learning: Augmented Reality vs. Cross-Sections. Anatomical Sciences Education*. 13. 10.1002/ase.1912.
6. Kamphuis, C., Barsom, E., Schijven, M., Christoph, L.H. (Noor). (2014). "Augmented Reality in Medical Education?." *Perspectives on Medical Education* 3. 10.1007/s40037-013-0107-7.
7. Irwansyah, F.S., Nur Asyiah, E., Maylawati, D.S., Farida, I., and Ramdhani, M.A. (2020) The Development of Augmented Reality Applications for Chemistry Learning. In: Geroimenko V. (eds) *Augmented Reality in Education (Springer Series on Cultural Computing)*. Springer, Cham. 10.1007/978-3-030-42156-4_9

8. Smith, K.C. and Nakhleh, M. (2011). "University Students' Conceptions of Bonding in Melting and Dissolving Phenomena.: *Chemistry Education Research and Practice* 12: 398–408. 10.1039/C1RP90048J.
9. Oh, S. and Byun, Y. (2012). The Design and Implementation of Augmented Reality Learning Systems. Proceedings – 2012 IEEE/ACIS 11th International Conference on Computer and Information Science, ICIS 2012. 10.1109/ICIS.2012.106.
10. Zan, N. (2015). The Effects of Smartphone Use on Organic Chemical Compound Learning. *US-China Education Review A* 5. 10.17265/2161-623X/2015.02.003.
11. Septi Irwansyah, F., Ramdani, I., and Ch, I. (2017). *The development of an Augmented Reality (AR) technology-based learning media in metal structure concept*. The Asian Education Symposium. 233–237. 10.1201/9781315166575-56.
12. Dalim, C. S. C., Piumsomboon, T., Dey, A., Billinghurst, M., and Sunar, S. (2016) TeachAR: An Interactive Augmented Reality Tool for Teaching Basic English to Non-native Children," 2016 IEEE International Symposium on Mixed and Augmented Reality (ISMAR-Adjunct), Merida (pp. 344–345). doi: 10.1109/ISMAR-Adjunct.2016.0113.
13. Kuang, Y. and Bai, X. (2019) "The Feasibility Study of Augmented Reality Technology in Early Childhood Education," 2019 14th International Conference on Computer Science & Education (ICCSE), Toronto, ON, Canada (pp. 172–175). doi: 10.1109/ICCSE.2019.8845339.
14. Kljun, M., Geroimenko, V., and Čopič Pucihar, K. (2020) Augmented Reality in Education: Current Status and Advancement of the Field. In: Geroimenko V. (eds) *Augmented Reality in Education* (*Springer Series on Cultural Computing*). Springer, Cham. 10.1007/978-3-030-42156-4_1
15. Wiley D. (2002) "Connecting learning objects to instructional design theory: a definition, a metaphor, and a taxonomy." In: *The Instructional Use of Learning Objects* (pp. 3–23). Bloomington, IN: The Agency for Instructional Technology.
16. Weerasinghe, M., Quigley, A., Ducasse, J., Pucihar, K., and Kljun, M. (2019). *Educational Augmented Reality Games*. 10.1007/978-3-030-15620-6_1.
17. Wolf, M., Söbke, H., and Baalsrud Hauge, J. (2020) Designing Augmented Reality Applications as Learning Activity. In: Geroimenko V. (eds) *Augmented Reality in Education* (*Springer Series on Cultural Computing*). Springer, Cham. 10.1007/978-3-030-42156-4_2
18. Mozelius, P., Jaldemark, J., Eriksson Bergström, S., Sundgren, M. (2020) The Concept of 'Bringing Your Own Device' in Scaffolded and Augmented Education. In: Geroimenko V. (eds) *Augmented Reality in Education* (*Springer Series on Cultural Computing*). Springer, Cham. 10.1007/978-3-030-42156-4_4
19. Naps, L., Eagan, J. R., and Norton, L. L. (2000) JHAVE: An environment to actively engage students in web-based algorithm visualizations. In Proceedings of the SIGCSE Session, pages 109–113, Austin, Texas, Mar. ACM Press, New York.
20. Bravo, C., Mendes, A.J., Marcelino, M.J., and Redondo, M.A. (2003) Animation and Synchronous Collaboration to Support Programming Learning (pp. 509–513). Proceedings of Second International Conference on Multimedia, Information and Communication Technologies in Education (m-ICTE'2003), Badajoz, Spain.
21. Halim, S. (n.d.) Visualgo – Visualising Data Structures And Algorithms Through Animation. [online] Visualgo.net. Available at: <https://visualgo.net>.
22. Klling, M., Quig, B., Patterson, A., and Rosenberg, J. (2003) "The BlueJ System and Its Pedagogy." *Journal of Computer Science Education, Special issue on Learning and Teaching Object Technology* 13(4).
23. Bednarik, R., Moreno, A., and Myller, N. (2005) Jeliot 3, an Extensible Tool for Program Visualization. 5th Annual Finnish / Baltic Sea Conference on Computer Science Education. November 17 – November 20. Turku Centre for Computer Science.
24. Ali, J. (2009). A Visualization Tool for Data Structures Course. Proceedings – 2009 2nd IEEE International Conference on Computer Science and Information Technology, ICCSIT 2009 (pp. 212–216). 10.1109/ICCSIT.2009.5234389.

14

Information Technology for Student Decision Support in College Planning

Robert R. Meyer[1], Athula D. A. Gunawardena[2], Sobitha Samaranayake[2], Vivek Deshpande[3], and Kirthi Premadasa[4]

[1]*University of Wisconsin-Madison, Madison, Wisconsin, USA*
[2]*University of Wisconsin-Whitewater, Whitewater, Wisconsin, USA*
[3]*University of Wisconsin, Madison, Wisconsin, USA*
[4]*University of Wisconsin-Platteville, Baraboo Sauk County, Wisconsin, USA*

14.1 Introduction

This chapter deals with the application of information technology (IT) in providing decision support to students facing the challenging problem of college degree planning (CDP), which is the process of determining, in order to satisfy requirements for a specific college degree at a given institution, a set of courses and a schedule for completing those courses. Although flexibility in the choice of subjects and majors to study has many benefits for students, CDP is especially important and difficult in the United States, where many students enter college without deciding on a focus for their studies and thus are faced with a myriad of potential *majors* (with possible majors often numbering in the hundreds at large universities). In order to select a major, students should carefully consider their interests and the requirements for that major. These requirements, especially in the United States at the current time, are often quite complex and may challenge the student to choose from a variety of rather different *major paths*, each of which corresponds to a set of courses that meets the requirements of the major. Even if a student has decided on a major, he or she must additionally determine how to deal with degree requirements regarding the courses needed to complete other required areas of study beyond the major. These complex U.S. higher education issues also affect a large number of students outside of the United States who wish to complete a college degree in the United States, often without the benefit of advice from peers or parents who have experience with U.S. colleges. While students outside the United States often have much less flexibility in choice and sequencing of courses, many countries are now changing their policies in this regard and moving toward greater latitude in course selection. For example, the National Education Policy 2020 [1],

DOI: 10.1201/9781003166702-14

which was approved by the Union Cabinet of India on 29 July 2020, envisions a new educational system to meet the needs of the 21st century. Some salient features of this new education policy include: (a) flexibility in the choice of subjects to study so that students can design their study and life plans; (b) emphasis on life skills such as communication, co-operation, teamwork, and resilience; (c) learning that is holistic, integrated, enjoyable, and engaging; (d) critical thinking and more holistic, inquiry-based, discovery-based, discussion-based, and analysis-based learning; and (e) development of the creative potential of each individual. These changes in the education policy of India will empower students to discover their own courses of study via decision support tools such as interest-aligned (IA) planning, which helps students understand and visualize degree paths based on their interests, skills, and aptitude.

This chapter focuses on the utility of IT for decision support in two closely related areas: (1) *major selection* (MS), a problem faced by pre-college students worldwide and also by the significant number of U.S. students (so-called *undeclared* students) who have enrolled in college without having chosen a focus for their studies, and (2) CDP, in which case it is assumed that a college major has been selected (or at least is being considered), but the set and sequencing of courses required to complete that major are yet to be determined, along with the courses needed to satisfy additional degree requirements beyond those for the major. These additional requirements often include (a) the selection and completion of a *minor* in a sub-area (which often may not need to be very closely related to the major) and (b) the completion of so-called *general education* (GE) requirements spread across a wide variety of topics. The technology that we discuss below focuses on two key aspects of college planning: (1) selection of areas and courses that are maximally aligned with a student's interests, which will be termed *interest-aligned* (IA) planning, and (2) completion of degree requirements in the shortest possible time, which will be termed *shortest-path* (SP) planning. For IA planning, we use a novel unified approach introduced in Gunawardena et al. [2] for decision support for MS and IA CDP. This approach is based on the concept of addressing IA planning by assigning college courses and college majors *Holland Profiles* (HP) (named after the psychologist John Holland – an HP is a numerical six-vector quantifying the level of certain characteristics, to be described below). Thus, recommendations for majors and courses are provided to a given student based on alignment of courses with that student's personal HP, where the latter is determined by a psychometric test. We then introduce tools for SP and show how IA and SP planning may be combined. For an alternative graph-based approach to CDP, we discuss visualization tools introduced in Samaranayake and Gunawardena [3] that allow students to easily visualize degree paths. Finally, we consider the utility of artificial intelligence (AI) tools for MS and CDP.

14.2 Challenges and Opportunities in College Planning

14.2.1 College Planning Is Hard for Undecided Students

To illustrate the complexity of the MS and CDP problems, let us consider a couple of typical institutions. The college catalog of the University of Wisconsin–Whitewater (UWW), a mid-sized U.S. public institution with about 11,000 students, has 574 pages of course information and degree requirements covering more than 50 possible majors and about twice that many possible minors. The University of Wisconsin–Madison, a large public institution with more than 43,000 students, has a 1,578-page catalog representing 232 majors and certificates. Even within a specific major at these institutions, a student generally has a choice of focus areas within the major. Thus, at UWW there are about 15,000 different computer science (CS) course sequences that meet the CS major requirements for a BS degree—as we will see below, some of these paths emphasize theory whereas others emphasize hands-on software development. In addition to deciding upon and meeting the requirements of both a major and minor, a student wishing to complete a degree at Wisconsin state colleges is faced with the challenge of selecting courses that meet GE requirements that specify the minimum number of credits needed in a variety of additional sub-areas. At UWW, for example, the GE requirements for a BS degree involve eight sub-areas: communication skills, calculation skills, quantitative and technical reasoning, cultural heritages, world of ideas, communities, physical health and well-being, and racial/ethnic diversity). These levels of complexity of choice are typical of public institutions, and, of course, a student who is considering applying to multiple institutions is faced with the additional problem of comparing the offerings and requirements presented in many college catalogs.

14.2.2 Information Technology (IT) Tools for College Planning

As noted above, a key technology that we describe here to assist with both MS and CDP is based on the comparison of HP data for persons and environments (in the application considered here, "environments" are college courses, college majors, college minors, etc.; Holland's original focus and the main current application of his approach is the matching of persons with careers). Holland's theory, introduced in Holland [4], classifies people and environments using rank ordering of six types – Realistic, Investigative, Artistic, Social, Enterprising, and Conventional (RIASEC) – and employs congruence measures to determine the degree of alignment between an individual's personality profile and the environment. In the approach presented in Gunawardena et al. [2], college courses are considered as environments in which students complete activities over a period of time. We show how the appropriate application of Holland's theory can help students select and plan completion of degrees through courses that are well aligned with their interests and thereby secure more satisfaction and success in their studies. Specifically, we adapt an optimization approach introduced in Gunawardena et al. [2] that has been implemented to provide an IA decision support system (DSS) for both MS and CDP, and we show how this methodology can be extended to a hybrid objective that addresses both IA and SP planning. Additional novel technologies are also described, including AI approaches and Dependency Evaluation and Visualization (DEV) [3], a graphical data visualization tool that enables students and advisors to see and easily understand course prerequisite structure and to readily determine paths that lead to the satisfaction of degree requirements.

14.3 Interest-aligned (IA) Planning: Extending Holland's Theory to College Courses for Decision Support

14.3.1 An Overview of Holland's Theory and Its Applications

As mentioned previously, Holland's theory, introduced by John Holland in 1959, classifies people and environments using rank ordering of six types: Realistic, Investigative,

Artistic, Social, Enterprising, and Conventional (RIASEC). These types are discussed in the two sections below. The theory goes on to employ congruence measures to find the degree of alignment between a profile of an individual's personality types (this profile is usually determined via psychometric tests) and the work (or educational) environment types (the environment profile may be determined by detailed assessments or other mechanisms as discussed below).

14.3.2 The Six Holland Personality Types and Their Assessment

14.3.2.1 Personality Types

Holland theorized that the six personality types that he considered were key to interactions with a person's environment and represented common personal traits that were the result of growing up in a particular setting. Each personality type definition that he formulated involved the acquisition of a particular set of competencies and, conversely, a deficit of a set of competencies associated with another personality type. These opposite personality types correspond to diagonally opposite trait pairs on what is called Holland's hexagon [5], as shown in Figure 14.1. We summarize the personality types associated with Holland's personality types given in Holland [5]:

Realistic: Doers who have mechanical ability and prefer to work with objects, machines, tools, plants, or animals. They prefer to work independently or outdoors and are frank, hands-on, and practical.

Investigative: Thinkers who like to observe, learn, investigate, analyze, research, or solve problems. They are generally reserved, independent, and scholarly.

Artistic: Creators who have artistic, innovative, or creative abilities. They like to work in unstructured environments, using their imagination or originality.

Social: Individuals who want to aid people by informing, training, or developing them. They are skilled in speech, concerned with the welfare of others, and compassionate.

Enterprising: Persuaders who like to interact with others by directing, influencing, or persuading. Their focus is generally on organization goals or economic gain. They are adventurous, outgoing, and energetic.

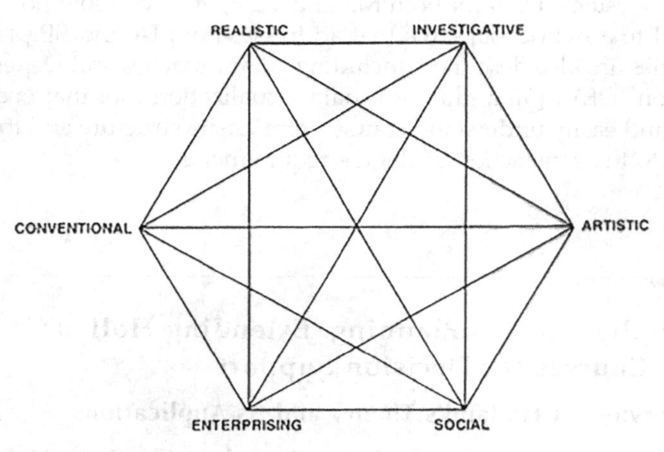

FIGURE 14.1
The six Holland traits displayed on the Holland hexagon.

Conventional: Well-organized persons who prefer to work with data and have corresponding clerical or numerical ability. They like structured environments and perform tasks in detail. They prefer working at a desk or office and are careful and conforming.

14.3.2.2 Personality Assessment and Holland Profile

The relative levels of these six traits in an individual are generally assessed by administering a psychometric test consisting of 60 or so questions. Another, but less preferable, option for profile assessment is direct assignment by an individual of relative values for the traits. In either case, the result is a numerical six-vector (with values typically in the 0–100 range for each trait) that we will call a *Holland Profile* (HP), indicating the relative strengths of the traits. As we will discuss below, an individual's HP may then be compared to HPs (also numerical six-vectors) associated with college courses in order to determine those courses that are most aligned with an individual's profile.

14.3.3 The Six Environmental Types and Their Assessment

14.3.3.1 Environmental Types

Holland's theory also defines six types of environments. In the application considered in this chapter, environments correspond to college courses or sets of college courses (we shall refer to these as *academic environments*). We restate the characterizations of environments used by the U.S. government on its O*NET career database site (https://www.onetonline.org/).

Realistic: Realistic environments frequently involve activities that include practical, hands-on problems and solutions. They often deal with plants, animals, or real-world objects such as tools and machinery. Many of the environments require working outside and do not focus on paperwork or working closely with others.

Investigative: Investigative environments frequently involve working with ideas and require an extensive amount of thinking. These environments can involve searching for explanations and figuring out problems mentally.

Artistic: Artistic environments frequently involve working with forms, designs, and patterns. They often require self-expression and assignments that can be accomplished without following a clear set of rules.

Social: Social environments frequently involve working with, communicating with, and teaching people. These environments often focus on helping or providing service to others.

Enterprising: Enterprising environments frequently involve starting up and carrying out projects. These environments can involve leading people and making many decisions. They often deal with business issues.

Conventional: Conventional environments frequently involve learning and following set procedures and routines. These environments generally include working with data and details rather than with ideas. Usually there is a clear line of authority to follow.

14.3.3.2 Assessing Holland Profiles for Courses and Sets of Courses

We consider college courses as environments in which students are expected to complete activities assigned by an instructor. The HP for a course can be assigned directly by domain experts via assessments of syllabi, course objectives, weekly assignments, etc., or it may be determined indirectly.

The Environmental Assessment Technique (EAT) [6,5] indirectly characterizes an environment by assessing the people in that environment. EAT assesses the most salient type of each person and uses the percentages of types as the Holland values for the environment. In Gunawardena et al. [2], the EAT concept is extended via an optimization model that can be solved to obtain an indirect assessment of a course. This model is a constrained least square problem based on student HPs and student course evaluations, and it is defined as follows:

Let $P = \{p_1, p_2, \ldots, p_m\}$ be a set of m students and let A be an $m \times 6$ matrix in which the jth row contains the transpose of the Holland personality vector, p_j^T. Let b be the vector in which the jth element contains the alignment score between −100 and 100 for the course assigned by person, p_j, $1 \le j \le m$. The solution vector, x, of the following constrained least squares minimization problem, where $\| \; \|_2$ denotes the 2-norm, is taken to represent the Holland vector for the course.

$$\text{Minimize} \|Ax - b\|_2^2$$

subject to

$$\sum_{i=1}^{6} x_i = 100$$

$$x_i \ge 0, \; 1 \le i \le 6.$$

The above quadratic programming model (see Nocedal and Wright [7] for a taxonomy of optimization models) can be solved quickly with a solver available in GAMS [8], IBM ILOG CPLEX Optimization Studio [9], or Gurobi [10], etc.

HPs for sets of courses corresponding to all the courses in a college major, or for a set of courses that satisfy the degree requirements of a major or minor, may be determined by a credit-weighted average of HPs of courses.

14.4 Congruence Measures for Determining Person–environment Alignment

Once a database of HPs has been established for both students and academic environments, a mechanism is needed to quantify the alignment between a particular student HP and an instance of an academic environment. More generally, algorithms for measuring alignment of people and environments are termed *congruence measures* (CM) and have been the subject of study and debate for more than 60 years.

In Gunawardena et al. [2], in addition to a new congruence measure, DP, 16 other congruence measures found in the literature are summarized. An earlier comparison of 12 of them can be found in Camp and Chartrand [11]. These existing measures have received mixed reviews in the literature. Although some of these measures have common features (i.e., Holland hexagon-related properties), these commonalities are not readily apparent due to the differing variable domains and structures of formulas used in the definitions of these measures. Gunawardena et al. [2] adapts and unifies these measures based on a vector–matrix framework that can be directly translated to a Matlab/Octave programming code, and also determines the correlation of these measures.

We utilize the measure DP in our system since it takes into account the fact that, in order to meet degree requirements, students must often select and complete courses in topics or areas that are not well-aligned with their personal HPs. The DP measure takes into account in a quantitative manner the interest types in a personal HP that are low ranked but which may be quite relevant to courses that must be completed to meet degree requirements. This feature enables a student to minimize the mismatch between his or her personal interests and courses selected to meet requirements. In this respect, academic environments differ from career environments (the most common area of application for Holland theory), which are generally more narrowly focused in terms of needed skills and responsibilities.

14.5 Decision Support for Major Selection and College Degree Planning

14.5.1 Major Selection (MS)

As noted above, MS is a problem faced by pre-college students worldwide and also by the significant portion of U.S. students who have enrolled in college without having chosen a focus for their studies. (Some studies have indicated that as many as 50% enter college as "undecided" students.) The MS problem is especially difficult for first-generation students and students who have little or no access to counseling services. The difficulty of the MS problem is further evidenced by the fact that an estimated 75% of U.S. college students change their major at least once before graduation [12]. The goal of our research is thus to provide students with a DSS that will help them to focus their consideration on those options that are well-aligned with their personalities, and thus empower them to make early and well-informed decisions that they are likely to adhere to throughout their college program.

Students are not only faced with the need to select a major from hundreds of options, but they often must take into account that for any given major there are generally a large number of possible course sets (called *major paths*) within the major that satisfy the requirements for that concentration. These major paths may correspond to very different foci, even in the case of technical majors. We have noted, for example, that even in the case of a medium-sized institution such as UWW, there are about 15,000 different CS course sequences that meet the CS major requirements for a BS degree, with some paths emphasizing theory (and thus requiring strong mathematical ability) while others emphasize team-based, hands-on code development. Thus, a DSS for MS should take into account the considerable variation that may exist within a major and respond to user queries with a ranked list of majors whose alignment scores are based on the most aligned major paths. The DSS user should also be provided with information about the course sets in the most-aligned paths when he or she requests further information about a particular major.

For a given college, a hashing grid structure for a database that may be efficiently accessed by a DSS is described in Gunawardena et al. [2]. This grid provides the user with clusters of major paths that are ranked according to the degree of alignment with the user's HP.

AI offers additional tools that could be applied to MS. From an AI viewpoint, we would consider HPs as input features that could be augmented by high school and college grades and aptitude scores to create student profiles. Data could then be collected on a per-student basis on majors of interest and majors selected. Conventional AI tools [13], such as k-Nearest-Neighbor, Artificial Neural Networks, or Support Vector Machines, could then be tested on this data to ascertain their utility in suggesting majors to new users based on their profiles.

This approach could be refined by collecting data from students who have already selected a major and have taken courses in the major. In that case, those students could be polled about their satisfaction with the major. This data, along with their profiles, could be used in AI models to predict student satisfaction with given majors. A similar approach could be used to predict satisfaction with respect to individual courses.

14.5.2 Optimization Models for College Degree Planning (CDP)

14.5.2.1 Shortest-path and Interest-aligned CDP

Recall that we assume for CDP that a particular college major has been selected. With an efficient technology for CDP, a student who has not yet decided on a course of study may investigate a collection of degree plans for a variety of majors and minors. The goal of CDP is to determine a set of courses that satisfies the requirements of the major (and minor, if required) and all other university requirements for the desired degree. CDP is even more complex than MS because overall degree requirements typically encompass a broad variety of diverse areas. In addition, previously completed courses (in both high school and college) must be taken into account for CDP as they may be used to at least partially satisfy some requirements, and course prerequisites must be recognized. Finally, if semester-by-semester plans are desired, which we will assume is the case, it is also necessary to know course rotation schedules that specify which courses are offered in which semesters.

We consider here two key aspects of CDP: shortest paths to degree completion and selection of courses that are maximally aligned with the student's HP. The SP/IA multiplan system, as described below, represents an integrated approach that addresses both of these aspects of CDP for a given university and is designed around five main components/tasks that in combination provide the user with the tools needed to easily investigate a broad variety of degree plan options:

The SP/IA Multiplan System

1. The user completes a *Holland interest assessment quiz* that provides an HP for the user.
2. If the user has not decided on a major or minor, a mapping is done, as described above, of the user HP vector into a recommended set of majors and minors available at the university that are well-aligned with that vector and from which the user selects preferred majors and minors.
3. An *automated degree planner (ADP)* that uses state-of-the-art optimization techniques and dynamic priority scheduling (discussed below) is applied to a database that includes courses labeled with their HPs. ADP efficiently generates a collection of SP semester-by-semester degree plans corresponding to the selected majors and minors. These plans are aligned with the user's HP. The system automatically varies a set of internal parameters that govern the course selection

process so that a large number of plans, e.g., 100, are generated and stored for review by the user.

4. A *clustering* method decomposes the collection of degree plans into small subsets (clusters) of related plans, labels each cluster with a six-letter Holland code interest vector corresponding to the cluster center, and presents these cluster labels to the user in the order of their proximity to the user's HP.
5. A degree plan *selection module* allows the user to view and compare degree plans, and then select and save one or more for further review with an advisor.

Having previously discussed components (1) and (2), we focus in this section on SP and IA technology for CDP. An informal statement of an IA optimization formulation of CDP for a particular user HP is as follows:

$$\text{Maximize} \sum_{\textit{courses } C_i \textit{ in an eligible course set}} DP(P, HP(C_i))$$

Here, DP represents the DP congruence measure, and the courses C_i appearing in the summation are the members of a set of courses that, together with courses already completed, satisfy all degree requirements and meet prerequisite and course rotation conditions as well as user-set semester credit limits. This is a combinatorial optimization problem that can be very large in size (taking into account that even a mid-sized college may offer thousands of courses and that a distinct binary variable is needed for each eligible course in each semester); hence, it presents a substantial challenge in terms of solvability. CDP is better addressed by approximating it by smaller problems. The following ADP algorithm takes into account the need to decompose CDP into more manageable semester-sized components as well as the desire to satisfy degree requirements as quickly as possible.

Automated degree planner (ADP) algorithm

1. According to the input selections (i.e., majors, minors, degree type (e.g., BA or BS)), set the corresponding requirements for the degree, and generate a set S^- of all the courses that appear in the requirements for the major and the prerequisites for these courses. (Note that the complexity of this step may be reduced by using generic courses for each area specified in the GE requirements, since these areas offer the greatest flexibility in choice of courses.)
2. Apply the user's completed courses to the requirements for the *major*, and generate a subset (S^*) of S^- that completes the major requirements. Based on recommendations from the major department, a limit (e.g., 2 or 3) may be set on the number of courses in the major allowed per semester. (This feature is intended to ensure that a student is not overwhelmed by taking too many demanding courses in one semester, but this value can be overridden in consultation with an advisor.)
3. A binary optimization problem is then solved to determine the minimum number of remaining semesters required to complete the major requirements (taking into account prerequisites and course rotation schedules). This number is then compared with the minimum number of semesters needed to complete the degree credit requirement (typically 60 credits for a two-year degree and 120 credits for a four-year degree), and the larger of these two numbers is used to set D, the semesters-to-degree value.

4. Apply the user's completed courses to the remaining set of degree requirements (these may include requirements for the *minor*, GE requirements, and minimum credit requirements), and generate a set T^- of all the courses that appear in the requirements and their prerequisites.
5. Generate a subset (T^*) of T^- that suffices to complete the remaining degree requirements while minimizing the total number of credits. This step requires the resolution of requirement conflicts (i.e., a course may be required by both a major and minor but it can be applied to one of those). ADP also selects a unique set of prerequisites for each course. The minimization is done with a heuristic algorithm that uses randomized sequences of T^- to generate potential T^*s.
6. A dynamic priority scheduling (DPS) algorithm is used to produce optimized semester plans. This involves the following three steps. For each future semester *i*:
   a. Construct the prerequisite-rotation graph (described below) for T^*.
   b. Calculate priority weights of the courses in T^*.
   c. Solve a knapsack problem [14] corresponding to maximizing a hybrid objective that includes priority weights and congruence measures. This produces an optimal set of courses (T^{**}) that are assigned to the current semester *i*. Remove the courses in T^{**} from T^* and start with step 5(a) for semester $i + 1$ if all degree requirements are not met.
7. Steps 5(a)–(c) are repeated until all courses needed to obtain a valid degree plan are determined. If more than D semesters are needed to obtain a valid degree plan, a warning will be attached to the plan advising the user that the plan does not have the desired number of semesters.

Each iteration of the DPS algorithm thus corresponds to a semester in the degree plan and selects the courses that will appear in the plan for that semester. DPS accomplishes this course selection by first constructing at each iteration a prerequisite rotation graph (PRG) and then using this graph to calculate priority weights for courses based on longest paths, prerequisite dependencies, and course rotation schedules. Details of this algorithm are complex and are given in Gunawardena et al. [15]. Figure 14.2 shows an example of a PRG for a set of 21 courses *C1-C21* spread over six semesters with completed courses, prerequisites, and course rotations as shown in Table 14.1. In the PRG, arcs indicate prerequisites, node labels correspond to priority weights, and arc labels show semester spacings.

The DPS algorithm then solves a knapsack problem based on these dynamic priority weights as follows:

Let P be a person and $C_1, C_2, \ldots, C_t$ be a collection of courses that are available in the semester S_i under consideration and $pw_1, pw_2, \ldots, pw_t$ be their priority weights, and let $c_1, c_2, \ldots, c_t$ be their corresponding credit hours. Let U_i be the maximum credit limit for the semester S_i. We assign courses to the ith semester by using dynamic programming to solve the following knapsack problem, where the binary variables *x[j]* have value 1 if course C_j is taken in the semester and 0 otherwise:

$$\text{Maximize } z = \mu \textstyle\sum_{j=1}^{t} pw_{j*}c_{j*}x[j] + \lambda \textstyle\sum_{j=1}^{t} DP(P,\ HP(C_j))_{*}x[j]$$

Subject to

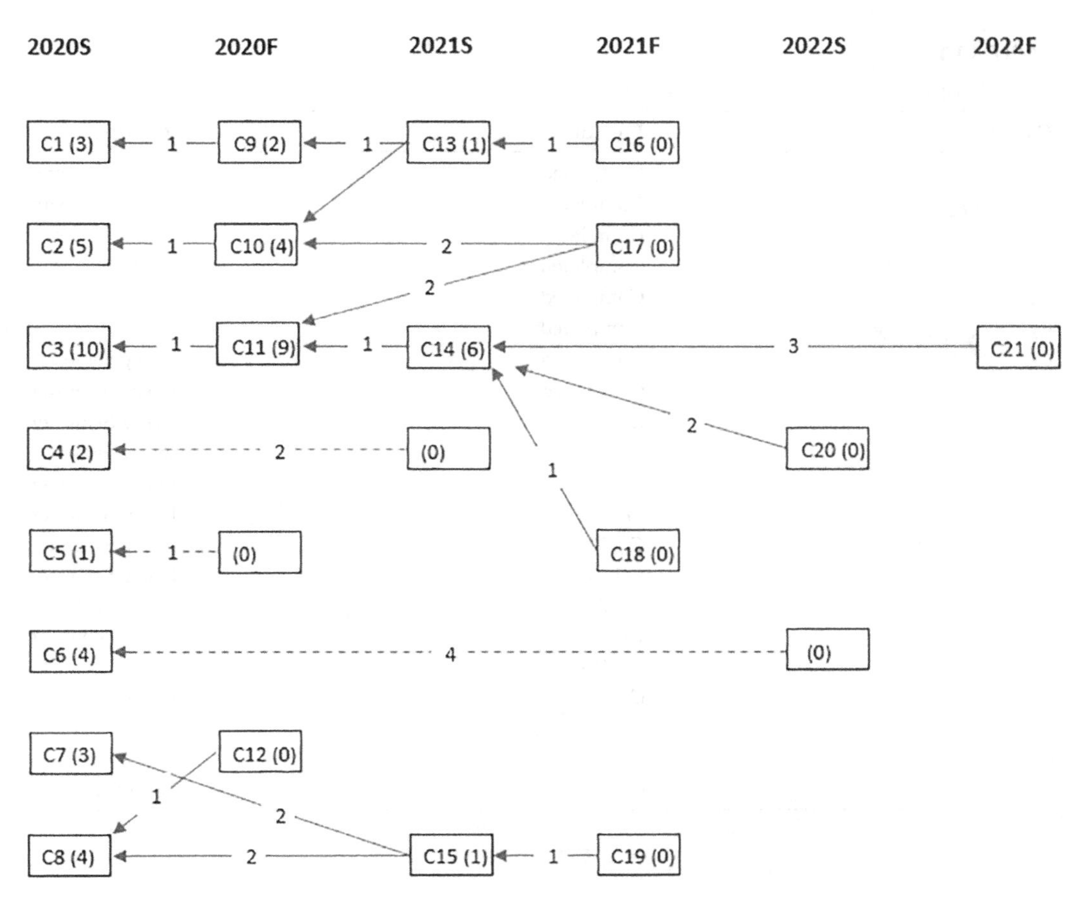

FIGURE 14.2
A Prerequisite Rotation Graph.

1. Semester credit limits: $\sum_{j=1}^{t} c_{j*}x[j] \leq U_i$
2. Other constraints based on user preferences such as limits on number of GE courses.

Here μ and λ are controlling parameters that can be used to generate different degree plans by adjusting the trade-off between the two competing objective components, SP and IA.

14.5.3 An Implementation of CDP via Visualization Models for IA and SP

We used C++/Linux to implement the ADP algorithm and custom-made tools to extract and insert degree requirements into a MySQL database. The Angular framework [16] was chosen as the platform for developing a DSS, and a PHP framework [17] was used for creating web services for the Angular framework. Course prerequisite structure for each major was stored in a database using an adjacency matrix of a directed graph *D(V, E)* where the set of nodes (*V*) represents courses and the set of edges (*E*) represents prerequisite relationships.

Figure 14.3 shows the course structure for the CS major at UWW, prior to completing any of the courses in the major. Nodes with a stack of courses represent prerequisite

TABLE 14.1

A Table of Prerequisites and Course Rotations

Course	Prerequisites	Rotation
C_1	Completed	Every Semester
C_2	Completed	Every Semester
C_3	Completed	Every Semester
C_4	Completed	Every Spring
C_5	Completed	Every Semester
C_6	Completed	Spring Even Years
C_7	Completed	Every Semester
C_8	Completed	Every Semester
C_9	C_1	Every Semester
C_{10}	C_2	Every Semester
C_{11}	C_3	Every Semester
C_{12}	C_8	Every Semester
C_{13}	C_9, C	Every Semester
C_{14}	$C_1 1$	Every Semester
C_{15}	C_7, C_8	Every Spring
C_{16}	C_{13}	Every Semester
C_{17}	C_{10},C_{11}	Every Fall
C_{18}	C_{14}	Every Semester
C_{19}	C_{15}	Every Semester
C_{20}	C_{14}	Every Spring
C_{21}	C_{14}	Fall Even Years

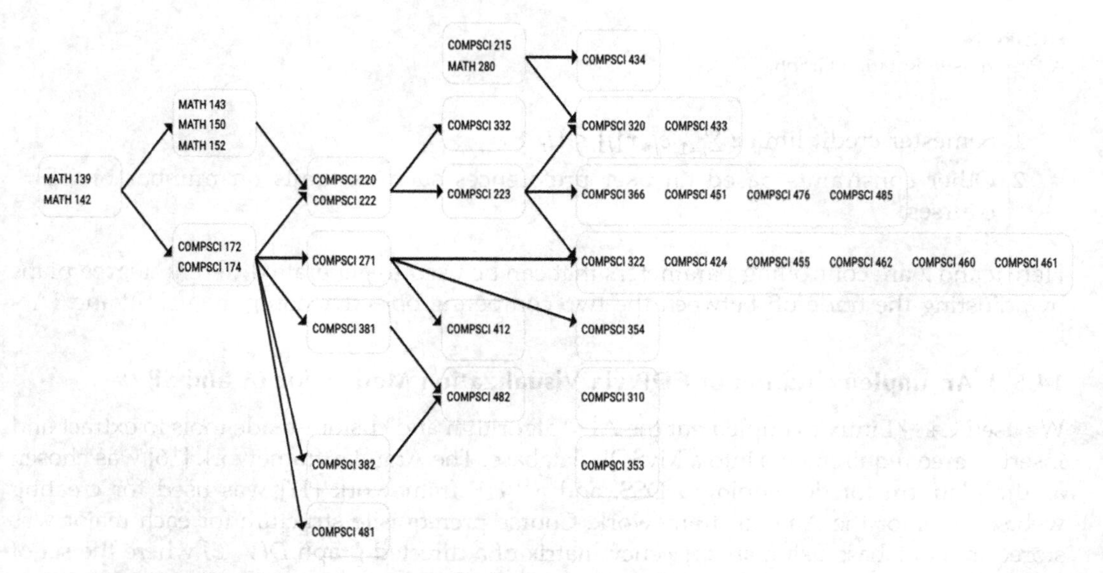

FIGURE 14.3
Course Structure for the Computer Science Major at UWW.

courses where only one of the courses needs to be taken to satisfy the prerequisite. If two or more arrows are pointing to the same child node, then each of the prerequisite relationships must be satisfied for the course list attached to the child node to be available.

A typical degree requirement belongs to one of the following categories:

- Type A: complete k courses from a set of p courses where $1 \le k \le p$
- Type B: complete at least m courses/units, but no more than n courses/units from a set of p courses where $0 \le m \le n \le p$
- Type C: complete k units from a set of p courses
- Type D: combination of Type A, Type B, and/or Type C requirements

There may be other requirements, such as GPA requirements, minimum number of credits/units needed to complete, internships, capstone projects, etc. Figure 14.4 shows the semester planning page of the DSS system implemented at UWW.

The semester planning page consists of a course prerequisite structure, list of the degree requirements, an indication of whether each requirement has been satisfied, and courses

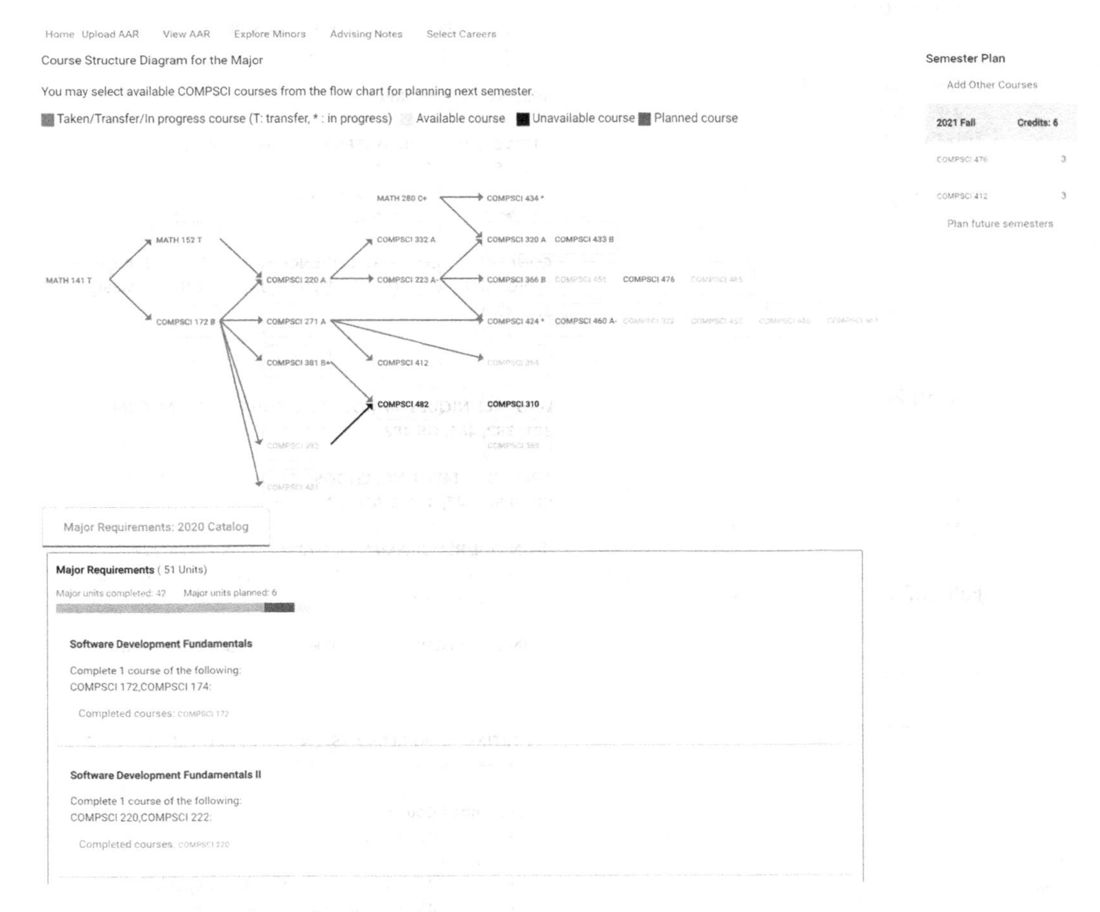

FIGURE 14.4
Semester Planning Page of the DSS System.

Semester Plan

Planned Semesters

Spring 2021

COMPSCI 215	3
COMPSCI 220	3
COMPSCI 271	3

Fall 2021

MATH 250	5
COMPSCI 223	3

Spring 2022

COMPSCI 460	3
COMPSCI 322	3
ENGLISH 090	3

Fall 2022

ENGLISH 101	3
ENGLISH 102	3
COMPSCI 310	3

Spring 2023

COMPSCI 461	3
COMPSCI 320	3
COMPSCI 412	3

Fall 2023

COMPSCI 433	3
ENGLISH 370	3
COMPSCI 476	3

Completed/Waived Courses: Show

Major Requirements

Core Courses: COMPSCI 172 OR COMPSCI 174
- COMPSCI 172 - Introduction To Java

Core Courses: COMPSCI 220 OR COMPSCI 222
- COMPSCI 220 - Intermediate Java

Core Courses: COMPSCI 223, COMPSCI 271, COMPSCI 366, COMPSCI 412, COMPSCI 433, COMPSCI 476
- COMPSCI 271 - Assembly Programming
- COMPSCI 223 - Data Structures
- COMPSCI 412 - Computer Organization And System Programming
- COMPSCI 433 - Theory Of Algorithms
- COMPSCI 476 - Software Engineering

SELECT ANY COMPSCI COURSE NUMBERED 300 OR ABOVE
- COMPSCI 310 - Intermediate Data Science

SPECIALIZATION AREAS

NETWORKING AND SYSTEMS: CHOOSE TWO COURSES FROM: COMPSCI 460, 461, OR 462
- COMPSCI 460 - Computer Networking
- COMPSCI 461 - Mobile Computing Architecture

CONCEPTUAL COMPUTER SCIENCE: COMPUTER SCIENCE FUNDAMENTALS: CHOOSE TWO COURSES FROM: COMPSCI 322, 332; MATH 450
- COMPSCI 322 - Computer Languages And Compilers
- COMPSCI 320 - Concepts Of Programming Languages

WEB TECHNIQUES: CHOOSE TWO COURSES FROM: COMPSCI 381, 382, 481, OR 482

APPLIED COMPUTING: CHOOSE TWO COURSES FROM: COMPSCI 347, 451; MATH 471

UNIQUE REQUIREMENTS: MATH 250 OR MATH 253
- MATH 250 - Applied Calculus Survey For Business And The Social Sciences

UNIQUE REQUIREMENT EITHER COMPSCI 215* OR MATH 280
- COMPSCI 215 - Discrete Structures

WRITING REQUIREMENTS: ENGLISH 370 OR ENGLISH 372
- ENGLISH 370 - Advanced Composition

Other Planned Courses
- MATH 040 - Pre-algebra
- MATH 041 - Beginning Algebra
- MATH 141 - Fundamentals Of College Algebra
- MATH 143 - Finite Mathematics For Business And Social Sciences

(Caption)

FIGURE 14.5
Automated Degree Plan using SP.

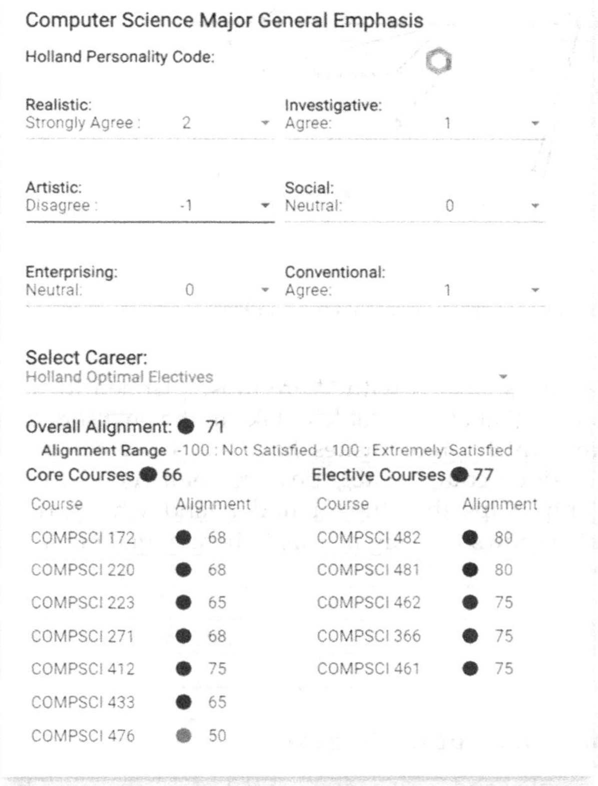

Computer Science Major General Emphasis

Holland Personality Code:

Realistic: Strongly Agree : 2
Investigative: Agree: 1
Artistic: Disagree : -1
Social: Neutral: 0
Enterprising: Neutral: 0
Conventional: Agree: 1

Select Career:
Holland Optimal Electives

Overall Alignment: ● 71
Alignment Range -100 : Not Satisfied 100 : Extremely Satisfied

Core Courses ● 66

Course	Alignment
COMPSCI 172	● 68
COMPSCI 220	● 68
COMPSCI 223	● 65
COMPSCI 271	● 68
COMPSCI 412	● 75
COMPSCI 433	● 65
COMPSCI 476	● 50

Elective Courses ● 77

Course	Alignment
COMPSCI 482	● 80
COMPSCI 481	● 80
COMPSCI 462	● 75
COMPSCI 366	● 75
COMPSCI 461	● 75

FIGURE 14.6
An Optimal Interest-aligned Major Path Selected by Our Approach.

credited toward satisfying each requirement. The course structure is updated dynamically to narrow down the major path choices, based on the completed and planned courses.

Figure 14.5 shows an automated degree plan generated using SP. Here, we have set $\mu = 1$ and $\lambda = 0$ in the above knapsack problem.

Figure 14.6 shows an optimal IA CS major path for a person with an HP vector, (2, 1, −1, 0, 0, 1), where the range for a Holland trait is taken to be [−2,2]. The optimal path shows good overall alignment with the courses selected (71 over a [−100, 100] interval) and generally good individual course alignments.

Figure 14.7 shows an IA major path flowchart that allows a student to see the hidden prerequisite courses needed for the selected major path in computer science (i.e., COMPSCI 381 and COMPSCI 382 are prerequisites for COMPSCI 482). In this example, the student has already taken (or is currently taking) the courses shown in bold. Course grades are displayed and * is a placeholder for grades for the courses that are in progress. The courses bracketed by underlines and overlines are those whose prerequisites are

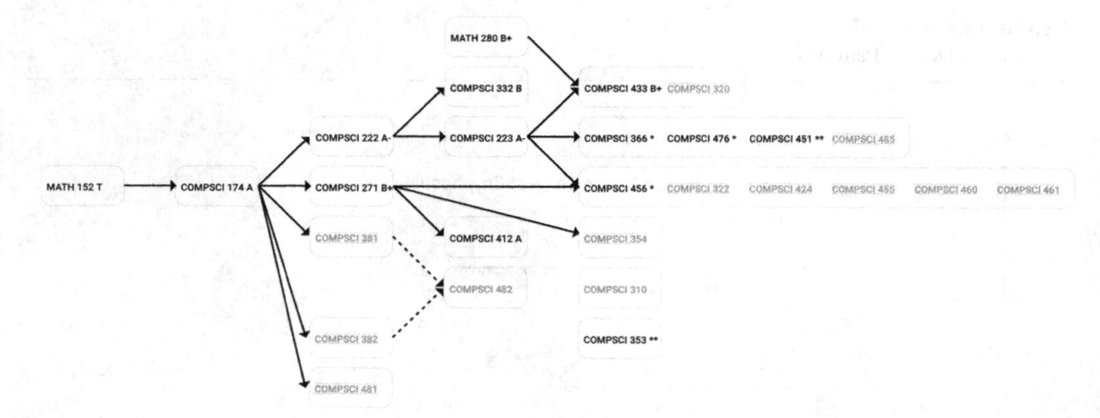

FIGURE 14.7
Interest-aligned Major Path Flowchart for a Student Who Has Completed Several Courses.

satisfied, and the courses labeled with ** are those planned for the next semester. Solid arrows point to courses that are available to take in the next semester, and dashed arrows indicate courses for which prerequisites have not been completed. In the actual implementation the various course categories are color-coded for easy recognition. This visualization not only helps the student understand why particular courses were recommended, but also shows the "big picture", thereby making it easier for the student to consider alternatives.

14.6 Conclusions and Future Research

MS and CDP are complex and very consequential problems faced by students worldwide. Appropriate use of tools from psychology and IT can provide students with the information that they need to make well-informed decisions regarding these key problems. With appropriate user interfaces, this information can be presented in a manner that is easy to understand and navigate. Key challenges for colleges in terms of providing the data needed for these tools include the development of appropriate profiles for courses and the encoding of degree requirements via standard structures. Most of the work associated with these data challenges for colleges can be automated, but care needs to be taken in terms of assessing and validating the results as well as performing updates whenever there are curricula changes. Long-term studies of student success and satisfaction with their education and careers are also needed in order to validate the utility of the DSS proposed here.

References

1. Ministry of Human Resource Development 2020. Final National Education Policy 2020. https://static.pib.gov.in/WriteReadData/userfiles/NEP_Final_English_0.pdf

2. Gunawardena, A. D. A., Meyer, R. R., Kularatna, T., Samaranayake, S., and Gunawardena, R. H. (2021). An Interest-Aligned System for Career and College Degree Planning. Private Communication. https:// cs.uww.edu/~athula/
3. Samaranayake, S. and Gunawardena, A. D. A. (2020) "Dependency Evaluation and Visualization Tool for Systems Represented by a Directed Acyclic Graph." *International Journal of Advanced Computer Science and Applications* 11(7): 1–8.
4. Holland J. L. (1959) "A Theory of Vocational Choice." *Journal of Counseling Psychology* 6: 35–45.
5. Holland J. L. (1997). *Making Vocational Choices*. Englewood Cliffs, NJ: Prentice-Hall.
6. Astin A.W. and Holland J. L. (1961) "The Environmental Assessment Technique: A Way to Measure College Environments." *Journal of Educational Psychology* 52(6): 308–316.
7. Nocedal, J. and Wright, S.J. (2006) *Numerical Optimization* (2nd ed.). Berlin, New York: Springer-Verlag.
8. Kallrath, J. (2004) *Modeling Languages in Mathematical Optimization* (First ed.). Norwell, USA: Kluer Academic Publishers.
9. IBM Corp. (2020) IBM ILOG CPLEX Optimization Studio 20.1.0 documentation. https:// www.ibm.com/support/knowledgecenter/SSSA5P_20.1.0/COS_KC_home.html
10. Gurobi Optimization, LLC 2021. *Gurobi Optimizer Reference Manual.* http:// www.gurobi.com.
11. Camp C. C. and Chartrand J. H. (1992) "A Comparison and Evaluation of Interest Congruence Indices." *Journal of Vocational Behavior* 41: 162–182.
12. Gordon, V. N. (1995) *The Undecided College Student: An Academic and Career Advising Challenge* (2nd. ed.). Springfield, IL: Charles C. Thomas.
13. Russell, S. and Norvig, P. (2020) *Artificial Intelligence: A Modern Approach* (4th Edition). London: Pearson Education, Inc.
14. Martello S. and Toth P. (1990) *Knapsack Problems: Algorithms and Computer Implementations.* New York, NY: John Wiley & Sons, Inc.
15. Gunawardena A. D. A., Meyer R. R., and Samaranayake S. 2017. Dynamic Priority Scheduling Based Automated Degree Planning Method and System. Private Communication. https:// cs.uww.edu/~athula/
16. Uluca, D. (2020) *Angular for Enterprise-Ready Web Applications* (Second Edition). Birmingham: Packt Publishing.
17. Welling, L. and Thomson, L. (2017) *PHP and MySQL Web Development*. London: Pearson Education, Inc.

15

Attention-based Image Captioning and Evaluation Methods

Khushboo Khurana
Department of Computer Science and Engineering, Shri Ramdeobaba College of Engineering and Management, Nagpur, Maharashtra, India

15.1 Introduction

Image understanding is a very critical topic in the field of artificial intelligence. Being able to generate a description after understanding image contents has attracted researchers from computer vision (CV) and natural language processing (NLP) communities. This task of generating a textual description expressing rich semantic information of an image is called *image captioning*. It involves the generation of a model that can recognize objects and the relationship between their attributes and activities, after which human-like image descriptions need to be generated. Despite these challenges, the problem has achieved significant performance over the past few years due to the advancement of deep neural networks.

The early works in image captioning used the encoder-decoder architecture [1–3]. It consists of two phases – encoding and decoding. The encoder extracts features from the image, and the decoder generates a sequence of words. Generally, image encoding is performed by convolutional neural network (CNN), and decoding is performed by a recurrent neural network (RNN). Figure 15.1 shows a general schematic of encoder-decoder based image caption frameworks.

However, encoder-decoder methods have limitations. These models cannot use information such as objects and other attributes from the image for caption generation. Also, these methods performed well on smaller datasets but failed to perform well on larger datasets.

To overcome these limitations, and inspired by the success of the encoder-decoder architecture with attention mechanism, attention was incorporated in the image captioning task. The attention mechanism is applied to extract image information for guiding the generation of descriptions in a multitude of works. *Attention* means to focus on the most relevant region of the image, corresponding to the text caption. For example, consider the image shown in Figure 15.2. For the generation of the word *operator,* the focus must be on the human present in the image; for the generation of the word *machine,* the focus must be on the machine in the image, and so on. This focus can be obtained by using the attention mechanism.

DOI: 10.1201/9781003166702-15

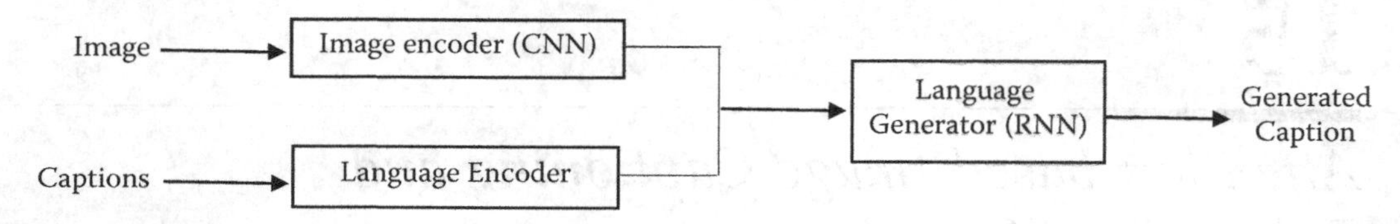

FIGURE 15.1
General Schematic of Encoder-Decoder Image-Captioning Framework.

FIGURE 15.2
Caption Corresponding to an Example Image.

Variants of pre-trained CNNs are used as the image encoder and RNNs with visual attention as the decoder [4–6] in image captioning methods with attention. CNN accepts a colored image as input. The lower layers detect the low-level visual cues like edges, whereas high-level semantic patterns like objects are detected by the higher layers. These methods utilize both features to better attend to the image details. The input to a language model for word prediction at the current time step is usually the linguistic word generated at the previous time step.

Two types of approaches are leveraged by researchers to apply attention: bottom-up and top-down. The bottom-up model is driven from data from the low-level features of the image, whereas the top-down model is driven by tasks and requires relevant prior knowledge. Various methods that utilize these approaches are discussed in this chapter.

In this chapter, we present a survey of various attention-based image captioning methods. The chapter is organized as follows. Section 15.2 presents the benchmark datasets widely used for image captioning. Attention-based image captioning techniques are discussed in section 15.3. Performance of these methods is compared and discussed in section 15.4. Various evaluation metrics used for evaluating the performance of captioning methods are discussed with examples in section 15.5. A case study on image captioning on industrial images is presented in section 15.6, followed by the conclusion in section 15.7.

15.2 Benchmark Datasets

The three most popular benchmark datasets used by image captioning researchers include Flickr8k, Flickr30k, and MSCOCO (Microsoft Common Objects in Context). Flickr8k [7] dataset comprises 8,000 images. Out of these, 6,000 images are used for training, 1,000 for validation, and 1,000 for testing. Flickr30k [8] contains 31,000 images. According to the widely used split presented by Karpathy and Fei-Fei [3], training is done on 29,000 images. The remaining 1,000 images are used for validation and 1,000 for testing. MSCOCO [9] is the largest dataset for image captioning, with 82,783 training images, 40,504 validation images, and 40,775 testing images. However, the test set does not contain the ground truth image descriptions. Hence, either the validation test is further split into validation and test set or a small subset of images are for validation and testing from the official validation set. In Karpathy and Fei-Fei [3], 5,000 images were used for validation and 5,000 images for testing. All these datasets have five reference captions per image.

15.3 Attention-based Image Captioning Techniques

Methods in attention-based image captioning literature can be broadly divided into three categories, namely spatial attention [10,11], channel-wise attention [11], and semantic attention [5,3,12]. Spatial attention, also referred to as visual attention, produces the feature of image regions corresponding to each word from the caption. Specifically, it conveys "where" information about the object with respect to the image. Figure 15.3 shows a general schematic of how attention is applied to the visual features from the image. Techniques utilize either the image features extracted from CNN or utilize RNN to extract region-based features.

Channel-wise attention recognizes "what" information about the objects. The semantic attention adopts some extra semantic concepts and attends to semantically important words. The semantic concepts can either be extracted from the image, as shown in Figure 15.4, or from the language model. Semantic attention involves combining visual attributes with the visual features in the language model.

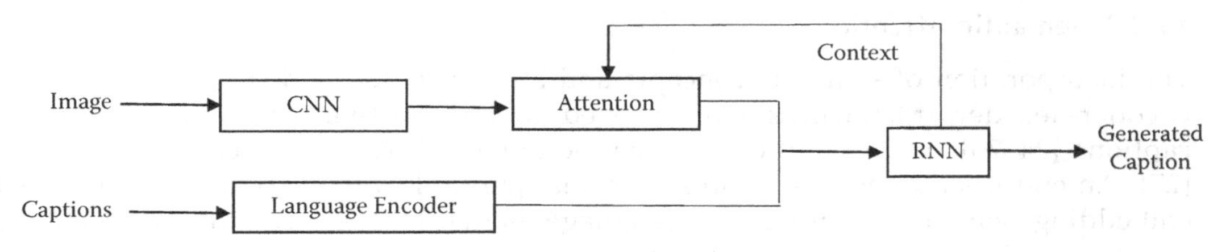

FIGURE 15.3
General Schematic of Image-captioning Framework with Spatial (Visual) Attention.

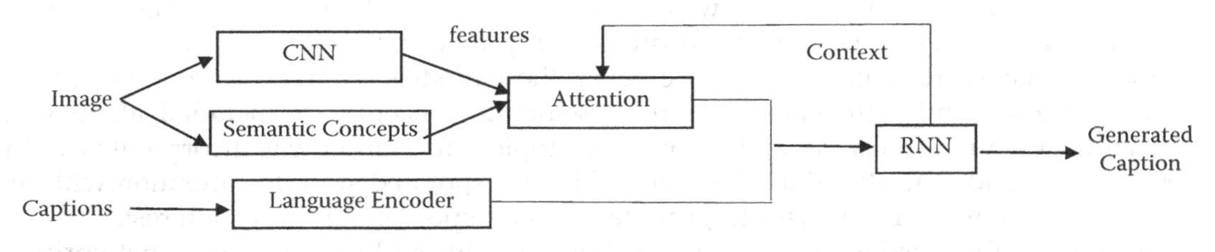

FIGURE 15.4
General Schematic of Image-Captioning Framework with Spatial and Semantic Attention.

These basic methods are fused to obtain better results and are called *fusion-based techniques*. Recently, attention-based methods that incorporate scene graphs and reinforcement learning (RL) have been proposed. We discuss various attention-based image captioning methods next. Table 15.1 shows an overview of selected impactful techniques.

15.3.1 Spatial Attention

Spatial attention models generally provide visual attention by considering spatial probabilities. The encoding for input image is obtained by considering the feature map from the last convolutional layer of a CNN. These feature maps are assigned weights based on the spatial probabilities. Xu et al. [11] proposed the first seminal work by application of a visual attention model for the task of image captioning. They combined the visual features extracted from CNN with the output of LSTM. Two types of attention mechanisms were proposed – hard and soft. Hard attention selects the most probable attentive region, and soft attention averages the spatial features with weights by allotting different weights at different spatial positions.

The methods based on soft attention proceed by establishing a link between visual representations in the image and semantic representations in the corresponding text captions. Recently, hard-up attention techniques [5,6] have also been proposed to find the most relevant regions based on bounding boxes.

In Fang et al. [23], a two-level word-level attention is applied. First, a bidirectional spatial embedding module is applied to obtain the visual representation of the image. The second module applies attention to extract word-level attention. The output word-level attention is utilized to predict the next word.

Anderson et al. [5] proposed an up-down model, which combines the bottom-up and top-down visual attention mechanism. The attention areas are given different weights and merged to be one attention feature at each time step. Thus, at one-time step, the attention mechanism can learn the spatial relationship of the selected areas.

15.3.2 Semantic Attention

The incorporation of semantic concepts and attributes detected from images in the encoder-decoder frameworks have proved to be effective for improving image captions [24–26]. Text-conditioned semantic attention was proposed in Yang et al. [27]. An end-to-end network is presented that jointly learns the image features, text embedding, semantic attention, and language model. This text-conditioned semantic information is fused with the attention mechanism to generate image descriptions. Attribute classifiers are used to capture the image features and to find the semantic concepts in You et al. [24].

A global semantic information was utilized in Jia et al. [10] to guide the LSTM for generating sentences. The global information is provided by extracting the correlation between images and their captions. However, these models require external resources to train these semantic attributes. Advanced semantic concepts are included along with semantic attention in Gan et al. [25], whereas topic information was incorporated in Lu et al. [6] to guide the attention. Zhou et al. [28] incorporated stimulus attention with the top-down attention mechanism to provide focus on specific regions of interest.

In most of the previous works, the attribute detector and the captioning network are independent. This may lead to the non-utilization of semantic information. Moreover, all the detected attributes are attended through the entire caption generation process.

TABLE 15.1

Overview of Existing Methods

Ref.	Method	Visual Feature Representation	Language Feature Representation	Dataset Used	Evaluation Parameter Used
FCN [13]	Fine-grained and semantic-guided visual attention	FCN encoder: VGG16	Long-short term memory (LSTM) encoder	MSCOCO	BLEU@N,METEOR, and CIDEr
HAF-REN [14]	Hierarchical attention fusion with reinforcement learning	Resnet-101	LSTM	MSCOCO	BLEU@N,METEOR, ROUGE, CIDEr
CGVRG [15]	Caption-guided visual relationship graphs	Faster R-CNN	Word embedding	MSCOCO and Visual Genome [16] for object detection	BLEU@N,METEOR, ROUGE, CIDEr, and SPICE
CSF [17]	Cascade semantic fusion architecture with three attentions: object-level, image-level, and spatial attention.	Faster R-CNN with CNN	LSTM	MSCOCO	BLEU@N,METEOR, and CIDEr
SCA-CNN [18]	Spatial and channel-wise attention and multi-layer visual attention	CNN: VGG-19 and ResNet-152	LSTM	Flickr8K, Flickr30K, and MSCOCO	BLEU@N,METEOR, ROUGE-L, and CIDEr.
DS-IC [19]	Injecting domain-specific ontology	Mask R-CNN	Stacked LSTMs	MSCOCO	BLEU@N,METEOR
R-AAM [16]	Reference-based on adaptive attention mechanism	ResNet-152	LSTM	MSCOCO	BLEU@N,METEOR, ROUGE-L, and CIDEr.
hLSTMat [20]	Temporal/Spatial attention with adaptive attention	ResNet101	LSTM	Flickr8K, Flickr30K, and MSCOCO	BLEU@N, METEOR, ROUGE-L, and CIDEr.
[21]	Semantic attention using attribute detectors	ResNet-101 based faster R-CNN	Two-layer LSTM	MSCOCO	BLEU@4, METEOR, ROUGE-L, and CIDEr.
[22]	Spatial and temporal	Faster R-CNN	LSTM	MSCOCO	BLEU@N,METEOR, ROUGE-L, and CIDEr.

This might lead to incorrect attention. To alleviate this, a deep model with a multimodal attribute detector (MAD) module is incorporated in Huang et al. [21]. It utilizes word embedding of attributes along with image features. This is followed by a subsequent attribute predictor (SAP) module that dynamically predicts a concise attribute subset at each time step to mitigate the diversity of image attributes.

Han et. al. [19] proposed an attention-based technique with object and attribute information. It reconstructs a generated caption to make it domain specific. This is done by specifying a domain-specific semantic ontology.

15.3.3 Adaptive Attention

Adaptive attention assists in deciding whether to use visual features or language model states in each decoding step. An approach based on adaptive attention was proposed in Lu et al. [4]. Another adaptive attention model was proposed in Chen et al. [18]. This method computes the context vector and assigns weights to visual features. It also utilizes an LSTM memory cell and assigns a weight to it. Based on this, it determines whether to use image features or text features for the generation of the current word.

Liu et al. proposed an image captioning system called the Reinforced Adaptive Attention Mechanism (R-AAM) [16], which incorporates additional information during attention. The textual information is added in the attention system as reference sentences. This helps to focus on the correct area in the image. The reference sentence is selected by computing the largest consensus score among the nearest images in the training data set. This reference sentence assists in selecting the highlight region for generating words in a time sequence. This method also expresses better semantic information.

Most of the techniques apply attention to both types of words – visual words like *man* and non-visual words like *the*. However, the prediction of non-visual words can be done without the application of attention. This issue is addressed in Gao et al. [20] by considering only visual words. A hierarchical LSTM with adaptive attention (hLSTMat) method presented in Gao et al. [20] utilizes either spatial or temporal attention. Adaptive attention is applied to decide whether to consider attention for visual information or language context information.

15.3.4 Cascade Attention

In Shi et al. [29], a cascade attention module is proposed, which sequentially processes different types of input. It allows higher priority input to affect the attention of other inputs. The global feature of the image, along with the region of interest (ROI) features from R-CNN, is used for captioning. Two interconnected LSTM layers are denoted as attention LSTM and language LSTM, respectively.

A cascade semantic fusion (CSF) architecture is proposed in Wang et al. [17]. Three types of visual attention semantics are utilized in a cascade manner. These include applying attention to objects in the image, the entire image, and spatial attention. The object-level features are extracted using a pre-trained detector. Next, a fusion module fuses object-level features with spatial features, resulting in image-level attention. The spatial features are obtained using Fast R-CNN [30] with CNN.

15.3.5 Fusion Techniques

Spatial and Channel-wise Attention: The spatial models apply weighted pooling on the attentive feature map. This may lead to the loss of spatial information. This was addressed in Chen et al. [18]. The Spatial and Channel-wise Attention in CNN (SCA-CNN) method extended the attention mechanism to multiple layers of CNN. Apart from spatial attention, channel-wise attention was also introduced to focus on different filters of the convolutional layer.

Spatial and Temporal: Along with spatial attention, temporal attention can prove to be more crucial for image captioning. In Ji et al. [22], a spatio-temporal memory attention (STMA) model is proposed. It also utilizes a memory mechanism incorporated in LSTM by introducing a cell for memorizing and propagating the attention information. The output gate is used to generate attention weights. Best results were obtained by combining STMA with up-down. STMA combines two classical attention-based image captioning methods: soft attention and adaptive attention.

Spatial and Semantic: In Wei et al. [31], StackVS, a multi-stage architecture, is proposed. It is a fine-gained image caption generation method that utilizes both visual and semantic information of an input image. Opposed to other methods where a single decoder is used, this method incorporates a stacked decoder model. It contains a sequence of decoder cells. These cells contain two LSTM layers. This enables assigning weights to both visual and semantic features. Both these features are input to the attention-based language model. Further, re-optimization is performed by the next decoder cell.

15.3.6 Scene Graph Based

Scene graphs have been employed in image encoders by the cited references [32–34]. Yao et al. used a unimodal setting, whereas Yang et al. and Guo et al. [33,34] used a multimodal approach. Relational semantics available in captions were utilized by She et al. [15]. Caption-guided visual relationship graphs (CGVRG) are constructed for image representation. First, the relationship triples are extracted from captions using a textual scene graph [35]. The graph is constructed with two nodes – object nodes and predicate nodes. The visual and textual features are integrated to enhance the CGVRG in the context of both modalities, as well as images and text, using graph convolution networks. Regional visual representations are used and extracted ROI pooling from Faster R-CNN. Textual encoding is obtained by word embeddings. Multi-task learning is employed for jointly predicting word and tag sequences. Each word in the caption is associated with a tag: object, predicate, or none. However, the images are segmented into a fixed-resolution grid at a coarse level. A model based on the Fully Convolutional Network (FCN)-LSTM network is presented in Zhang et al. [13] that addresses this problem. This method is capable of generating fine-grained attention maps. Grid-wise labels are allocated and co-related to other grids in the image. Also, the grid resolution of the attention module can be adjusted. It helps to capture the objects in the images as well as larger scene areas such as the sky or beach.

15.3.7 Reinforcement Learning Based

Image captioning methods that utilize RL have gained significant performance. Mostly these methods consider CIDEr as the reward function. However, other measures can be considered. A Hierarchical Attention Fusion (HAF) model was proposed in Wu et al. [14]. HAF was used along with RL by integrating multi-level feature maps. A pre-trained Resnet model was used with hierarchical attention for obtaining the feature maps. The CIDEr scores were re-evaluated using the Revaluation network, by assigning different weights for each word while generating the caption. Further, Scoring Network

(SN) scored the generated sentences based on the ground truth captions. Such sentence-level reward also considered benefits from unmatched ground truth sentences. The classical structure of Zhang et al. [13] was used as the baseline network, which generated a normalized attention weight α_t according to LSTM hidden state h_t at each time step t. And α_t was used to attend a different spatial of the image features Att as the final representation A.

In self-critical sequence training (SCST) [20], CIDEr was used as the reward along with the policy gradient method. The same reward was given to each word as the gradient weight. However, each word may have different importance in the caption. Yu et al. [36] computed word-level importance by computing Monte Carlo rollouts. However, the method was computationally expensive. In Wu et al. [14], Revaluation network was exploited to obtain the word-level rewards and allot different weights to each word.

15.4 Performance Comparison and Discussions

The performance comparison of various attention-based image captioning techniques for the MSCOCO dataset based on the standard evaluation metrics is discussed in Table 15.2. NIC v1 [2], DeepVS [3], and m-RNN [12] are end-to-end multimodal networks. These utilize CNN as an image encoder and sequence modeling via RNN. These models apply spatial attention only. emb-gLSTM [10] incorporates semantic attention, by utilizing the correlation between the image and its description as global semantic information. Compared to these pure semantic and spatial attention methods, the application of channel-wise attention with spatial attention is in SCA-CNN. However, its results surpass only a few methods and not hard attention, the reason being SCA-CNN is a single model as opposed to an ensemble model, which obtains better results. The FCN [13] method with attention Resolution 27 × 27 with Joint Context Feature performs better in comparison with the SCA-CNN method. In R-AAM [16], the reference statement information is incorporated. It guides the image to describe a more correct statement. It also helps to better describe the correct position information.

Up-down [5], HAF-REN [14], STMA + up-Down [22], and MAD + SAP [21] are the best performing models and have competitive results on the image captioning task. Up-down is a spatial attention model that utilizes both a bottom-up and a top-down approach. HAF + REN is a reinforcement-based technique, and it utilizes word-level reward and sentence-level reward. Both rewards lead to improvements across most metrics.

The STMA network combined with the up-down strategy gave results comparable to those obtained by RL methods. This is attributed to the additional attention from LSTM (A-LSTM) of the up-down model and the spatio-temporal learning strategy of STMA.

Moreover, compared with up-down, the MAD + SAP [21] model utilizes high-level semantic information to generate more fine-grained captions. Thereby, it outperforms the up-down model by incorporating appropriate semantic information into CIDEr-D optimized models. Image attribute words are more important than non-attributes such

TABLE 15.2

Performance Comparison of Existing Attention-Based Image Captioning Methods on MSCOCO Dataset (S-Spatial Attention, A-Adaptive Attention, Se-Semantic Attention, C- Chanel-wise, T-temporal, R-Reinforcement Learning)

Ref	Type of Attention	B-1	B-2	B-3	B-4	METEOR	CIDEr	ROUGE-L
NIC v1 [2]	S	0.666	0.461	0.329	0.246	–	–	–
DeepVS [3]	S	0.625	0.450	0.321	0.230	0.195	0.660	–
emb-gLSTM [10]	Se	0.670	0.491	0.358	0.264	0.227	–	–
m-RNN [12]	S	0.670	0.490	0.350	0.250	–	–	–
Soft-Attention [11]	S	0.707	0.492	0.344	0.243	0.239	0.773	–
Hard-Attention [11]	S	0.718	0.504	0.357	0.250	0.230	–	–
FCN [13]	Se	0.712	0.514	0.368	0.265	0.247	0.882	–
SCA-CNN [18]	S, C	0.712	0.542	0.404	0.302	0.244	0.912	0.524
R-AAM [16]	A	0.731	0.568	0.431	0.328	0.258	0.990	0.538
DS-IC[19]	S	0.791	0.624	0.471	0.359	0.271	–	–
hLSTMat [20]	T,S, A	0.794	0.635	0.487	0.368	0.282	1.205	0.577
Up-Down [5]	S	0.802	0.641	0.491	0.369	0.276	1.179	0.571
HAF-REN [14]	RL	0.803	0.629	0.478	0.355	0.272	1.162	–
STMA + up-Down [22]	S	0.803	0.646	0.498	0.377	0.283	1.231	0.581
MAD + SAP [21]	Se	0.805	0.651	0.504	0.384	0.286	1.251	0.587

as *the* and *is*. Hence, the attribute words are assigned higher weights compared to non-attributes in the TF-IDF-based CIDEr-D metric. It achieves the best results.

15.5 Evaluation Metrics and Discussion

15.5.1 BiLingual Evaluation Understudy (BLEU)

BLEU score [9] is computed by finding the similarity between the test caption (referred to as *candidate*) and the reference sentences using N-grams of words. Mostly BLEU@1, BLEU@2, BLEU@3, and BLEU@4 are computed for evaluating the generated caption with respect to the reference captions. The BLEU score is given by,

$$\text{BLEU@N} = \text{BP}.\ e^{\left(\sum_{n=1}^{N} w_n \log p_n\right)}$$

where N is the N-gram under consideration, P_n is the modified n-gram precision, w_n is the weight associated with n-gram, and BP is brevity penalty given by,

$$BP = \begin{cases} 1, & \text{if } c > r \\ e^{\left(1-\frac{r}{c}\right)}, & \text{if } c \leq r \end{cases}$$

c is the length of the candidate caption, and r is the effective reference corpus length. For multiple reference translations, the closest reference sentence length is considered to be r.

The modified n-gram precision is computed by,

$$p_n = \frac{\sum_{C \in \{candidates\}} \sum_{n-gram \in C} Count_{clip}(n-gram)}{\sum_{C' \in \{candidates\}} \sum_{n-gram' \in C'} count(n-gram')}$$

That is, by dividing clipped counts, the total number of n-gram candidate words $count_{clip}$ is given by,

$$Count_{clip} = \min(Count, Max_{ref})$$

Next, we discuss an example of how BLEU scores can be computed. Consider the following reference and candidate statements for computing the BLEU Score.

Reference-1: The roll machine operators are working

Reference-2: The operators of the roll machine are rolling the plates

Candidate: The operators of the roll machine are working

The values for count, $Count_{Ref-1}$, $Count_{Ref-2}$, Max_{Ref} and $Count_{clip}$ for unigram, bi-gram, 3-gram and 4-gram are presented in Tables 15.3 to 15.6, respectively. Unique n-gram words are considered from the candidate statements, and *count* is the maximum number of times a word occurs in any single reference translation. $Count_{Ref}$ is the frequency of occurrence of candidate n-gram in the reference caption. Max_{Ref} is

TABLE 15.3

Count Table for Unigrams

Unigrams in Candidate	Count	$Count_{Ref-1}$	$Count_{Ref-2}$	Max_{Ref}	$Count_{clip}$
the	2	1	3	3	2
operators	1	1	1	1	1
of	1	0	1	1	1
roll	1	1	1	1	1
machine	1	1	1	1	1
are	1	1	1	1	1
working	1	1	0	1	1
				Σ	8

TABLE 15.4

Count Table for Bigrams

Bi-grams in Candidate	Count	$Count_{Ref-1}$	$Count_{Ref-2}$	Max_{Ref}	$Count_{clip}$
the operators	1	0	1	1	1
operators of	1	0	1	1	1
of the	1	0	1	1	1
the roll	1	1	1	1	1
roll machine	1	1	1	1	1
machine are	1	0	1	1	1
are working	1	1	0	1	1
				Σ	7

TABLE 15.5
Count table for 3-grams

3-grams in Candidate	Count	$Count_{Ref-1}$	$Count_{Ref-2}$	Max_{Ref}	$Count_{clip}$
the operators of	1	0	1	1	1
operators of the	1	0	1	1	1
of the roll	1	0	1	1	1
the roll machine	1	1	1	1	1
roll machine are	1	0	1	1	1
machine are working	0	0	0	0	0
				Σ	5

TABLE 15.6
Count table for 4-grams

4-grams in Candidate	Count	$Count_{Ref-1}$	$Count_{Ref-2}$	Max_{Ref}	$Count_{clip}$
the operators of the	1	0	1	1	1
operators of the roll	1	0	1	1	1
of the roll machine	1	0	1	1	1
the roll machine are	1	0	1	1	1
roll machine are working	1	0	0	0	0
				Σ	4

the maximum number of n-grams occurrences in any reference count. $Count_{clip}$ is the minimum of count and Max_{Ref}.

For the example under consideration, c = 8 and r = 6. Thereby, the value of BP is 1, as c > r.

Computation of BLEU-1

Referring to Table 15.3, the modified precision for unigram (n = 1) for the assumed reference captions and candidate caption will be,

$$P_1 = \sum Count_{clip} / Count_{candidate} = 8/8 = 1.$$

Since the number of words (1-gram) in the candidate caption ($Count_{candidate}$) is 8. BLEU-1 will be,

$$\text{BLEU-1} = \text{BP}.\ e^{(\sum_{n=1}^{N=1} w_1 log\ p_1)}$$

The value of weight w_1 will be 1/n = 1

$$\text{BLEU-1} = \text{BP}.\ e^{1\ x\ log\ 1} = 1$$

Computation of BLEU-2

For BLEU-2, the value of n = 2 and weights will be ½. Therefore, w_1 = 0.5 and w_2 = 0.5. The number of 2-grams is 7 (refer to Table 15.4).

$P_2 = 7/7 = 1$

$$\text{BLEU-2} = BP.\ e^{(\Sigma_{n=1}^{N=2} w_n \log n)}$$
$$= 1\ x\ e^{(0.5\ x \log 1 + 0.5\ x \log 1)}$$
$$= 1xe^0 = 1x1 = 1$$

Computation of BLEU-3

For BLEU-3, the value of n = 3 and weights will be 1/3. Therefore, w_1 = 0.33, w_2 = 0.33, and w_3 = 0.33. (refer to Table 15.5).

$P_3 = 5/6 = 0.8333$

$$\text{BLEU3} = \text{BP}.\ e^{(\Sigma_{n=1}^{N=3} w_n \log n}$$
$$= 1\ x\ e^{(0.33x \log 1 + 0.33x \log 1 + 0.33x \log 0.8333)}$$
$$= 1\ x\ e^{-0.02613} = 0.9742$$

Computation of BLEU-4

For BLEU-4, the value of n = 4 and weights will be 1/4. Therefore, w_1 = 0.25, w_2 = 0.25, w_3 = 0.25, and w_4 = 0.25 (refer to Table 15.6).

$P_4 = 4/5 = 0.8$

$$\text{BLEU4} = \text{BP}.\ e^{(\Sigma_{n=1}^{N=4} w_n \log n)}$$
$$= 1\ x\ e^{(0.25x \log 1 + 0.25x \log 1 + 0.25x \log 0.8333 + 0.25x \log 0.8)}$$
$$= 1\ x\ e^{(-0.0440)} = 0.9569$$

15.5.2 METEOR

METEOR score [37] is based on unigrams in their original form, stemmed forms, and meanings. It utilizes recall along with precision, as recall helps in obtaining high levels of correlation with human judgment. METEOR (M) score is computed using the formula,

$$M = F_{mean}(1 - Pn)$$

where Pn is the penalty given by,

$$Pn = 0.5\left(\frac{c}{v_m}\right)^3$$

c is the number of chunks, and v_m is the number of unigrams that are mapped or matched. This factor is referred to as fragmentation.

$$F_{mean} = \frac{10x\ P\ x\ R}{R + 9P}$$

The precision (P) and recall (R) value are given by,

$$P = \frac{m}{w_t} \text{ and } R = \frac{m}{w_r}$$

Here, m is the number of unigrams in the candidate that are found in the reference caption, and w_t is the number of unigrams in the candidate caption. w_r is the number of unigrams in the reference caption.

Consider the following reference and candidate statements for computing the M score.
Reference: The operator is working on press machine
Candidate: The operator at press machine is absent.
P = 5/7 = 0.7142 and R = 5/7 = 0.7142

$$F_{mean} = \frac{10x\ P\ x\ R}{R + 9P} = 0.7141$$

Fragmentation = chunks / matches = 3 / 5 = 0.6
Pn = 0.5 $(0.6)^3$ = 0.108
M = F_{mean} (1 – Pn) = 0.7141(1 – 0.108) = 0.6369

15.5.3 ROUGE (Recall-Oriented Understudy for Gisting Evaluation)

We discuss ROUGE-L (Longest Common Subsequence), variant of the ROUGE score [38], in this section. Two natural language sentences are considered as a sequence of words, and LCS is computed between them. LCS is the longest common subsequence length between any two sentences. It is expected that if the LCS is longer, the sentences are more similar. If X is the reference caption and Y is the candidate caption with lengths m and n, respectively.

$$R = LCS(X, Y)/m$$

$$P = LCS(X, Y)/n$$

The F-measure to compute the similarity based between X and Y based on R and P is given by,

$$F - measure = \frac{(1 + \beta^2)RP}{R + \beta^2 P}$$

where is β controls the relative importance of the precision and recall. It is usually set to a very high value. The benefit of using LCS is that it does not require matching of adjacent words and requires only sequential matching of words that may or may not be consecutive.

15.5.4 CIDEr (Consensus-based Image Description Evaluation)

The CIDEr [39] score efficiently captures sentence similarity, grammar, importance, precision, and recall. It is the number of times n-gram w_k occurs in a reference sentence S_{ij}, referred to as $h_k(S_{ij})$ for candidate sentence C_i.

TF_IDF weighting gk(Sij) for each n-gram wk using

$$g_k(S_{ij}) = \frac{h_k(S_{ij})}{\Sigma_{w_l \in \omega} h_l(S_{ij})} \log\left(\frac{|I|}{\Sigma_{I_p \in I} \min(1, \Sigma_q h_u(S_{pq}))}\right)$$

where ω is the vocabulary of all n-grams. The first part is TF, and the second part is IDF. It was shown that CIDEr achieves higher accuracy over existing metrics for measuring consensus.

15.6 Image Captioning on Industrial Images: Case Study

A small dataset comprising 1,024 images was collected from an aluminum rolling mill that manufactures utensils. Captions were allotted to each of these images by the industry managers. Each image has two ground truth captions. Further, data augmentation was applied to generate five images corresponding to each image by rotation, shearing, zoom, brightness reduction, and enhancement. The encoder-decoder based technique proposed by Xu et al. [11] for image captioning utilizing the local attention mechanism of Bahdanau attention was used. VGG16 pre-trained model was used as the CNN encoder, which acts as the image encoder, and GRU was used as the RNN decoder, which performs language generation. A batch size of 64 was used for training, and the model was trained for 20 epochs. The train-test split used was 75%-25%. However, the predicted sentences contained repeated words, which were identified and removed before producing the image caption.

The results obtained for the test data based on the evaluation metrics are in Table 15.7. Although the BLEU score values are not very competitive, compared to the BLEU scores that are obtained on large datasets as discussed in Table 15.2, the CIDEr value is competitive. CIDEr is a more precise measure to evaluate the correctness of the generated sentences. Consequently, we consider that the attention-based models have the capability to be applied in images from an industrial environment. The results can be further improved by increasing the training data.

Figure 15.5 shows sample results obtained on test images with one ground truth caption mentioned to show the effectiveness of the predicted caption. The generated

TABLE 15.7

Results on Industrial images

B-1	B-2	B-3	B-4	METEOR	CIDEr	ROUGE-L
0.586	0.507	0.448	0.399	0.336	3.419	0.582

(a)

Predicted Caption: the hand press machine is switched off with no operator

Actual Caption: the hand press machine is not being operated

(b)

Predicted Caption: the circle cutting machine is switched off and plates are available

Actual Caption: there is no operator on the circle cutting machine and the cutting of the machine is occupying space

(c)

Predicted Caption: there is no operator on the circle cutting machine but near huge amount of aluminium plates are available for and visible

Actual Caption: circle cutting machine is loaded but switched off there is no operator and cutting is occupying space

(d)

Predicted Caption: the operator of aluminium tope are there near spinning machine

Actual Caption: spinning machine operator is working on the machine the raw material is available in the white bag and processed aluminium tope are there near the machine

FIGURE 15.5
Sample Results on Images from an Industrial Environment. (a) Correct Prediction (b) Partially Correct Prediction (c) Partially Correct with Grammatical Errors in the Predicted Caption (d) Correct Caption with Incorrect Grammar

sentences are grammatically correct for test images in Figure 15.5a,b. A correct caption is generated for the image in Figure 15.5a. The circle cutting machine is identified correctly; however, the aluminum cuttings are incorrectly identified as plates for the image in Figure 15.5b. However, there are grammatical errors in the generated captions for images in Figure 15.5c,d. Considering the obtained results, incorporation of grammar rules along with the decoder can significantly improve the semantics of the generated captions. The image captions from industrial images can be utilized for summary generation and can be further extended for videos. This can facilitate the work summarization in a factory.

15.7 Conclusion

This chapter presented a survey on image captioning, which is a challenging task involving an understanding of vision and language. We presented a categorization of attention-based techniques. The techniques that utilize only spatial attention may not fully utilize the attention mechanism capabilities. The multi-layer CNNs are naturally spatial. It is evident from the survey that techniques that fuse different attention mechanisms or utilize RL are capable of handling the challenges of the image captioning task. Moreover, the evaluation metrics that are used to evaluate the performance of the captions generated are also discussed. Although considerable performance is achieved by the existing techniques, there is still much scope for improvement. Particularly, it will be interesting to cater to various application domains. The advancement in attention-based methods opens up enormous opportunities in many application domains. Industries and military surveillance will be interesting areas for the application of image and video captioning. The task of image captioning is further extended to videos and question-answering, which are the emerging research topics related to image captioning.

References

1. Donahue, J., Anne Hendricks, L., Guadarrama, S., Rohrbach, M., Venugopalan, S. Saenko, K., and Darrell, T. (2015) Long-term recurrent convolutional networks for visual recognition and description. In *Proceedings of the IEEE conference on computer vision and pattern recognition* (pp. 2625–2634). IEEE.
2. Vinyals, O., Toshev, A., Bengio, S. and Erhan, D., (2015) Show and tell: A neural image caption generator. In Proceedings of the IEEE conference on computer vision and pattern recognition (pp. 3156–3164). IEEE.
3. Karpathy, A. and Fei-Fei, L. (2015) Deep visual-semantic alignments for generating image descriptions. In *Proceedings of the IEEE conference on computer vision and pattern recognition* (pp. 3128–3137). IEEE.

4. Lu, J., Xiong, C., Parikh, D. and Socher, R. (2017) Knowing when to look: Adaptive attention via a visual sentinel for image captioning. In Proceedings of the *IEEE conference on computer vision and pattern recognition* (pp. 375–383). IEEE.
5. Anderson, P., He, X., Buehler, C., Teney, D., Johnson, M., Gould, S. and Zhang, L. (2018) Bottom-up and top-down attention for image captioning and visual question answering. In Proceedings of the *IEEE conference on computer vision and pattern recognition* (pp. 6077–6086). IEEE.
6. Lu, J., Yang, J., Batra, D. and Parikh, D., 2018. Neural baby talk. In Proceedings of the *IEEE conference on computer vision and pattern recognition* (pp. 7219–7228). IEEE.
7. Hodosh, M., Young, P. and Hockenmaier, J. (2013) "Framing Image Description as a Ranking Task: Data, Models and Evaluation Metrics." *Journal of Artificial Intelligence Research* 47: 853–899.
8. Young, P., Lai, A., Hodosh, M. and Hockenmaier, J. (2014) "From Image Descriptions to Visual Denotations: New Similarity Metrics for Semantic Inference over Event Descriptions." *Transactions of the Association for Computational Linguistics* 2: 67–78.
9. Papineni, K., Roukos, S., Ward, T., and Zhu, W.J. (2002, July) Bleu: a method for automatic evaluation of machine translation. In *Proceedings of the 40th annual meeting of the Association for Computational Linguistics* (pp. 311–318). ACM.
10. Jia, X., Gavves, E., Fernando, B., and Tuytelaars, T. (2015) Guiding the long-short term memory model for image caption generation. In Proceedings of the IEEE international conference on computer vision (pp. 2407–2415). IEEE.
11. Xu, K., Ba, J., Kiros, R., Cho, K., Courville, A., Salakhudinov, R., Zemel, R., and Bengio, Y. (2015, June). Show, attend and tell: Neural image caption generation with visual attention. In International conference on machine learning (pp. 2048–2057). PMLR.
12. Mao, J., Xu, W., Yang, Y., Wang, J., Huang, Z. and Yuille, A. (2014) "Deep Captioning with Multimodal Recurrent Neural Networks (m-rnn)." *arXiv preprint arXiv* 1412: 6632.
13. Zhang, Z., Wu, Q., Wang, Y., and Chen, F. (2018) "High-Quality Image Captioning with Fine-Grained and Semantic-Guided Visual Attention." *IEEE Transactions on Multimedia* 21(7): 1681–1693.
14. Wu, C., Yuan, S., Cao, H., Wei, Y., and Wang, L. (2020) "Hierarchical Attention-Based Fusion for Image Caption With Multi-Grained Rewards." *IEEE Access* 8: 57943–57951.
15. Shi, Z., Zhou, X., Qiu, X. and Zhu, X. (2020) "Improving Image Captioning with Better Use of Captions." *arXiv preprint arXiv* 2006: 11807.
16. Liu, S., Bai, L., Guo, Y. and Wang, H. (2018, September). Reference Based on Adaptive Attention Mechanism for Image Captioning. In *2018 IEEE Fourth International Conference on Multimedia Big Data (BigMM)* (pp. 1–8). IEEE.
17. Wang, S., Lan, L., Zhang, X., Dong, G., and Luo, Z. (2019) "Cascade Semantic Fusion for Image Captioning." *IEEE Access*, 7: 66680–66688.
18. Chen, L., Zhang, H., Xiao, J., Nie, L., Shao, J., Liu, W., and Chua, T.S. (2017) "Sca-cnn: Spatial and Channel-Wise Attention in Convolutional Networks for Image Captioning." In Proceedings of the IEEE conference on computer vision and pattern recognition (pp. 5659–5667). IEEE.
19. Han, S. H. and Choi, H. J. (2020, February) Domain-Specific Image Caption Generator with Semantic Ontology. In *2020 IEEE International Conference on Big Data and Smart Computing (BigComp)* (pp. 526–530). IEEE.
20. Gao, L., Li, X., Song, J., and Shen, H. T. (2019) "Hierarchical LSTMs with Adaptive Attention for Visual Captioning." *IEEE Transactions on Pattern Analysis and Machine Intelligence* 42(5): 1112–1131.

21. Huang, Y., Chen, J., Ouyang, W., Wan, W., and Xue, Y. (2020) "Image Captioning with End-to-End Attribute Detection and Subsequent Attributes Prediction." *IEEE Transactions on Image Processing* 29: 4013–4026.
22. Ji, J., Xu, C., Zhang, X., Wang, B., and Song, X. (2020) "Spatio-Temporal Memory Attention for Image Captioning." *IEEE Transactions on Image Processing* 29: 7615–7628.
23. Fang, F., Wang, H. and Tang, P. (2018, October) Image captioning with word level attention. In *2018 25th IEEE International Conference on Image Processing (ICIP)* (pp. 1278–1282). IEEE.
24. You, Q., Jin, H., Wang, Z., Fang, C. and Luo, J. (2016) Image captioning with semantic attention. In *Proceedings of the IEEE conference on computer vision and pattern recognition* (pp. 4651–4659). IEEE.
25. Gan, Z., Gan, C., He, X., Pu, Y., Tran, K., Gao, J., Carin, L. and Deng, L. (2017) Semantic compositional networks for visual captioning. In *Proceedings of the IEEE conference on computer vision and pattern recognition* (pp. 5630–5639).
26. Yao, T., Pan, Y., Li, Y., Qiu, Z. and Mei, T. (2017) Boosting image captioning with attributes. In *Proceedings of the IEEE International Conference on Computer Vision* (pp. 4894–4902). IEEE.
27. Yang, Z., He, X., Gao, J., Deng, L. and Smola, A. (2016) Stacked attention networks for image question answering. In *Proceedings of the IEEE conference on computer vision and pattern recognition* (pp. 21–29). IEEE.
28. Zhou, L., Xu, C., Koch, P., and Corso, J.J. (2017, October) Watch what you just said: Image captioning with text-conditional attention. In Proceedings of the on Thematic Workshops of ACM Multimedia 2017 (pp. 305–313). ACM.
29. Shi, J., Li, Y., and Wang, S. (2019, September) Cascade attention: multiple feature based learning for image captioning. In *2019 IEEE International Conference on Image Processing (ICIP)* (pp. 1970–1974). IEEE.
30. Ren, S., He, K., Girshick, R., and Sun, J. (2016) Faster R-CNN: towards real-time object detection with region proposal networks. *IEEE Transactions on Pattern Analysis and Machine Intelligence* 39(6): 1137–1149.
31. Wei, W., Cheng, L., Mao, X., Zhou, G. and Zhu, F. 1909 Stack-vs: Stacked visual-semantic attention for image caption generation. *arXiv preprint arXiv* (2019): 02489.
32. Yao, T., Pan, Y., Li, Y. and Mei, T. (2018) Exploring visual relationship for image captioning. In *Proceedings of the European conference on computer vision (ECCV)* (pp. 684–699). Springer.
33. Yang, X., Tang, K., Zhang, H., and Cai, J. (2019) Auto-encoding scene graphs for image captioning. In *Proceedings of the IEEE/CVF Conference on Computer Vision and Pattern Recognition* (pp. 10685–10694). IEEE.
34. Guo, L., Liu, J., Tang, J., Li, J., Luo, W. and Lu, H. (2019, October) Aligning linguistic words and visual semantic units for image captioning. In *Proceedings of the 27th ACM International Conference on Multimedia* (pp. 765–773). ACM.
35. Schuster, S., Krishna, R., Chang, A., Fei-Fei, L., and Manning, C. D. (2015, September) Generating semantically precise scene graphs from textual descriptions for improved image retrieval. In *Proceedings of the fourth workshop on vision and language* (pp. 70–80). Association for Computational Linguistics.
36. Yu, L., Zhang, W., Wang, J., and Yu, Y. (2017, February) Seqgan: Sequence generative adversarial nets with policy gradient. In *Proceedings of the AAAI conference on artificial intelligence* (Vol. 31, No. 1). Association for the Advancement of Artificial Intelligence.
37. Banerjee, S. and Lavie, A. (2005, June) METEOR: An automatic metric for MT evaluation with improved correlation with human judgments. In *Proceedings of the acl workshop on intrinsic*

and extrinsic evaluation measures for machine translation and/or summarization (pp. 65–72). Association for Computational Linguistics.

38. Lin, C. Y. (2004, July) Rouge: A package for automatic evaluation of summaries. In *Text summarization branches out* (pp. 74–81). Association for Computational Linguistics.
39. Vedantam, R., Lawrence Zitnick, C. and Parikh, D. (2015) Cider: Consensus-based image description evaluation. In *Proceedings of the IEEE conference on computer vision and pattern recognition* (pp. 4566–4575). IEEE.

16

Smart Vehicle Monitoring and Control System using Arduino and Speed Gun: A Case Study

Reena Thakur, Muskan Qureshi, Sakshi Sarile, Shreya Pandit, and Laveena Tahilani
Jhulelal Institute of Technology, Nagpur

16.1 Introduction

In this massively populated environment, the entire population relies heavily on automobiles and the road transport network in order to travel from one location to another. From the beginning of civilization, transport has been a component of human growth. As the population increases, there is also a significant rise in the number of cars. According to an estimate by the U.S. Publisher Ward [1,2], approximately 1.05 billion vehicles were on roads, except off-road and heavy construction vehicles. Road traffic injuries are set to be the fifth major cause of death, according to World Health Statistics 2008. According to the WHO Global Status Report on Road Safety 2015 [3,4], there are approximately 1.25 million deaths each year due to road accidents. This report also shows that, during accidents, the major casualties are due to lack of immediate medical attention. Although technology has evolved so much, the accident detection and prevention mechanisms used now are the same ones implemented years ago and are all static measures, such as road signs, speed breakers, etc. Extensive studies have been conducted in the field of transport, particularly in the field of accident identification and avoidance. Methods based on the Vehicular Ad-hoc Network (VANET) [5,6] propose using information collected from neighboring vehicles to predict accidents through VANET communication and machine learning approaches. Accidents are predicted from collected data by observing traffic behavior. The road accident survey [7,8] recommends the use of Internet of Things (IoT) technology in the identification of accidents. Integrated systems designed using micro controller boards, such as Arduino, equipped only with piezoelectric sensors were primarily used by existing IoT systems. In this chapter, the smart vehicle monitoring and control system using Arduino and speed gun also proposes the use of IoT technology for early accident detection using the sensor-acquainted Raspberry Pi. The smart vehicle system also uses an image classification model based on machine learning to find the severity of the accident. The system is also familiar with GPS and the GSM module to find the location of the vehicle and communicate via cellular network.

DOI: 10.1201/9781003166702-16

16.2 Motivation

In India, since there is no method to handle or monitor the speed of vehicles on the roads, the traffic population has grown tremendously. This has contributed to a growing number of collisions as well. It is not important whether such incidents are the result of driving while intoxicated, because even a person who has not consumed alcohol can drive recklessly. The implementation of modern and advanced speed enforcement technologies is important to solve this issue and reduce death rates due to accidents. In order to encourage people to lead happier lives, information and communication technologies (ICT) have contributed greatly to society by solving many emerging problems. Vehicle monitoring while driving to guarantee that the vehicle and the people inside are safe is one such technology.

A smart vehicle speed control system using Arduino and speed sensors has been suggested. To help with road safety, this new approach will identify and penalize speeding vehicles for violations of the rules or infringements of authority.

Several schemes for the detection of rash driving on highways have been developed in the past. Many of these methods need human attention and take a lot of time, which is difficult to implement. For example, the speed monitoring systems are hand-held guns by police officers that permit them to test the speed of a vehicle.

16.3 Existing Technology

Traditional vehicle monitoring and control systems are inefficient and ineffective due to the frequency with which they sound, providing little safety or security to vehicle owners as well as others. This has resulted in the proliferation of *Smart Vehicle Monitoring and Control Systems*. This chapter proposes a control device that is connected to the vehicle and offers a range of protection and safety features. The proposed scheme is planned and implemented on the basis of Arduino and speed gun use.

16.4 Related Work

In recent years, smart cards have been commonly used to build different automated systems since the IoT has increased in popularity. This portion of the chapter addresses some of the notable efforts in which these cards have been used to support important applications as well as some algorithms to address parking and traffic control.

An algorithm for the treatment of real-time parking was suggested by Zhao et al. [9]. This algorithm was used to move a parking system from online to offline, and then a mathematical model was set up to explain the system offline based on a linear problem. Finally, an algorithm was created in order to compute this linear problem. With the aid of some simulations of the final method, the proposed algorithm was then assessed. The work to automate the entire car as well as the car parking was presented in Bonde et al. [10]. This chapter also explores a prototype designed for the automation of a car parking system that is capable of monitoring and controlling the number of cars that can be parked in a given time slot and, on the basis of availability, in a specific parking area. An Android program is used on the device.

In Kwan and Moghavyemi [11], the authors used smart card technology to include a credit mechanism for a prepaid metering system. Because of the visibility of the credits left on the card, the customer can monitor the usage of electricity, gas, and water in households as necessary. The credit available on the smart card will be deducted continuously for each usage, and the card can be recharged using an application designed to add balance to the card. This method acts as an improvement over conventional systems for billing water, electricity, and gas.

In Omar and Djuhari [12], the authors created a card scheme for learners in educational institutions.

As many as 98% of respondents in one study said they have seen ambulances on public roads in Dearborn Heights, but 82.9% said they have failed to respond adequately when emergency vehicles approached [13]. The speed adaptive traffic control system, the pollution adaptive traffic control system (PATC), the weather information system, and the master control center (MCC) are the key components discussed by Al-Dweik et al. [14]. The goal of the scalable enhanced roadside unit is to improve traffic flow by using PATC and MCC to monitor weather, road repair facilities, and driving during inclement weather to open or close particular routes.

Smart cards can serve various purposes, such as retail, transportation, education, and banking. The use of chips instead of magnetic strips in smart cards will ensure improved security. The data stored on the smart card can also be modified using some features. After a significant amount of work on smart cards was checked, researchers noted that while smart cards have been widely used in different arenas, there is still no such system that uses only one single card [13,15] for fully automated parking, including toll systems.

16.5 Technologies Important for Smart Cities

Smart cities are cities in which everything is interconnected, and this relies heavily on technology. Let us look at six innovations that are key to smart cities. Technological literacy is a key to turning a city into a well-connected, safe, and resilient smart city where knowledge is not only open but also discoverable.

It is not a new phenomenon that smart cities are all about providing their people with intelligent services that can save their time and ease their lives. It is also about linking them to the government, where they can give the government their input as to how they want their city to be. And without technology, this goal cannot be transformed into reality. Using technology, when combined with activities, authorities are able to collect city data, and this intelligence, makes the cities smarter and stronger.

16.5.1 Information and Communication Technology (ICT)

For a city to be smart, it must develop a two-way communication channel. This is where ICT comes in (Figure 16.1). ICT creates a connection between citizens and the government, where people can engage with the government, and the government can create a city that its citizens choose in exchange. ICT enables the government to determine the

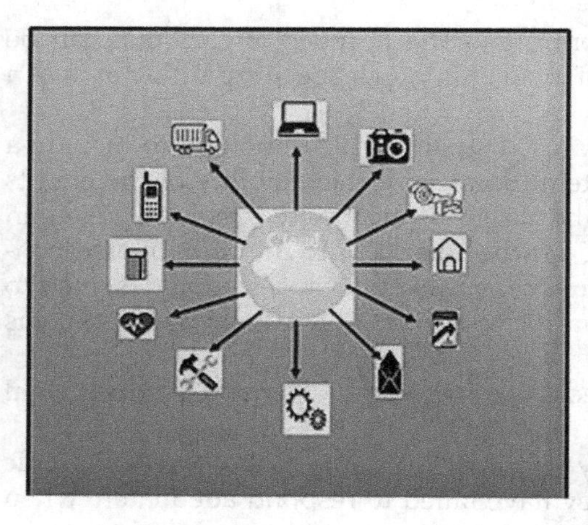

FIGURE 16.1
ICT.

state's demand trend and thereby create a pool of resources to manage the same online. In a culture, the electronic means of communication helps to establish a collective wisdom that can be used to leverage resources through analytics and deep learning.

16.5.2 Internet of Things (IoT)

The IoT is like the city's veins distributed throughout and linking each point (Figure 16.2). Every system that is part of a smart city needs to be linked to other systems so that they can speak to each other and make decisions for themselves that allow the megacity population to manage resources in return. This is where the IoT comes in, offering a body of interacting devices with the ideal model that offers intelligent solutions to daily problems. In smart cities, all smart solutions are based on the IoT in which they are linked and intelligent enough to determine their actions.

16.5.3 Sensors

Sensors are hidden yet pervasive elements of the urban environment (Figure 16.3). They are a vital element of any smart control device. These devices improve based on their

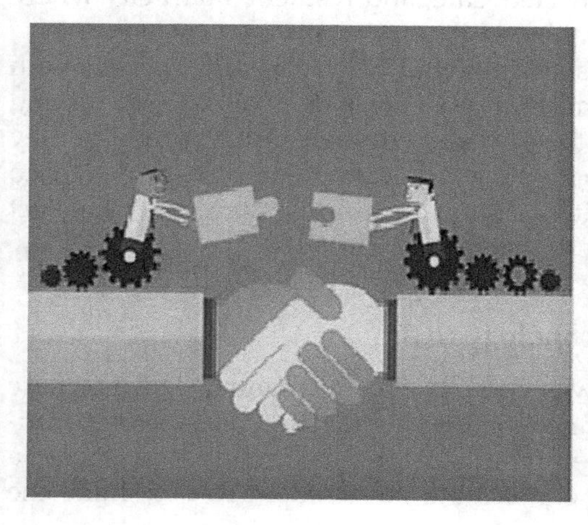

FIGURE 16.2
Example of IoT.

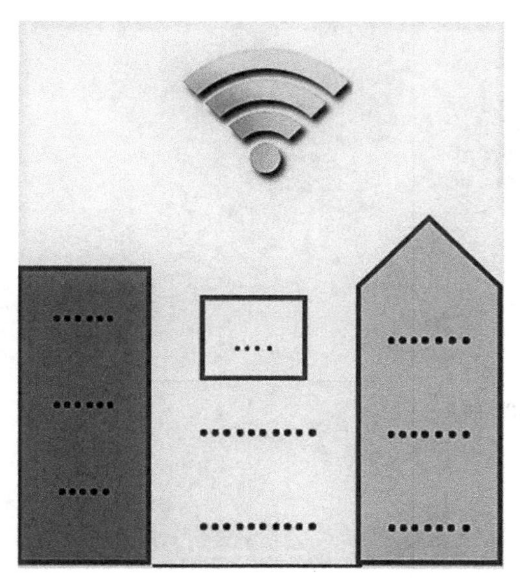

FIGURE 16.3
Example of Sensors.

environment and usually are equipped with a set of sensors to collect the necessary data in order for their control systems to be aware of their environment. To define their settings, smart control devices then use the necessary variables and adjust their operations accordingly.

The availability of a multitude of different sensors as well as technology that is continually changing has made applications that have previously been unworkable due to limited availability and high costs available to many users. Sensors are like converters that transform physical parameters to an electronic signal, which can be represented or fed into an autonomous device by humans. Among other things, these signals for traditional sensors include humidity, temperature, air pressure, light, and a number of other parameters.

16.5.4 Geospatial Technology (GT)

Whatever is designed in a smart city has to be accurate. A sustainable plan is needed to collect reliable, precise, and comprehensive information. This is the position of geospatial technology (GT), which provides the fundamental basis and ultimately the fabric on which to create the solution (Figure 16.4).

GT helps define the need precisely so that a better approach can be applied to it. In order to promote software-based solutions around intelligent infrastructure, GT provides the required framework for data collection and transformation of observations in these collections.

16.5.5 Artificial Intelligence (AI)

The smart city is a new paradigm that generates a massive amount of knowledge. These data are of no use until they are processed, which, in return, produces data. This enormous amount of data requires artificial intelligence (AI) that can make sense of the knowledge (Figure 16.5).

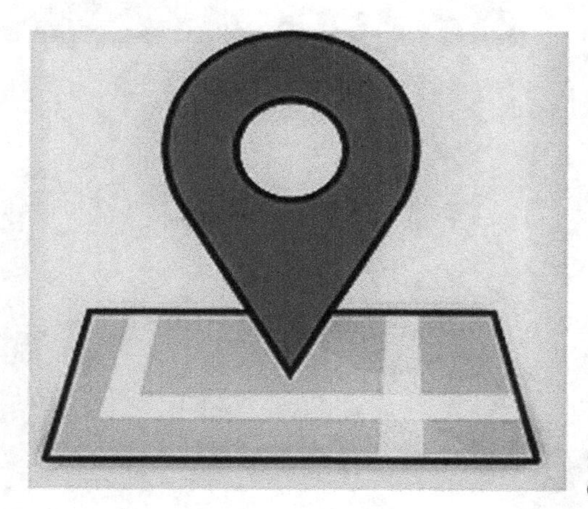

FIGURE 16.4
GT.

FIGURE 16.5
Artificial Intelligence.

AI facilitates machine-to-machine interaction through the processing and sensing of data. Let us look at an example to understand the fascinating use of AI in smart cities. In a system where energy spikes appear to occur, AI can learn where they normally occur and under what conditions, and this data can be used for improve power grid management. Likewise, AI can play a part in intelligent traffic management and health services.

16.6 Benefits of Smart Cities and Technologies

This section describes the benefits of smart cities and technologies.

1. *Environmental Impact*

The key force behind building smart and sustainable cities is reducing the CO2 footprint. One of the biggest benefits is enhancing energy production and storage, reforming waste management, and improving traffic conditions.

2. *Optimized Energy and Water Management*

Smart grids and smart water management are recurring themes in smart cities. Energy use and control of potable water ensure the energy supply and tap water quality throughout the region.

3. *Transportation*

It is important to transport goods, services, and people cleanly and efficiently. Many cities are turning to smart technology to alleviate traffic congestion and provide users with real-time alerts, in the hope of maximizing mobility.

4. *Security*

For all cities, protection is a priority. Accelerated smart city growth could allow communities to better track their residents, thanks to CCTV cameras with facial recognition. In addition, state-of-the-art CCTV cameras, as well as fire alarms, are now fitted with motion and smoke detectors.

16.7 Smart City Solutions

1. *Digital Water Management*

Hydraulics is one of the industries that has begun applying the innovations of the moment in its work processes. Some of the technologies that make up the Water 4.0 model are IoT, big data, and AI. These technologies aim to promote sustainability and achieving better industry performance, but also achieving reuse and combating scarcity.

As a whole, the impacts of climate change are minimized, and all processes in this field are transformed to help shape smart cities. "The trend in the water sector is marked by the disruption of automation, connectivity and intelligence," according to Xavier Torra, President of the Eurecat Technology Centre.

One of the smartest city strategies proposed at CES (CES is the only place to experience the entire connected ecosystem that brings together the technologies, solutions, players and audiences in the smart city) was specifically linked to the management of water. Specifically, it is a system built by Protectionist with consumption and leak detection sensors. According to this company, one out of every three liters of water in the world is lost via leaks, and as a solution, they offered this technology to serve both homes and entire cities. In real time, a mobile application is responsible for informing whether there is a problem, so it immediately senses the leakage and alerts to the cost of water consumption.

2. *Citizen Emergency Response Devices*

In order to help transform cities, there are several areas in which technical advances and new digital models are required. Some factors are more important than others, but in any condition, protection must not be ignored. Owing to inadequate preparation and lack of mitigation criteria, among other factors, protection is one of the worst controlled variables.

It is important to incorporate security when turning a city into a smart one that allows for a global and cross-cutting view of the potential problems and risks from data collection and analysis. In this respect, the Emergency Response Community presented two gadgets whose purpose is, among other things, to improve security and deal with emergency situations far more effectively.

FITT360 LIVES is a *wearable*, a kind of necklace that has three integrated cameras that allow it to make 360° video calls and immediately share the videos with up to four people so that they know what is happening around the person by means of audio and video. For cities, these devices allow people to act faster in cases where, for example, health or police services are required.

A further highlight was Cepton, a technology created by industry veterans in the United States that is able to identify, monitor, and organize with high accuracy both object and behavioral data. It can not only optimize traffic and plan for residents, but it can also ensure public safety and provide rapid responses to emergency situations.Cepton's proprietary, MMT™-based technology, according to the company, allows accurate, scalable, and cost-effective solutions that provide intelligent applications with long-range, high-resolution 3D perception.

3. *IoT and Traffic Management*

Transport and its infrastructure are other common concerns when talking about the growth of smart cities. That is why devices such as IoT RADAR came to CES with the goal of meeting the great challenge faced by smart cities. The role IoT RADAR plays in this case is to track not only the number of vehicles on the road, but also their speeds and accidents in real time – and all with 95% precision. It can be of great benefit to navigation systems to collect this data, which can be used for monitoring, protection, or collision alerts.

We also find projects in this same region by companies such as Toyota, which exhibited their proposal to create a city of the future in Japan, in particular on Mount Fuji. Woven City is a 100% intelligent city that will be characterized by being fully connected and controlled by resources such as solar power and hydrogen fuel cells. The town was introduced as a research facility to be used as a model for potential ventures, if all goes well. Both permanent residents and researchers can live in this environment that replicates reality to test and continue to evolve innovations such as robots, personal mobility, smart homes, etc.

4. *Smart Vehicles*

People cannot foresee all accidents using the current methods. Therefore, they must at least be able to recognize injuries as early as possible. Medical care must be given immediately to any person or animal who is involved in an accident. Sometimes, individuals injured in incidents cannot travel for urgent medical attention, which leads to their injuries being misjudged or to accident-related legal procedures. Even if an individual feels okay, with any accident, there is no risk in being evaluated. The main purpose of this chapter's case study is to quickly diagnose an accident, determine the condition of the driver, and notify the authorities of it.

The entire system is divided into different sets of modules. Each is designed to perform specific tasks or operations, as shown in Figure 16.6.

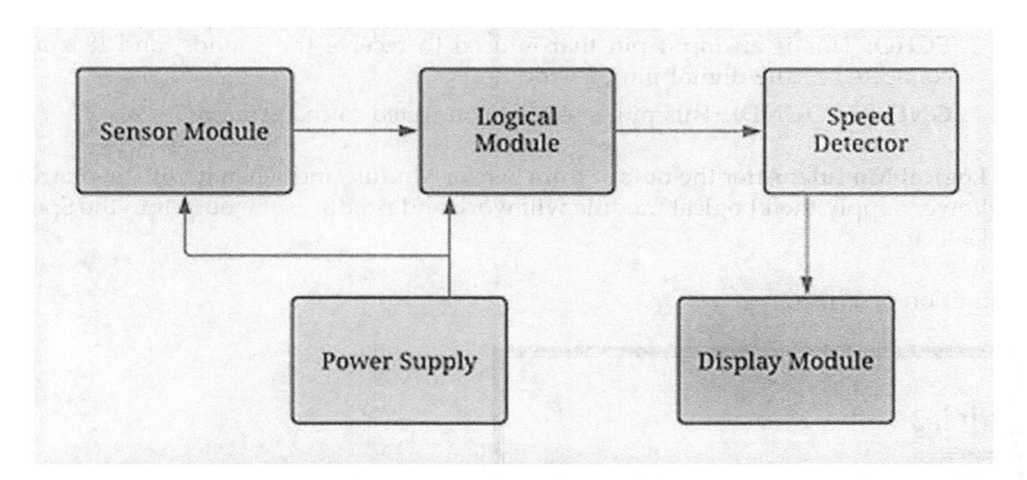

FIGURE 16.6
Block Diagram of Proposed System.

- **Sensor Module**: The task of this module is to acquire the speed of the vehicle using ultrasonic sensors, which usually work on radio waves. Through the radio waves, the ultrasonic sensor takes the speed of the vehicle. The ultrasonic sensor captures distance and time, and based on this distance and time, the ultrasonic sensor calculates the speed: speed = distance /time.
- **Speed Detector:** As mentioned above, speed is detected and calculated using an ultrasonic sensor. The ultrasonic sensor usually works on radio waves. It passes an ultrasonic pulse that moves through the air and bounces back to the sensor if there is an obstacle.
- **Display Module**: The calculated speed means the exceeded speed limit passes to the liquid-crystal display (LCD), which displays the calculated speed using Raspberry Pi cameras. If the speed of the vehicle has exceeded the limit, then the camera captures the image of the driver with his of her vehicle using Raspberry Pi cameras. The captured information is passed through the GSM module using a Sim of 2G, which is used because there might be an area where a 3G or 4G network will not be available; 2G is easily available all over India.
- **Power Supply:** There are four pins of an ultrasonic sensor:
    1. **VCC**
    2. **TRIG (trigger)**
    3. **ECHO**
    4. **GND (ground)**

1. **VCC:** The VCC pin of the ultrasonic sensor is connected through the 5v of Arduino.
2. **TRIG (trigger):** This is an output pin that is used to transmit sounds from the ultrasonic sensor. It is connected to the digital pin of Arduino.

3. **ECHO:** This is an input pin that is used to receive the sounds, and it is also connected to the digital pin of Arduino.
4. **GND (GROUND)**: This pin is simply connected to the ground.

Logical Module: After the output from Sensor Module and when it will get ok from Power supply then Logical module will work and produces the output to the Speed Detector.

Definition of Wiring:

```
Wiring
======
Ultrasonic sensor - Arduino Nano:
VCC  --- 5V
Trig --- D3
Echo --- D4
GND  --- GND
```

16.8 Case Study

A. *Objective*

There have been improvements and advancements on how to avoid road accidents in the past few years. Traffic congestion presents many issues for individuals. Many things happen because of this. A new approach to estimating vehicle speed has been established to reduce this issue. We suggest a smart vehicle speed control system using Arduino and speed sensors. In view of road safety, we describe a new technique for detecting and fining a speeding vehicle for ignoring the laws or forcing the consulted authority to take action. Many devices have been developed in the past to identify rash driving on highways. Most of the techniques require human attention and entail a great deal of effort, which is hard to execute.

The latest speed-monitoring devices are police officers' hand-held guns that allow them to verify the speed of a vehicle. Vehicle speed detection is based on the use of Doppler radar to detect the velocity of moving vehicles. The Doppler effect can be used to calculate the speed of vehicles and detect those crossing speed limits. The major factor used for speed measurement is the shift in frequency between transmitted and reflected high-frequency waves. The Doppler radar-based speed detector can communicate with a microprocessor-based system for calculation and comparison. The person may use the HD camera connected to the device to provide a view of the road in real time. It is possible to link the device through the Internet to the server.

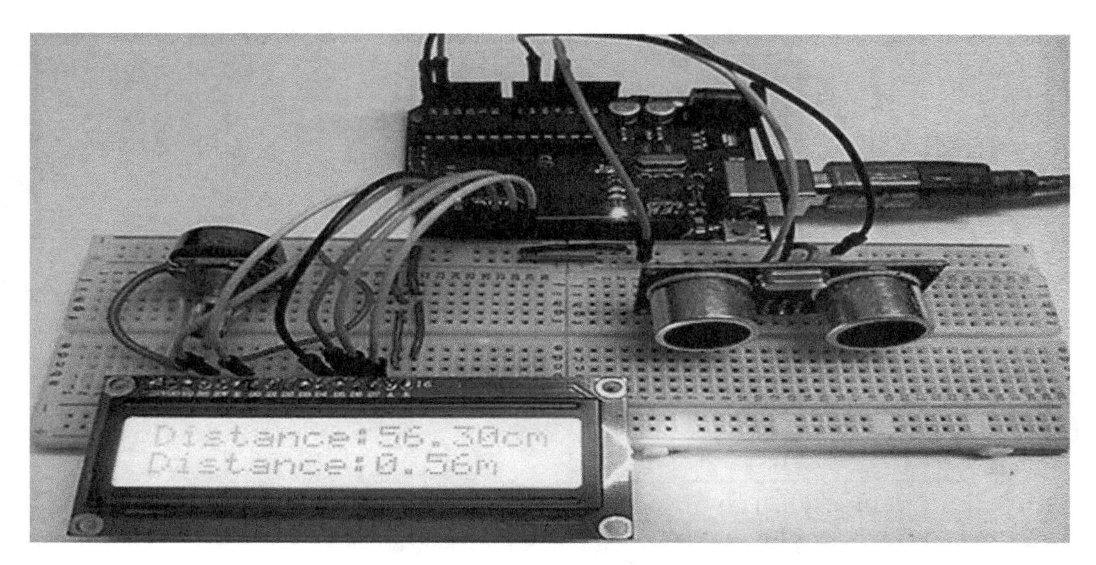

FIGURE 16.7
Smart Vehicle (Proposed System).

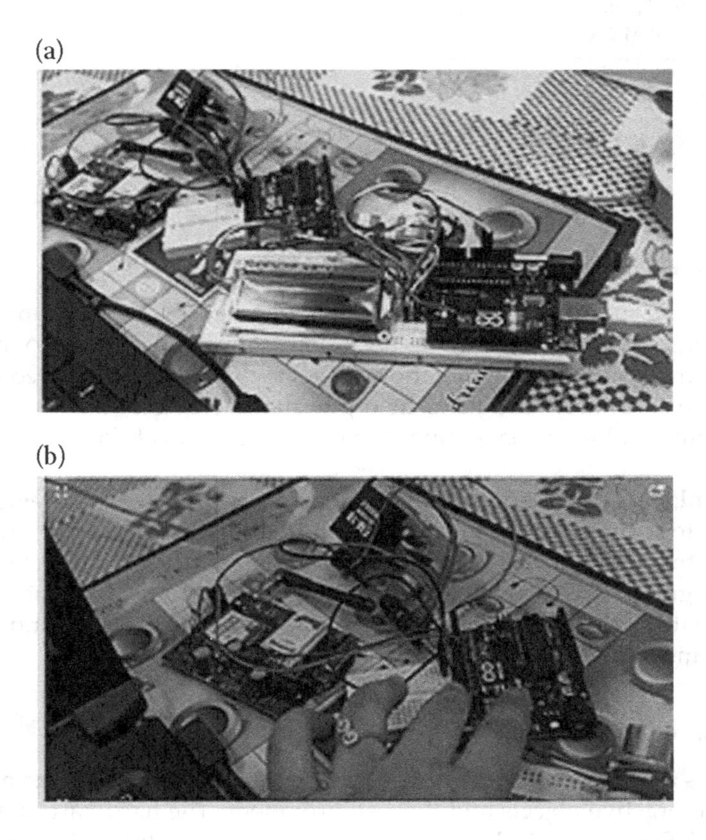

FIGURE 16.8
(a) and (b) Arduino Uno Board.

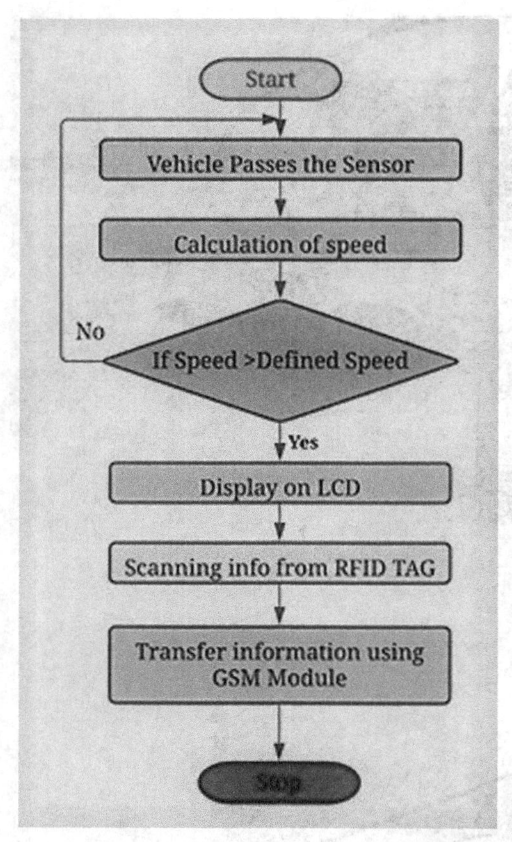

FIGURE 16.9
Schema of Approach.

B. *Vehicle Speed Detection Using Arduino Speed Gun*

There are specific guidelines for driving cars on roads. In any country, the most recommended practice is driving a safe speed on certain roads, i.e., if your vehicle's speed exceeds this limit, you will be in violation of the law. To detect the speed of a moving car, patrol officers usually rely on a hand-held gun that works on Radar technology or Lidar technology. This is a long process since the officer must check for speeding manually for each vehicle.

Automatic plate number identification is an image processing device that uses the plate number to identify the vehicle. The aim is to distinguish the plate number of the vehicle from the picture and to identify the plate number characters using optical character recognition. Moreover, the digitized plate number will be sent to the next station, where it will be displayed on the LCD screen (Figure 16.7), and thus the test for speed can be measured.

C. *Arduino and Sensor Device*

The Arduino is a microcontroller with a single board built to make the application more accessible to interactive objects and their environment. The hardware is dependent on an 8-bit Atmel AVR microcontroller with an open-source software board, a 32-bit Atmel

ARM. Current models consist of a USB interface, 6 analog input pins, and 14 digital I/O pins that allow the user to connect to various extension boards.

The Arduino Uno Board (Figure 16.8) is a microcontroller based on the ATmega328. These have 14 digital input or output pins (6 are for PWM outputs), a 16 MHz ceramic resonator, a USB link, an ICSP header, 6 analog inputs, a power jack, and a reset button. This includes all the necessary support for the microcontroller. To get started, the board is simply connected to a computer with a USB cable or with an AC-to-DC adapter or battery. The Arduino Uno Board varies from all other boards, and it does not have the FTDI USB-to-serial driver chip. The Atmega16U2 programmed as a USB-to-serial converter is included (Atmega8U2 up to version R2).

D. *Methodology*

Schema of Approach

As shown in the flow diagram (Figure 16.9), once the vehicle is on the road, it passes an ultrasonic sensor. Then, the speed of the vehicle is calculated. If the speed is not greater than the defined speed, then the vehicle is not violating the law. However, if the speed is greater than the defined speed, the number will be displayed on the LCD screen. Then, the RFID Tag attached on the vehicle will be scanned to collect all the details of the vehicle, such as registration number. This information is transferred using GSM module in which a 2G/3G Sim is attached.

16.9 Algorithm

The following table shows the pseudo-code of the algorithm.

Algorithm 1 UltrasonicRadar

```
// Arduino firmware for implementation of traffic radar with ultra-
sonic distance sensor.
// Wiring
// Ultrasonic sensor - Arduino Nano
// Assign- VCC, Trig, Echo,GND
SoftwareSerialgsm(ON, OFF);
LiquidCrystal_I2C lcd(0x27, 16, 2);
// trigger pin from ultrasonic sensor
// echo pin from ultrasonic sensor
// Define duration, distance, time and speed variables:
Start setup
// initialize the LCD
// Turn on the blacklight and print a message
// set the trigger_pin as an output
// set the trigger_pin as an input
endsetup
Do
```

```
{
// First ultrasonic measurement... // start the serial communication
// clear the trigger_pin
// Set the trigger_pin on HIGH state for 10 micro seconds:
// Read the echo_pin and return the sound wave travel time in
microseconds:
// Calculate the distance:
distance1 euqal to duration1 multiply 0.034 by 2;
use delay
in continuation with algorithm 2.
```

Algorithm 2 Second ultrasonic measurement

```
time2 is equal tomillis();
// Set the trigger_pin on HIGH state for 10 micro seconds:
// Read the echo_pin and return the sound wave travel time in
microseconds
// Calculate the distance:
distance2 is equal to duration2 multiply 0.034 by 2;
// Compute traveled distance, elapsed time and speed:
traveled_distanceis equal to distance1 minus distance2;
elapsed_timeis equal to time2 minus time1;
calculate measured_speed
// Print speed only if object final distance is less than 0.5 m away
Check if distance2 is less than 50 then
{
// Print speed only if object moves for at least 1 cm.
{
// Print if object is approaching or moving away
if (measured_speed> 20)
{
lcd.print("Speed:");
lcd.setCursor(0, 1);
lcd.print(fabs(measured_speed));
lcd.print(" km/h");
delay(5000);
}
}
}
}
```

16.10 Implementation

The step-by-step implementation of the proposed case study is shown below.

Step 1. Input through Sensor: The task of this module is to acquire the speed of the vehicle using ultrasonic sensors (Figure 16.10). The ultrasonic sensor usually

(a)

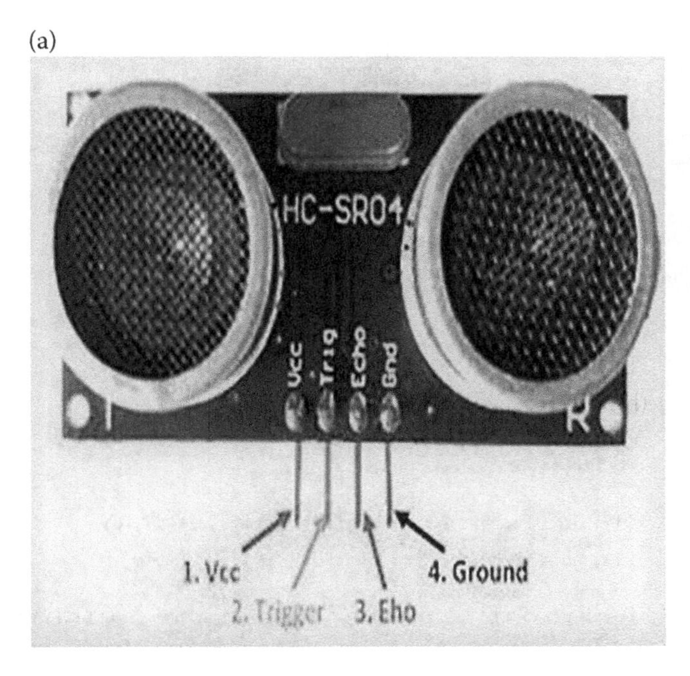

(b)

```
softwareserial gsm(0,1);

LiquidCrystal_I2C lcd(0x27, 16, 2);

const int trigger_pin = 3;
const int echo_pin = 4;

long duration1;
long duration2;
int distance1;
int distance2;
float traveled_distance;
unsigned long time1;
unsigned long time2;
float elapsed_time;
float measured_speed;

void setup() {

   lcd.begin();
   gsm.begin(9600);
   Serial.begin (9600, SERIAL_8N2);
   pinMode (0, OUTPUT);
   pinMode (1, OUTPUT);

   lcd.backlight();
   pinMode(trigger_pin, OUTPUT);
}
```

FIGURE 16.10
(a) Ultrasonic Range Finder, (b) Initializing Devices.

works on radio waves to determine the speed of the vehicle by collecting distance and time. Then, the sensor calculates the speed: speed = distance/time.

Step 2. Calculation of Speed: The speed taken by ultrasonic sensor passes to the Arduino. The Arduino (Figure 16.11), having 13 analog and 7 digital pins,

(a)

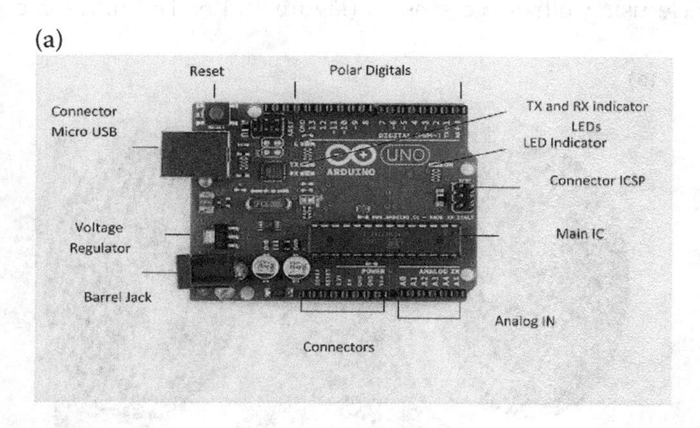

(b)

```
time1 = millis();
digitalwrite(trigger_pin, LOW);
delayMicroseconds(2);

digitalwrite(trigger_pin, HIGH);
delayMicroseconds(10);
digitalwrite(trigger_pin, LOW);

duration1 = pulseIn(echo_pin, HIGH);
distance1 = duration1*0.034/2;
```

FIGURE 16.11
(a) Arduino – UNO R3 Development Board, (b) Start Timer on First Sensor Trigger.

(a)

(b)

```
time2 = millis();
digitalwrite(trigger_pin, HIGH);
delayMicroseconds(10);
digitalwrite(trigger_pin, LOW);
duration2 = pulseIn(echo_pin, HIGH);
distance2 = duration2*0.034/2;
```

FIGURE 16.12
(a) LCD Display, (b) Stop Timer at Second Sensor Trigger.

(a)

(b)

```
traveled_distance = distance1-distance2;
elapsed_time = time2-time1;
measured_speed = traveled_distance/elapsed_time*36;
```

FIGURE 16.13
(a) Camera – Image Processing, (b) Calculate Speed Using Time Difference.

takes the speed from the ultrasonic sensor and calculates the speed as per the limit. If the number exceeds the limit, then it passes to the LCD.

Step 3 **Displaying of Speed in LCD:** The exceeded speed limit passes to the LCD (Figure 16.12), which displays the calculated speed.

Step 4 **Capture of Image of Limit-Exceeding Vehicle Using Camera:** The camera (Figure 16.13) captures an image of the driver with his or her vehicle using Raspberry Pi cameras.

Step 5 **Scanning of Information Using RFID Scanner from RFID Tag:** The RFID Tag (Figure 16.14) will be built in the vehicle, such as in the front of the vehicle. It can be included in all vehicles in the future.

Step 6 **The Captured Image Is Included with the Information Passed Using a GSM Module:** The captured information is passed through a GSM module (Figure 16.15) using a Sim of 2G because there might be an area where a 3G or 4G network will not be available; 2G is easily available all over India.

16.11 Summary

This chapter demonstrated how smart vehicle tracking systems are implemented using Arduino and speed sensors. In view of road safety, we described a new technique for detecting and fining a speeding vehicle for violating the rules or forcing the consulted authority to take action. This chapter also described various technologies that are important for smart cities, which will have many practical and economic benefits. The detailed working and flow diagram gave a better understanding of the proposed system.

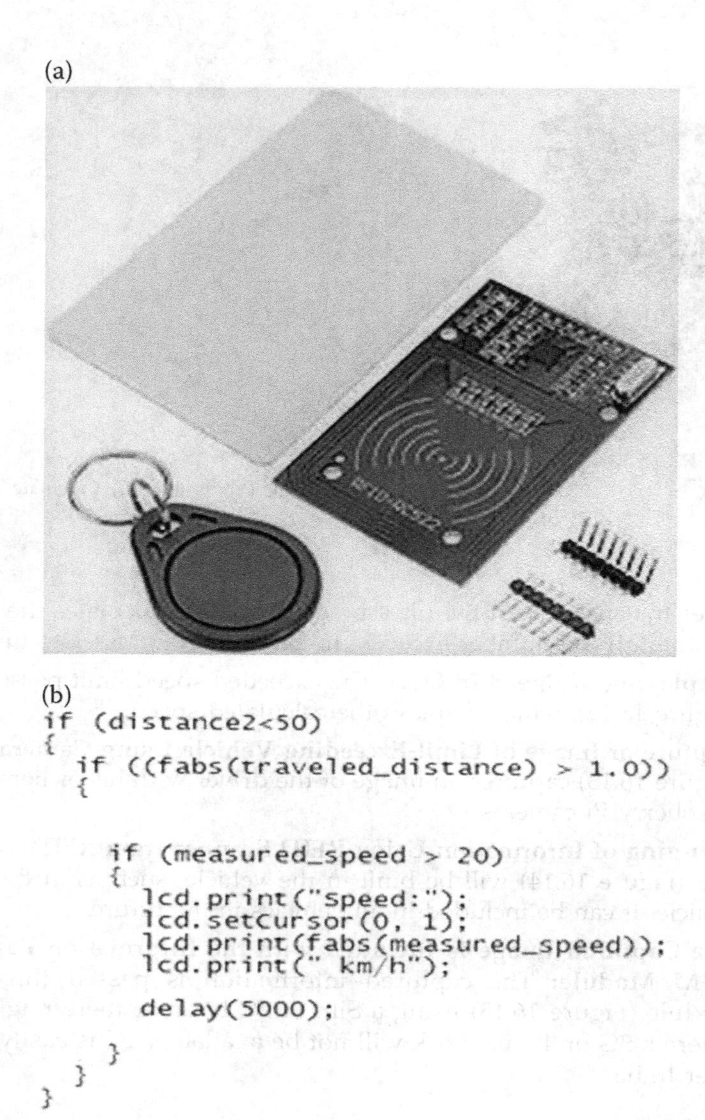
(a)
(b)
if (distance2<50)
{
if ((fabs(traveled_distance) > 1.0))
{
if (measured_speed > 20)
{
lcd.print("speed:");
lcd.setCursor(0, 1);
lcd.print(fabs(measured_speed));
lcd.print(" km/h");
delay(5000);
}
}
}

FIGURE 16.14
(a) RFID, (b) Check if Speed Limit Crossed or Not Crossed

1. Alerts the parked vehicle's registered driver if it is opened or picked up by generating a ringing tone and sending a vehicle location SMS message. Such SMS messages allow the registered driver to locate the vehicle at any time.
2. Sends an automated appeal message to the ambulance service, police, relatives, and/or family in the event of a major collision.
3. In the case of a less urgent situation, allows the driver to send appeal messages with a single click for accidents.

FIGURE 16.15
GPRS Module.

References

1. Motor vehicle, Wikipedia, https://en.wikipedia.org/wiki/Motor_vehicle.
2. John Sousanis, WardsAuto, World Vehicle Population Tops 1 Billion Units, http://wardsauto.com/news-analysis/worldvehicle-population-tops-1-billion-units
3. List of countries by traffic-related death rate, "https://en.wikipedia.org/wiki/List_of_countries_by_trafficrelated_death_rate"
4. World Health Organization (2015) Global status report on road safety.
5. Nejdet, D., and Abdulhamit, S. (2012, March) "Traffic Accident Detection By Using Machine Learning Methods", International Symposium on Sustainable Development, At Bosnia and Herzegovina, Volumemac_mac 2 (pp. 468–474). https://www.researchgate.net/publication/288511790
6. Najada, H. A. and Mahgoub, I. (2016) "Anticipation and alert system of congestion and accidents in VANET using Big Data analysis for Intelligent Transportation Systems," 2016 IEEE Symposium Series on Computational Intelligence (SSCI), Athens (pp. 1–8). IEEE Xplore.
7. Patel R., Dabhi V. K., and Prajapati H. B. (2017) "A survey on IoT based road traffic surveillance and accident detection system (A smart way to handle traffic and concerned problems)," 2017 Innovations in Power and Advanced Computing Technologies (i-PACT), Vellore (pp. 1–7). IEEE Xplore.
8. Singhal, A., Sarishma, and Tomar, R. (2016) "Intelligent accident management system using IoT and cloud computing," 2016 2nd International Conference on Next Generation Computing Technologies (NGCT), Dehradun (pp. 89–92). IEEE Xplore.
9. Zhao X., Zhao K., and Hai F. (2014) "An algorithm of parking planning for smart parking system,'" in Proc. 11th World Congr. Intell. Control Autom. (WCICA) (pp. 4965–4969). doi: 10.1109/WCICA.2014.7053556

10. Bonde, D. J., Shende, R. S., Gaikwad, K. S., Kedari, A. S., and Bhokre, A. U. (2014) "Automated car parking system commanded by Android application," in Proc. Int. Conf. Comput. Commun. Inform. (ICCCI), Coimbatore, India, Jan (pp. 1–4). doi:10.1109/ICCCI.2014.6921729
11. Kwan, B. H. and Moghavvyemi, M. (2002) "PIC Based Smart Card Prepayment System." In Proc. of 2002 Student Conf. on Research and Development (pp. 440–443). IEEE Xplore.
12. Omar, S. and Djuhari, H. (2004) "Multi-Purpose Student Card System Using Smart Card Technology."Information Technology Based Proceedings of the Fifth International Conference on Higher Education and Training, 2004. ISBN: 0-7803-8596-9. IEEE.
13. Savolainen, P., Datta, T., Ghosh, I., and Gates, T. (2010) "Effects of Dynamically Activated Emergency Vehicle Warning Sign on Driver Behavior at Urban Intersections." *Transportation Research Record* 2149: 77–83. [CrossRef].
14. Al-Dweik, A., Muresan, R., Mayhew, M., and Lieberman, M. (2017) "IoT-based multi-functional scalable real-time enhanced road side unit for intelligent transportation systems," In Proceedings of the 2017 IEEE 30th Canadian Conference on Electrical and Computer Engineering (CCECE), Windsor, ON, Canada, 30 April–3 May (pp. 1–6). IEEE.
15. https://www.pparke.in/, last accessed on5/11/2018 [9] http://www.fastag.org/, last accessed on1/11/2018.

17

Deep Learning Approaches to Pedestrian Detection: State of the Art

Kamal Hajari[1], Dr. Ujwalla Gawande[2], and Prof. Yogesh Golhar[3]
[1]*Yeshwantrao Chavan College of Engineering*
[2]*Yeshwantrao Chavan College of Engineering*
[3]*G H Raisoni Institute of Engineering and Technology*

17.1 Introduction

Pedestrian detection is an essential method of automated video surveillance. It is an obligatory step for eliciting exact information from real-world scenes about objects in the environment [1][2]. Nowadays, pedestrian detection is utilized in numerous computer vision applications, such as elder fall detection; hospital employee and patient monitoring, such as newborn infants and patients with physical impairments; prisoner surveillance; pedestrian detection in high-crowded areas, such as railway stations, vegetable markets, shopping malls, airports, weddings, and birthday parties; vehicle collision detection on highways; pedestrian counting in a crowded area; and pedestrian traffic surveillance in transportation. However, pedestrian detection accuracy is affected by diverse variants in human presence, trajectory, and posture. Non-uniformed uncertain illumination, complex background scenarios, pedestrian deformation, object occlusion, and object shadow continue to be unsolved problems in this area [3].

In recent times, the COVID-19 pandemic led to circumstances where crowded areas were monitored using an active pedestrian detection system. It aids in identifying hot spots, where the probably of COVID-19 virus transmission would increase due to more human interactions in a highly dense region. Therefore, the requirement for efficient pedestrian surveillance systems has increased. Many researchers' focus is on enhancing the performance of existing pedestrian detection systems by employing deep learning approaches.

State-of-the-art convolutional neural network (CNN) based architectures have been adopted for pedestrian detection. In recent times, one of the most utilized object detection models is You Only Look Once (YOLO) [4]. It is the most reliable, accurate, and fast deep vision architecture compared to other CNN-based object detection models. Xiang Long et al. [5] proposed PP-YOLO, or PaddlePaddle You Only Look Once, for an efficient and fast object detection model in real time. YOLO has different variants, such as YOLO v1, YOLO v2, YOLO v3, YOLO v4, and YOLO v5 [6].

DOI: 10.1201/9781003166702-17

Other CNN architecture specifically used for instance-level and semantic segmentation and classification frameworks are Mask Region based-CNN [7][8], Faster Region based-CNN [9][10], Fast Recurrent-CNN [11], Region-based Fully Convolutional Network (R-FCN) [12], Single Shot MultiBox Detector (SSD) [13], Fully Convolutional Network (FCN) [14], Deep Convolutional Generative Adversarial Network (DCGAN) [15], Residual Neural Network (ResNet) [16], GoogLeNet [17], Visual Geometry Group (VGG Net) [18], ZFNet [19], AlexNet [20], Deep Belief Network (DBN) [21], and LeNet [22], which are utilized for human detection in real-time environments. The aforementioned deep learning architecture is significantly more reliable than the traditional neural network models such as artificial neural network (ANN) [23], Support Vector Machine (SVM) [24], AdaBoost [25], Radial Basis Neural Network (RBN) [26], and Probabilistic Neural Network (PNN) [27].

LeNet was the first CNN-based architecture proposed by Kangming et al. [22] in 1998. It is a two-layered architecture consisting of a convolutional and a max pooling layer along with the backpropagation network for classification of desired object category. LeNet utilizes the tanh activation function for determining whether an input layer neuron should be activated or not through calculating cumulative summation of associated weights and bias. It is less accurate and has low computational capabilities, because it consists of only two layers to process the input. LeNet has trained and tested on the Modified National Institute of Standards and Technology (MNIST) database with 50,000 images classified into 10 different categories. It was utilized for handwritten signature detection. It had an error rate of 26.2%.

A similar architecture, AlexNet, was proposed by Krizhevsky et al. [20] in 2012. It employs the ReLU activation function with the cross entropy loss function. It has a five-layer architecture trained and tested on the ImageNet database with 1 million images divvied into 1,000 categories. It is more accurate than LeNet. It reduces error rate to 15.4%. The modified version of antecedent architectures was ZFNet proposed by Zeiler et al. [19] in 2013. ZFNet employs 7 x 7 filters for convolving the input image in the first layer. It reduces error rate further to 11.2%. Next, Liu et al. [21] introduced Deep Belief Network (DBN) in 2009. DBN consists of multiple layer-by-layer architectures along with a feature detector. The greedy training algorithm increases the run-time complexity. However, this architecture was not adaptive to new or complex data.

Simonyan et al. [18] proposed the VGGNet in 2014 that reduces the error rate up to 7.2%. It increased the convolutional layers from 11 to 19 layers, along with 16 times smaller filter size, 3 x 3. A similar variant of this architecture, introduced by Vanhoucke et al. [17][18], is ResNet and GoogLeNet in 2014. It reduces the error rate to 6.7%. Further, He et al. [16] in 2015 proposed ResNet, which reduces the error rate to 3.57%. ResNet contains 152 convolutional layers and a backpropagation algorithm for training. Afterward, deep learning architectures are utilized for semantic segmentation. In 2015, Jonathan et al. [19] proposed the first semantic segmentation model, i.e., FCN, that divides the pixel of the image into a different class.

For instance, Ross Girshick et al. [28] introduced R-CNN. This architecture uses a selective search algorithm for identifying the region of interest (RoI) in an input image, sometimes called the region proposals. Other variants of the R-CNN are Faster R-CNN [9][10], Fast R-CNN [11], and Mask R-CNN [6][7]. These are all different types of deep learning architectures used for pedestrian detection [1][2][3]. However, despite the fact that deep learning architecture gives outstanding results in terms of object detection, it is computationally expensive for unknown data. Hence, it is a complex task to design and employ deep learning architecture for real-world applications.

The rest of the chapter is organized as follows: Section 17.2 presents a proposed pedestrian database in academic institutions along with the most commonly used pedestrian database used by researchers. Section 17.3 outlines the issues and challenges of deep learning architectures used for pedestrian detection. Section 17.4 discusses the recent growth and deep learning architecture innovation. Section 17.5 describes open-source libraries and platforms for the development of deep learning frameworks based on real-world computer vision systems. Section 17.6 describes the proposed case study for scale-invariant pedestrian detection using Mask R-CNN. Section 17.7 concludes with future research directions.

17.2 Pedestrian Datasets

State-of-the-art techniques for pedestrian detection include Adaptive Local Binary Pattern (LBP), Histogram of Oriented Gradient (HOG), and the spatio-temporal context information-based method. These methods were tested on pedestrian benchmark databases [1]. In this section, we describe the commonly used pedestrian datasets used by researchers. Figure 17.1 shows images from each of the datasets. Next, we compare each dataset with its size, environmental condition, and behavior captured by the dataset.

17.2.1 Caltech Pedestrian Dataset

The Caltech Pedestrian dataset consists of 2,300 pedestrians along with 350,000 annotated bounding boxes. It was recorded in daylight conditions on an urban road in regular traffic. The video was recorded at 640 × 480 resolution from a camera mounting on a vehicle for capturing on-road pedestrians walking on the street. An annotation of pedestrians is utilized for validating the performance and accuracy of different pedestrian detection algorithms [29].

17.2.2 MIT Pedestrian Dataset

The MIT Pedestrian dataset was the first pedestrian dataset; it is reasonably small and comparatively high quality. This dataset comprises 709 pedestrian sample images, either in front or a back view, with a relatively limited range of poses images taken in city streets [30].

17.2.3 Tsinghua-Daimler Pedestrian Dataset

The Tsinghua-Daimler Pedestrian dataset captured humans walking on the street from a camera mounted on a vehicle in a daylight urban environment. This dataset includes tracking attributes of pedestrians, annotated labeled bounding boxes, ground truth images, and float disparity map files. It is divided into a training set that comprises 15,560 pedestrian images along with 6,744 annotated pedestrian images. The testing set comprises more than 21,790 sample pedestrian images with 56,492 annotated pedestrian images [31].

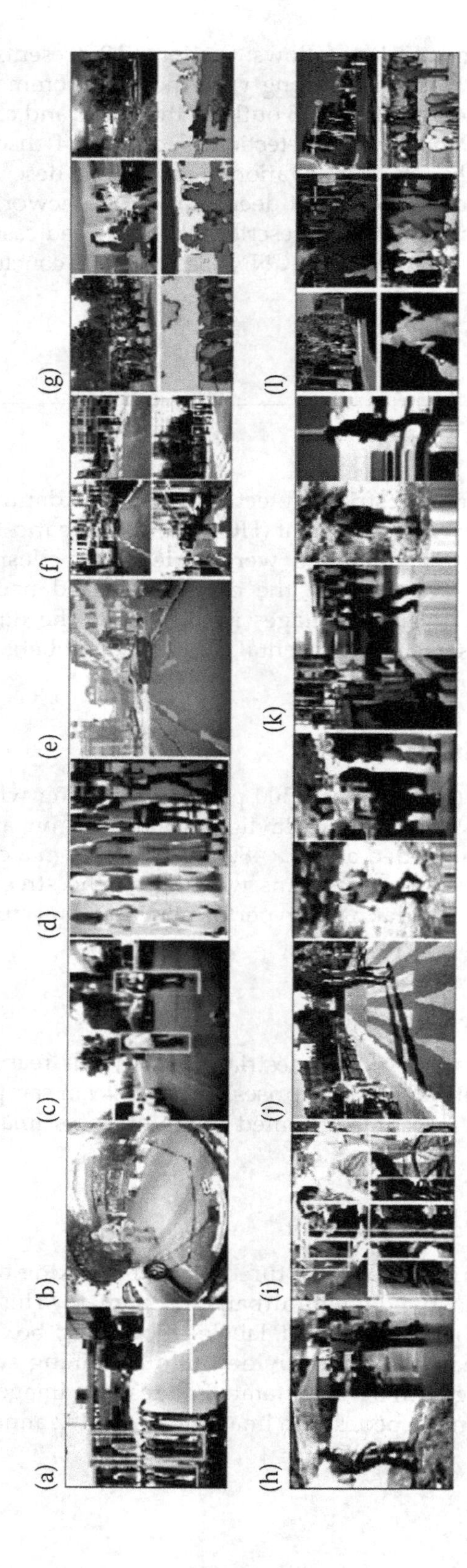

FIGURE 17.1
Sample Images of Pedestrian Datasets. (a) Unique Annotated Pedestrian of Caltech Dataset. (b) Sample Images of GM-ATCI Rear-View Dataset. (c) Sample Images of Cyclist Pedestrian in Tsinghua-Daimler Datasets. (d) Sample Images of NICTA Dataset. (e) Sample Images of ETH Urban Dataset. (f) Sample Images of TUD-Brussels Dataset. (g) Sample Images of Microsoft COCO Pedestrian Dataset (h) Sample Images of Static Pedestrian INRIA Datasets. (i) Sample Images of PASCAL Object Dataset. (j) Sample Images of CVC-ADAS Collection of Pedestrian Datasets. (k) Sample Images of MIT Pedestrian Dataset. (l) Sample Images of Mapillary Vistas Research Dataset.

17.2.4 Advanced Technical Center (ATCI) Pedestrian Dataset

The Advanced Technical Center (ATCI) Pedestrian dataset is a rear-view pedestrian database captured using a vehicle-mounted standard automotive rear-view camera for assessing rear-view pedestrian detection at different locations such as in-house and outside parking lots, city roads, and private driveways. The dataset comprises 250 video clips of 76 minutes each, along with 200K labeled pedestrian bounding boxes. This dataset was captured in both day and night scenarios, with varying weather conditions [32].

17.2.5 National Information and Communication Technology Australia (NICTA) Pedestrian Dataset

The National Information and Communication Technology Australia (NICTA) dataset was created in different cities and countries. The dataset comprises 25,551 unique pedestrians. The 50K annotated pedestrians are utilized for validating the performance and accuracy of different pedestrian detection algorithms [33].

17.2.6 ETH Pedestrian Dataset

The ETH Pedestrian dataset was utilized to observe traffic scenes from inside vehicles. The pedestrian behavior was captured from stereo equipment fixed on a stroller attached to a car. The database can be used for pedestrian detection and tracking from movable platforms in an urban scenario. The dataset includes different traffic agents such as cars and pedestrians. [34].

17.2.7 TUD-Brussels Pedestrian Dataset

The TUD-Brussels Pedestrian dataset was created with a moving platform in an urban environment. Crowded urban street behavior was recorded using a camera mounted on the front side of the vehicle. It can be useful in automotive safety scenarios in urban environments [35].

17.2.8 National Institute for Research in Computer Science and Automation (INRIA) Pedestrian Dataset

The National Institute for Research in Computer Science and Automation (INRIA) Pedestrian dataset is one of the popular static pedestrian detection datasets. It comprises moving human behavior with significant variation in pose, appearance, clothing, background, illumination, contrast, etc., coupled with movable cameras and complex background scenarios. [36].

17.2.9 PASCAL Visual Object Classes (VOC) 2017 and 2007 Dataset

The PASCAL Visual Object Classes (VOC) 2017 and 2007 dataset consists of static objects with different views and poses in an urban environment. The aim behind the creation of this dataset was to recognize the visual object classes in real-world scenes. The 20 different classes in this dataset include animal, tree, road sign notation board, vehicle, and pedestrian [37].

17.2.10 Microsoft Common Object in Context (COCO) 2018 Dataset

Microsoft [20] created the Common Object in Context (COCO) 2018 dataset recently for spur object detection along with a focus on detecting different objects in context. The annotations comprise different instances of objects related to 80 classes of objects and human segmentations for 91 different categories. There are key point annotations for pedestrian instances and five image labels per sample image. The COCO 2018 dataset challenges include (1) real-scene object detection with segmentation mask, (2) panoptic segmentation, (3) pedestrian key point evaluation, and (4) DensePose estimation in a crowded scene [38].

17.2.11 Mapillary Vistas Research Dataset

The Mapillary Vistas Research dataset was utilized for real-world scene segmentation on street images [21]. Panoptic segmentation resolves both pedestrian and different non-living classes, consolidating the concept of semantic and instance segmentation tasks efficiently.

Table 17.1 shows a comparative analysis of the pedestrian databases with their purposes for video surveillance. We have also included our proposed dataset, discussed later in the chapter. The association is performed in terms of the use of the dataset, size of the dataset, real-world environment scenarios, type of labeling, and annotation. These details are used for validating the object detection and tracking algorithm performance [39].

17.2.12 Proposed Pedestrian Dataset in Academic Institution

We propose a pedestrian database consisting of different behaviors of students under different conditions in academic activities such as students studying in a practical lab, examination hall scenarios, classrooms, a student cheating in an exam hall, a student taking an answer book outside the exam hall, a student stealing a mobile phone or other electronic devices such as a mouse or keyboard, a student stealing lab equipment, a student dispute on the college premises, a student disturbing another student, a student threatening another student, etc. [41].

Students' behavior on college premises is recorded using a high-quality DSLR camera from different viewing angles. The video is recorded at 30 f/s. The database includes 100 sample videos of approx. 20- to 30-minute duration for each sample video. Figure 17.2 shows sample frames of the video sequence available in the dataset.

17.3 Pedestrian Detection – Issues and Challenges

Moving object features are constantly changing as they move over time in video captured by the camera. In a video surveillance system, the RoI is a pedestrian that must be identified and tracked in the video [22]. However, this is tedious because of the many challenges and difficulties involved in real time. These difficulties happen at three distinct levels of the pedestrian detection life cycle: problems related to the camera, problems related to video acquisition, and problems related to detecting persons. Camera problems include camera motion and non-rigid object deformation.

TABLE 17.1
Comparative Analysis of Pedestrian Datasets

Data Source	Purpose	Image or Video Clips	Annotation	Environment	Ref.	Year
Caltech Pedestrian dataset	Detection and tracking of pedestrians walking on the street	250,000 frames (in 137 approximately minute-long segments)	350,000 bounding boxes and 2,300 unique pedestrians were annotated	Urban environment	[29]	2012
MIT	City street pedestrian segmentation, detection and tracking	709 pedestrian images, 509 training and 200 test images	No annotated pedestrian	Daylight scenario	[30]	2000, 2005
Daimler	Detection and tracking of pedestrians	15,560 pedestrian samples, 6,744 negative samples	2D bounding box overlaps criterion and float disparity map and a ground truth shape image	Urban environment	[31]	2016
ATCI	Rear-view pedestrian segmentation, detection, and tracking	250 video sequences	200K annotated pedestrian bounding boxes	Dataset was collected in both day and night scenarios, with different weather and lighting conditions	[32]	2015
NICTA 2016	Segmentation, pose estimation, learning of pedestrians	25,551 unique pedestrians, 50,000 images	2D ground truth image	Urban environment	[33]	2016
ETH	Segmentation, detection, tracking	Videos	Dataset consists of other traffic agents such as different cars and pedestrians	Urban environment	[34]	2010
TUD-Brussels	Detection, tracking	1,092 image pairs	1,776 annotated pedestrian	Urban environment	[35]	2009

(Continued)

TABLE 17.1 (Continued)

Comparative Analysis of Pedestrian Datasets

Data Source	Purpose	Image or Video Clips	Annotation	Environment	Ref.	Year
INRIA	Detection, segmentation	498 images	Annotations are marked manually	Urban environment	[36]	2005
PASCAL VOC 2012	Detection, classification, segmentation	11,530 images, 20 object classes	27,450 ROI annotated 6,929 segmentations	Urban environment	[37]	2012
MS COCO 2018	Object detection, segmentation, keypoint detection, DensePose detection	300,000 images, 2 million instances, 80 object categories	5 captions per image	Urban environment	[21]	2018
Mapillary Vistas dataset 2017	Semantic understanding of street scenes	25,000 images,152 object categories	Pixel-accurate and instance-specific human annotations for understanding street scenes.	Urban environment	[38]	2017
Proposed Pedestrian Dataset in Academic Institution	Detection, tracking	3,60,000 frames/sec	2000 annotated pedestrian	Urban environment daylight conditions	[40]	2021
MS COCO 2017	Recognition, segmentation, captioning	328,124 images, 1.5 million object instances	Segmented people and objects	Urban environment	[37]	2017
MS COCO 2015	Recognition, segmentation, captioning	328,124 images, 80 object categories	Segmented people and objects	Urban environment	[37]	2015

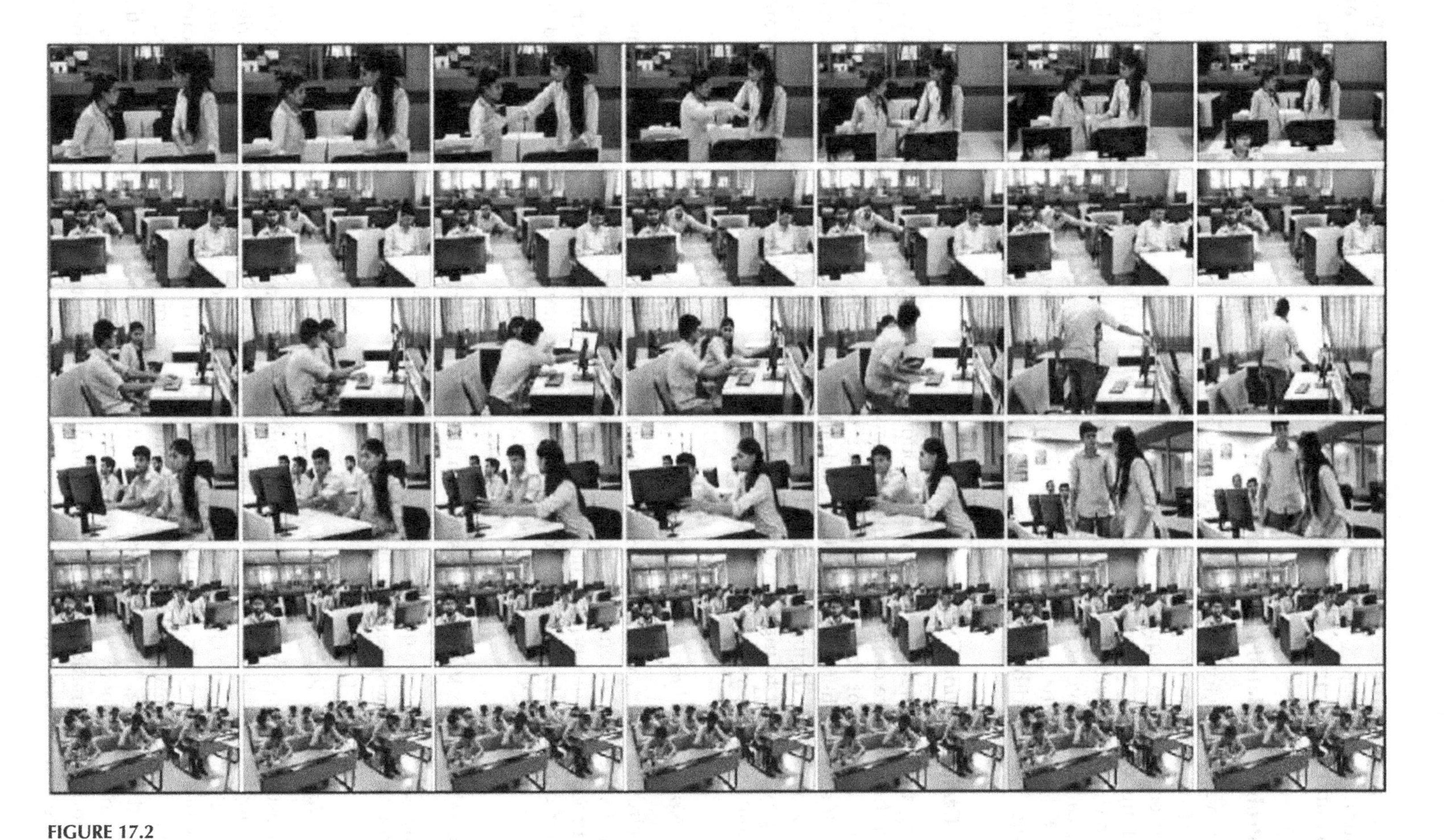

FIGURE 17.2
Sample Images of Proposed Database. The First Row Illustrates Two Girls Disputing in the Lab. The Second Row Illustrates the Scenario of Students Stealing the Mobile Phone in a Lab. The Third Row Illustrates a Scenario of a Student Threatening Another Student. The Fourth Row Shows the Same Threatening Scenario with the Front View. The Fifth Row Shows the Scenario of Students Stealing the Lab Equipment. The Sixth Row Shows the Scenario of Cheating in the Exam Hall.

Video acquisition problems involve illumination changes, abrupt changes, complex backgrounds, and object shadows. Human detection and tracking challenges are occlusion and pose variation. Each issue and challenge is described in this section.

17.3.1 Problems Related to the Camera

There are various factors associated with video acquisition methods. For example, the stability of sensors can immediately affect the quality of the video. In some instances, the method used for video acquisition might create boundaries for designing object detection and tracking systems. Moreover, blur reduce and block affects the quality of video [23]. Noise is a different factor that can seriously deteriorate the quality of the image. Besides, multiple cameras have sensors, frame rates, and dropping features that affect image quality. A low-quality or low-resolution image can influence the accuracy of moving object detection algorithms. Figure 17.3 shows examples of these challenges.

17.3.1.1 Camera Motion

Measuring and remunerating the camera motion is necessary when working with moving objects in the presence of moving cameras. However, these are not straightforward tasks to perform because of the potential camera's depth variations and complicated movements. In literature, there are several methods that use simple movement cameras such as Pan Tilt Zoom (PTZ) digital cameras. This controlled head movement facilitates using a homography in order to manage camera motions, which results in creating a panorama background complete video [2].

17.3.1.2 Non-rigid Object Deformation

In some circumstances, various parts of a moving object might have complex movements in terms of speed and orientation. For instance, a walking dog wags its tail and rotates its body. It produces a tremendous challenge, especially for non-rigid objects in the presence of moving cameras. In Artacho and Savakis [41], articular models were presented for moving non-rigid objects to handle non-rigid object deformation.

17.3.2 Challenges in Video Acquisition in the Real World

The challenges of video acquisition are generally related to the environmental conditions and surrounding atmosphere lighting and visibility.

17.3.2.1 Illumination Variation Scenarios

The contrast lighting scenarios of the scene and the target might change due to motion, the view angle of the light source, different times of day or night, reflection from bright planes, whether the scene is indoors or outdoors, partial or complete occlusion of the light source by other objects, etc. The direct impact of these differences results in background appearance changes, which cause false positive detections for the methods based on background modeling. Thus, it is essential for these methods to adapt their model to this illumination variation in the real world. Meanwhile, because the object's appearance changes under illumination variation, appearance-based detection and tracking methods may not be able to track the object in the video sequence [42].

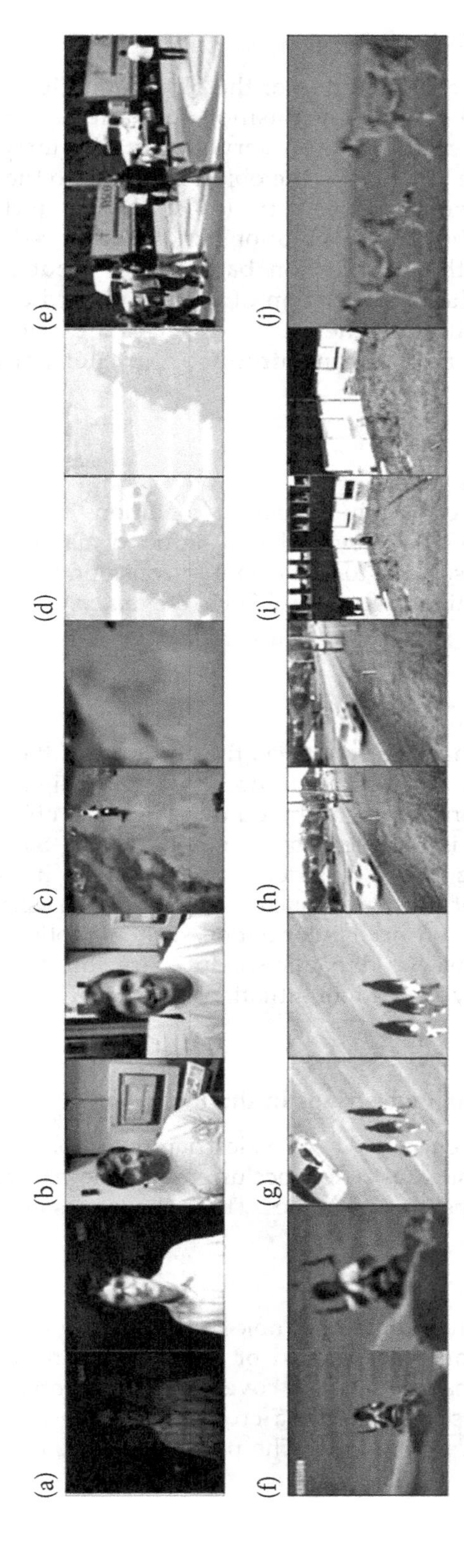

FIGURE 17.3
Issues and Challenges of Pedestrian Detection. (a) Illumination Variation Example (Ross dataset [2]). (b) Appearance Change Example (Ross Dataset [1]). (c) Abrupt Motion Example (COCO Dataset [38]). (d) An Example of Occlusion Challenge (Kalal Dataset [1]). (e) An Example of Free Motion of Camera in the Michigan University Dataset [2]. (f) An Example of Dynamic Background Challenge (Zhang Dataset [2]). (g) An Example of Shadow Challenge (Pedestrian 4 in the Kalal Dataset [2]). (h) Panning in Camera Example in the CDNET Database [2]. (i) Zooming in Camera Example in the CDNET Database [2]. (j) Non-rigid Moving Object Example in a Video Sequence [2].

17.3.2.2 Presence of Abrupt Motion

Sudden changes in the speed and direction of the object's motion or sudden camera motion are another challenge of video acquisition that affects object detection and tracking. If the object or the camera moves very slowly, the temporal differencing methods may fail to detect the portions of the object coherent to the background [43]. Meanwhile, a very fast motion produces a trail of a ghost detected in the region. So, if this object's motion or the camera's motion is not considered, the object cannot be detected correctly by methods based on background modeling. On the other hand, for tracking-based methods, prediction of motion is hard or even impossible. As a result, the tracker might lose the target. Even if the tracker does not lose the target, the unpredictable motion can introduce a greater amount of error into algorithms [44].

17.3.2.3 Complex Background

The background may change, especially in natural conditions with high variability of textures, such as outdoor scenes. Moreover, the background may be dynamic or static; it may contain movement. These need to be taken into consideration in the background of many moving object detection algorithms. Such movements can be sequential or non-sequential [45].

17.3.2.4 Shadows

The appearance of shadows on an object affects the accuracy of the moving object detection algorithm. Shadows are formed due to occlusion of the light source by another object. Static shadow is generated as background when the object is not moving. However, a dynamic shadow is produced by a moving object. Since it has the same motion features as the moving object, it is deeply connected to it. For the removal of shadows, one can use different properties such as texture, color, shape, or applying a shadow detection model based on prior information such as motion, illumination conditions, and variation [46]. However, dynamic shadows are still difficult to distinguish from moving objects, especially in outdoor situations where the background is usually complex.

17.3.3 Challenges of Detecting Pedestrians in the Real World

There are several real-world scenarios where pedestrian detection is a challenging task due to natural circumstances such as object occlusion and the natural behavior of pedestrians as well as pose or appearance changes. These challenges are described in brief.

17.3.3.1 Object Occlusion

Objects may be blocked or occluded by other objects in a real-world scene. In this case, some portions of the object can be concealed or hidden behind other objects, called *partial occlusion*, or the object can be entirely covered by other objects, called *complete occlusion*. For example, suppose you are in a crowded area recoding video of a pedestrian walking on the sidewalk or street. The pedestrian may be occluded by other pedestrians, cars, trees, etc.

17.3.3.2 Moving Object View Angle Changes – Pose Variation

In real-world situations, most objects occur in 3D and 2D planes. Hence, each rotation in the direction of the third axis may change the object's view and appearance [28]. Pedestrian detection and tracking algorithm accuracy gets impaired due to fluctuations in the pose. The same pedestrian may look different in sequential frames. The appearance can vary continuously. Moreover, objects may have unusual variations in their appearance, such as facial expressions, light variation, different color clothes, hats, masks on the face, etc., that the tracking algorithm does not recognize. Again, problems include non-rigid objects and changes in appearance over time [47–49].

17.4 Recent Developments and Architectural Innovations in CNN

There are several methods for enhancing pedestrian detection accuracy using deep learning architecture [58–51], as shown in Figure 17.4. However, pedestrian detectors need to overcome several issues produced by complex backgrounds, varying scales of pedestrians in the captured image, occlusion, illumination changes, etc. These issues are partially addressed, which significantly influences the performance of pedestrian detectors. Most of the pedestrian detectors utilize HOG [2]. In this approach, gradient information is applied for detecting the object in various directions. The weakness of this approach is 1) somewhat occluded pedestrians are not detected accurately, and 2) the computation process is time-consuming. Hence, it is not a proper fit for a real-time video surveillance system.

Moreover, Minkesh et al. [47] exhibited a mixed approach of LBP and HOG to manage partial occlusion of pedestrians. Piotr Dollar et al. [1] introduced a consolidated approach of HOG descriptors and LUV where L represented luminance, and U and V components represented a chromaticity value of color in an image. For fast computation and increasing, the accuracy of pedestrian detection researcher goes toward the region-based deep learning strategies to overcome the existing issues of the hand-crafted feature-based approach [39,41]. Nathan Silberman et al. [49] exhibited the segmentation approach by employing a coverage loss method for object segmentation in an indoor environment. In this technique, CNN and scale-invariant feature transform (SIFT) are used in combination to extract meaningful features. The limitations of this approach are 1) due to tree structure description, non-neighboring sections are not connected to form a segment when the objects in a scene are blocked or occluded; 2) the coverage loss function does not resolve the problem of false positives; 3) the object detection and instance segmentation process are slow due to heavy computation involvement; and 4) these methods' performance is poor without depth information of the image. Artacho and Savakis [41] presented an instance of segmentation with deep, densely correlated Markov Random Fields (MRF). The architecture blends patch-based CNN prediction and global MRF reasoning. In this approach, input image instances have been extracted using different size patches (270 x 432, 180 x 288, and 120 x 192). CNN has been used for assigning instance labels to each extracted patch. MRF consist of the unidirectional graphical model. Each vertex represents an instance label of each pixel. This method has several drawbacks, such as 1) it is suitable only for a single type of object, 2) interconnected object consideration fails in the case of an occluded object, and 3) object

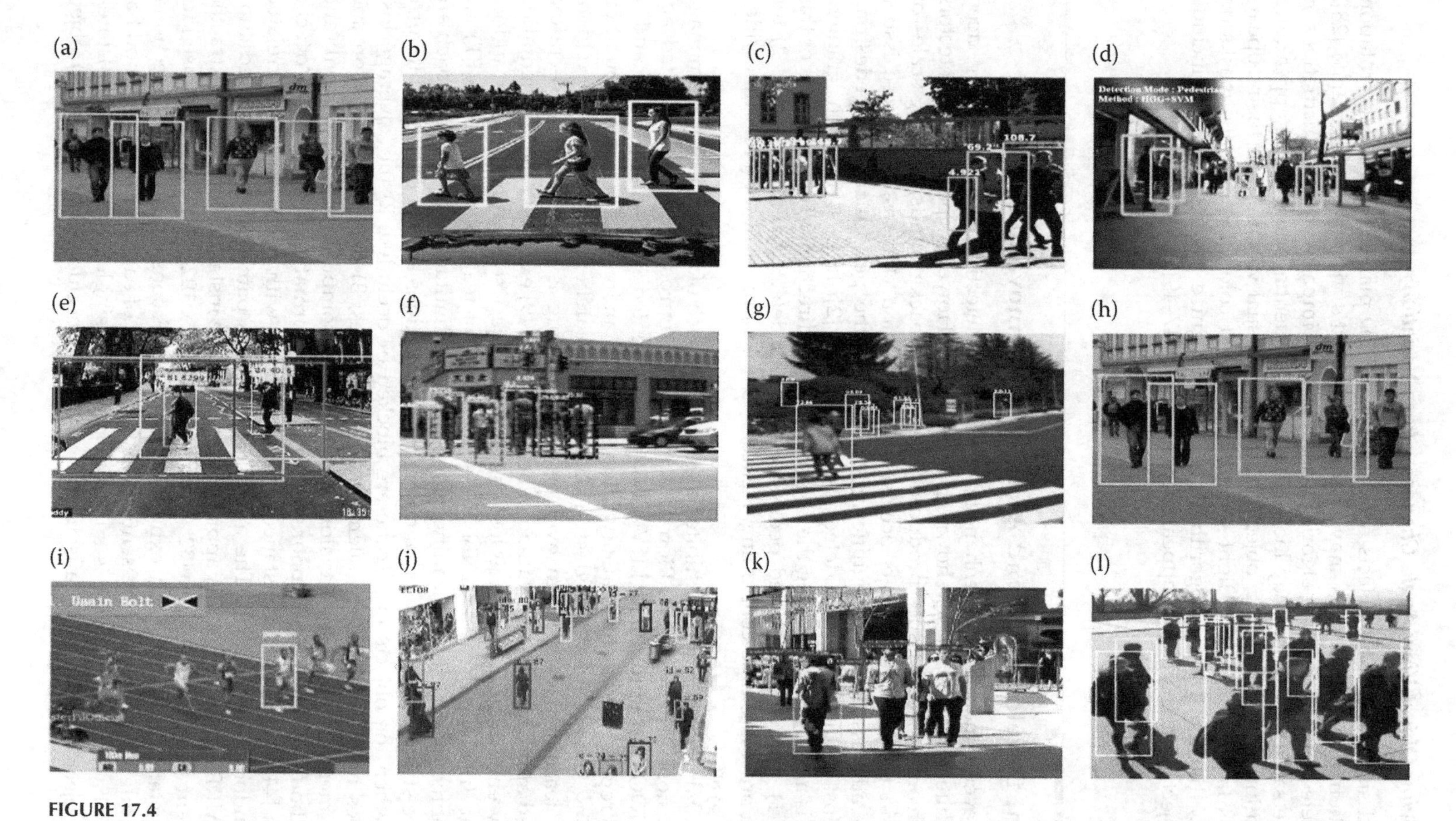

FIGURE 17.4
Example of Pedestrian Detection and Tracking. (a) Detecting Pedestrians Outdoors, Walking along the Street. (b) ADAS Pedestrian Detection. (c) Pedestrian Detector Is Based on the Aggregate Channel Feature Detector. (d) Real-Time Vehicle and Pedestrian Detection of Road Scenes. (e) Pedestrian Action Prediction Is Based on the Analysis of Human Postures in the Context of Traffic. (f) Pedestrian Detection Based on the Hierarchical Co-occurrence Model (g) Cross-modal Deep Representations for Robust Pedestrian Detection. (h) Pedestrian Detection OpenCV. (i) Object Tracking with dlib C++ library. (j) Multiple Object Tracking with Kalman Tracker. (k) Multi-class Multi-object Tracking using Changing Point Detection. (l) Pedestrian Tracking using Deep-occlusion Reasoning Method.

detection fails in the case of small object size. This method achieves an average false positive rate of 83.9% and an average false negative rate of 0.375%. If a predicted instance does not overlap with any ground truth instance, then it is a false positive instance. Similarly, if a ground truth instance does not overlap with any prediction, then it is a false negative instance.

Gawande et al. [48] presented a Relief R^2-CNN for pedestrian extraction in real-time environments. In this approach, the main focus was on faster RoI generation from the convolutional features of trained CNN. Here, input normalized feature maps were used to produce the RoI. A normalized feature map was generated by splitting each feature map element by its maximum feature map value. Object detection and classification are similar to R-CNN. This approach can be beneficial in person re-identification. Drawbacks of this approach are 1) the R^2-CNN is not tested for real-time application, and 2) the classification task requires more time because R^2-CNN requires repetitive fine-tuning of object localization. Kelong Wang et al. [32] suggested a unified joint detection structure for cyclist pedestrian detection. The method uses Fast R-CNN. In this approach, multi-layer feature fusion and multi-target candidate selection is composed to improve detection accuracy and to determine the difficulties of frequent false detection and missed detection of pedestrian and cyclist targets.

17.5 Recent Open-Source Libraries and Platforms for Deep Learning

Many useful open-source libraries and platforms are available for designing computer vision applications that utilize deep learning. These libraries are free to users for experimentation and research. The main use of the following libraries and platforms is to provide a flexible, scalable, platform-independent, and user-friendly development environment for machine learning applications.

17.5.1 TensorFlow

TensorFlow is an extensively used open-source library developed using JavaScript. Its main feature is fast development of machine learning models and efficiently performing deep learning computation. It can operate smoothly on Grpahics Processing Unit (GPU) as well as multi-core Central Processing Unit (CPU). TensorFlow uses a lightweight embedded device called TensorFlow Lite. It supports different neural network operations such as training, testing, validation, and deployment for large products.

17.5.2 Keras

Keras is another efficient open-source neural network library developed in Python language. It runs on top of Microsoft Cognitive Toolkit, TensorFlow, PlaidML, and Theano. It is a scalable, extensible, and modular library. Keras provides fast experimentation, parallelism, and deep neural network support for large-volume data processing with efficiency. It is mostly utilized in artificial intelligence and deep learning models.

17.5.3 PyTorch

PyTorch is another open-source Python machine learning framework-based library. It accelerates the deployment and prototyping of machine learning algorithms. PyTorch was designed using object-oriented language, i.e., C++ for security. It is a combination of a deep neural network and a tensor framework for storing and processing features, graph traversing, GPU task handling, etc. It is mostly used for Natural Language Processing (NLP) and several computer vision applications.

17.5.4 Scikit-Learn

Scikit-learn is an actively used machine learning library in Python. It incorporates easy alliance with other ML libraries like Pandas and NumPy. Scikit-learn facilitates various algorithms for classification, clustering, regression, model selection, dimension reduction, etc. It is easy to use the focused library for data modeling. It handles complex data-relevant tasks such as storing, manipulation, handling, and visualization.

17.5.5 Pandas

Pandas is one of the most popular open-source data analysis libraries. It is utilized for the manipulation and interpretation of complex equations and formulas. Pandas is effective in multi-dimensional series computation for machine learning programmers. The important features of Pandas consist of dataset normalization, pivoting, merging and integration of datasets, handling of false data with alignment, and various indexing options for improving searching and sorting functionality, such as hierarchical axis indexing, data filtration options, and fancy indexing. Pandas utilizes data frames, which is a two-dimensional description of data that allows programmers to create customized data frame objects.

17.5.6 Spark MLlib

Apache Spark MLlib is an open-source machine learning library that facilitates easy scaling of your complex computations. It is quick, simple, and easy to configure the library. It allows fluid integration with other tools. Spark MLlib immediately became a suitable tool for developing machine learning algorithms and applications. Some of the important algorithms and APIs that programmers working on machine learning using Spark MLlib employ are regression, clustering, optimization, dimensional reduction, classification, basic statistics, and feature extraction.

17.5.7 Theano

Theano is an open-source Python library that aids you to define, optimize, and evaluate mathematical formulations comprising multi-dimensional arrays. Theano's main feature is that it provides excellent integration with NumPy and uses GPU to deliver fast data-intensive calculations. Furthermore, it has an effective symbolic differentiation and allows dynamic code generation in the C programming language. Theano is mainly designed to perform complex mathematical neural network algorithms used in deep learning. Hence, it is a convincing tool for developing deep learning projects. It can handle complex structures and convert them into effective code that utilizes NumPy and other native fundamental libraries.

17.5.8 MXNet

MXNet is used to train and deploy deep neural network-based architecture. It is highly scalable and faster when building a trained model. Apache's MXNet is compatible with programming languages such as Python, C, C++, Perl, Julia, R, Scala, Go, etc. MXNet's key feature is it is platform-independent and adaptive per a project's requirements. Companies such as Amazon's AWS, Intel, Microsoft, and MIT support and use MXNet to develop deep learning framework-based applications.

17.5.9 Google Colab

Google Colaboratory is an open-source platform provided by Google for performing deep learning-based experiments on a cloud platform. This platform facilitates the in-built GPU, command-line execution of arbitrary Python script, installation, and downloading facility of open-source packages of Python libraries. The GUI is user friendly for ease of user experience compared to the other platforms available on the market. The Google Colab platform is fast and efficient, with a required feature for deep learning-based computer vision applications in real time.

17.6 Case Study

Ross Girshick et al. proposed Fast R-CNN [11] to solve the problem of scale variance by scaling the image using a brute-force technique. Yunchao Gong et al. [30] presented a multi-scale filtered approach on all objects of various sizes in an image. However, this is a difficult task because the same object may appear to be different sizes in different instances or may appear to be similar to other objects in the image. The available approach in the literature is less accurate as well as computationally complex and requires more time to detect objects of different sizes. The state-of-the-art method for handling scale-variance problems is Scale-Aware Fast R-CNN (SAF R-CNN), which was proposed by Jianan Li et al. [43].

The divide-and-conquer technique was employed with Fast R-CNN to resolve the scale variance issue. SAF R-CNN utilized a large-size and small-size sub-network combination for each instance in the image. The confidence cumulative score was computed for each instance of the object. The scale-aware layer was allocated to the higher weight in the large-size sub-network and a lower size weight in the small sub-network for robust detection of the instances at various scales. The limitation of this approach was 1) irrespective of the scale of instances in the proposal, both the sub-networks computed the cumulative confidence score for each instance, which increased the time needed for the detection and identification of the object in the image, and 2) the small- and large-size instances of an object require additional time for training and testing.

The proposed scale-invariant, Mask R-CNN, addresses the existing method issues by using the scale-invariant feature map generation algorithm. The main idea of scale-invariant feature map generation is a scale-invariant feature map with a two-stage backbone network built into a unified CNN architecture. SIM R-CNN in the first phase takes the original input image through the bottom shared convolutional layers to extract all its feature maps. These feature maps are convolved with different-size filtered masks to generate the confidence score, which is later used to generate the final detection results, defined over the proposal size. The conclusive results can always be enhanced by the scale-invariant feature map generation process and two-stage backbone deep network, appropriate for the current input of different scales. Therefore, SIM R-CNN can detect performance in a wide range of input scales of an image. This framework can be utilized in crowded areas to accurately detect different-size pedestrians.

17.7 Conclusions and Future Directions

Deep learning framework usage in real-time object detection is an active research area in computer vision. This chapter provided a comprehensive overview of the deep learning framework, its application to pedestrian detection, and recent deep learning-based pedestrian detection methods. We can say from analysis of state-of-the-art literature that deep learning frameworks are more accurate than CNNs but require higher computation power and more time for processing a large volume of data. In terms of pedestrian detection, all the issues available in the literature are not address by a single method. There is still scope for improvement in deep learning frameworks. In the proposed case study, we addressed the scaling issue of images. Different-size pedestrians were not detected accurately by the deep learning framework. We proposed a novel approach for scale-invariant Mask R-CNN that detects small- and large-size pedestrians effectively. Finally, we proposed several encouraging future directions to gain a thorough knowledge of object detection. There is room for improvement in terms of multiple pedestrian detection in real time using a deep learning framework, reduction of time and computational complexity of the deep vision system using unsupervised learning techniques, etc. This review is also meaningful for the development of deep neural networks based on computer vision applications using open-source libraries.

References

1. Dollar, P., Wojek, C., Schiele, B., and Perona, P. (2012, April) "Pedestrian Detection: An Evaluation of the State of the Art." *IEEE Transactions on Pattern Analysis and Machine Intelligence (TPAMI)* 34 (4): 743–761.
2. Dalal, N. and Triggs, B. (2005) "Histograms of Oriented Gradients for Human Detection", Proceeding of IEEE Conference on Computer Vision and Pattern Recognition (CVPR) (pp. 886–893), San Diego, CA, USA, 20th-25th June 2005.
3. Ess, A., Leibe, B., and Gool, L. (2017) "Depth and appearance for mobile scene analysis", Proceeding of IEEE International Conference on Computer Vision (ICCV) (pp. 1–8), Venice, Italy, 22nd-29th Oct. 2017.
4. Bochkovskiy, A., Wang, C.-Y., Liao, H.-Y. (2020) "YOLOv4: Optimal Speed and Accuracy of Object Detection", Proceeding of International Conference of Computer Vision and Pattern Recognition (pp. 1–17), Seattle, Washington, 16th–19th June 2020.
5. Long, X., Deng, K., Wang, G., Zhang, Y., Dang, Q., Gao, Y., Shen, H., Ren, J., Han, S., Ding, E., Wen, S. (2020) "PP-YOLO: An Effective and Efficient Implementation of Object Detector",

Proceeding of International Conference of Computer Vision and Pattern Recognition (pp. 1–8), Seattle, Washington, 16th–19th June 2020.

6. Redmon, J., Divvala, S., Girshick, R., and Farhadi, A. (2016) "You Only Look Once: Unified, Real Time Object Detection", Proceeding of IEEE Conference on Computer Vision and Pattern Recognition (CVPR) (pp. 1–10), Las Vegas, Nevada, USA, 26th June–1st July 2016.
7. He, K., Gkioxari, G., Dollar, P., and Girshick, R.(February, 2020) "Mask R-CNN." *IEEE Transactions on Pattern Analysis and Machine Intelligence* 42(2): 386–397.
8. He, K., Gkioxari, G., Dollar, P., and Girshick, R. (2017) "Mask R-CNN", Proceeding of IEEE International Conference on Computer Vision (ICCV) (pp. 2980–2988), Venice, Italy, 22nd–29th October 2017.
9. Ren, S., He, K., Girshick, R., and Sun, J. (2017) "Faster R-CNN: Towards Real-Time Object Detection with Region Proposal Networks." *IEEE Transactions on Pattern Analysis and Machine Intelligence (TPAMI)* 39(6): 1137–1149.
10. Ren, S., He, K., Girshick, R., and Sun, J. (2015) "Faster R-CNN: Towards Real-Time Object Detection with Region Proposal Networks", International Conference on Neural Information Processing Systems (NIPS), Montreal (pp. 1–9), Quebec, Canada, 7th–12th Dec. 2015.
11. Girshick, R. (2015) "Fast R-CNN", International Conference on Computer Vision (ICCV) (pp. 1441–1448), Santiago, Chile 7th–13th Dec. 2015.
12. Dai, J., Li, Y., He, K., and Sun, J. (2016) "R-FCN: Object Detection via Region-based Fully Convolutional Networks", Proceeding of EEE Conference on Computer Vision and Pattern Recognition (CVPR) (pp. 1–11), Las Vegas, Nevada, USA, 26th June–1st July 2016.
13. Dai, J., Li, Y., He, K., and Sun J. (2016) "R-FCN: Object Detection via Region-Based Fully Convolutional Networks", Proceeding of IEEE Conference on Computer Vision and Pattern Recognition (CVPR) (pp. 1–11), Las Vegas, Nevada, USA, 26th June–1st July 2016.
14. Anguelov, L., Erhan, D., Szegedy, C., Reed, S., Fu, C.-Y., and Berg, A. (2016) "SSD: Single Shot MultiBox Detector", European Conference on Computer Vision (ECCV) (pp. 1–17), Amsterdam, Netherlands, 11th–14th October 2016.
15. Anguelov, J., Shelhamer, E., and Darrell, T. (2015) "Fully Convolutional Networks for Semantic Segmentation", Proceeding of IEEE Conference Computer Vision and Pattern Recognition (CVPR) (pp. 1–10), Boston, Massachusetts, 8th–10th June 2015.
16. Radford, A., Metz, L., and Chintala, S. (2015) "Unsupervised Representation Learning with Deep Convolutional Generative Adversarial Networks", Proceeding of IEEE Conference Computer Vision and Pattern Recognition (CVPR) (pp. 1–10), Boston, Massachusetts, 8th–10th June 2015.
17. He, K., Zhang, X., Ren, S. and Sun, J. (2015) "Deep Residual Learning for Image Recognition", Computer Vision and Pattern Recognition (CVPR) (pp. 1–10), Boston, Massachusetts, 8th–10th June 2015.
18. Szegedy, C., Ioffe, S., Vanhoucke V., and Alemi A. (2016) "Inception-v4, Inception-ResNet and the Impact of Residual Connections on Learning", Proceeding of IEEE Conference Computer Vision and Pattern Recognition (CVPR) (pp. 1–10), Las Vegas, Nevada, USA, 26th June–1st July 2016.
19. Simonyan, K. and Zisserman, A. (2015) "Very Deep Convolutional Networks for Large-Scale Image Recognition", Proceeding of IEEE Conference Computer Vision and Pattern Recognition (CVPR) (pp. 1–10), Boston, Massachusetts, 8th–10th June 2015.
20. Zeiler, M. and Fergus, R. (2013) "Visualizing and Understanding Convolutional Networks", Proceeding of IEEE Conference Computer Vision and Pattern Recognition (CVPR) (pp. 1–11), Portland, Oregon, USA, 23rd–28th June 2013.
21. Alex, K., Sutskever, I., and Hinton, G. (2012) "ImageNet Classification with Deep Convolutional Neural Networks", International Conference on Neural Information Processing Systems (NIPS) (pp. 1–9), Lake Tahoe, Nevada, United States, 3rd - 6th Dec. 2012.
22. Kangming, L. (March, 2016) "Research on an Improved Pedestrian Detection Method Based on Deep Belief Network (DBN) Classification Algorithm." *International Journal of Information Systems and Technologies (RISTI)* 17(3): 7–87.

23. Lecun, Y., Bottou, L., Bengio, Y., and Haffner, P. (Nov, 1998) "Gradient-Based Learning Applied to Document Recognition." *Proceedings of the IEEE* 86(11): 2278–2324.
24. Kang, S., Byun, H., and Lee, S. (2002) "Real-Time Pedestrian Detection Using Support Vector Machines", First International Workshop on SVM: Pattern Recognition with Support Vector Machines (pp 268–277), Niagara Falls, Canada, 10th August 2002.
25. Geronimo, D., Sappa, A. D., Lopez, A. and Ponsa, D. (2006) "Pedestrian Detection using AdaBoost Learning of Features and Vehicle Pitch Estimation", International Conference on Visualization, Imaging, and Image Processing (pp. 1–8), Spain, 28th–30th August 2006.
26. Alireza, A., Mollaie, K., Reza, M., Yasser, B., Andi, S., and Hosein (2011) "Improved Object Tracking Using Radial Basis Function Neural Networks", International Conference on Machine Vision and Image Processing (MVIP), Tehran, Iran (pp. 1–5), Nov 16th–17th Nov. 2011.
27. Wu, C., Yue, J., Wang, L., and Lyu, F. (March, 2019) "Detection and Classification of Recessive Weakness in Superbuck Converter Based on WPD-PCA and Probabilistic Neural Network." *International Journal MDPI Electronics* 8(290): 1–17
28. Victor Emil, N., Cristian, T., and Mihai, N. (2009) A neural network approach to pedestrian detection, WSEAS International Conference on Computers (pp. 374–379), Wisconsin, United States, 23rd July 2009.
29. Caltech Pedestrian dataset, Computer Vision Lab, Available at: http://www.vision.caltech.edu/Image_Datasets/CaltechPedestrians/ [Accessed: 22nd September 2020]
30. MIT Pedestrian Dataset. Center for Biological and Computational Learning at MIT. Available at: http://cbcl.mit.edu/software-datasets/PedestrianData.html [Accessed: 3rd September 2020]
31. Daimler Pedestrian Detection Benchmark Dataset, Computer Vision Lab, Available at: http://www.gavrila.net/Research/Pedestrian_Detection/Daimler_Pedestrian_Benchmark_D/Daimler_Mono_Ped__Detection_Be/daimler_mono_ped__detection_be.html [Accessed: 3rd September 2020].
32. Advanced Technical Center (ATCI) Pedestrian dataset, Computer Vision Lab, Available at: https://sites.google.com/site/rearviewpeds1/ [Accessed: 3rd September 2020].
33. NICTA Pedestrian dataset, Available at: www.nicta.com.au/category/research/computer-vision/tools/automap-datasets/ [Accessed: 3rd September 2020].
34. ETH Pedestrian dataset, Available at: www.vision.ee.ethz.ch/~aess/dataset/ [Accessed: 3rd September 2020].
35. TUD-Brussels Pedestrian dataset, Computer Vision Lab, Available at: https://www.mpi-inf.mpg.de/departments/computer-vision-and-machine-learning/publications [Accessed: 3rd September 2020].
36. INRIA Pedestrian dataset, Available at: http://pascal.inrialpes.fr/data/human/ [Accessed: 3rd September 2020].
37. PASCAL Pedestrian dataset, Computer Vision Lab, Available at: http://pascallin.ecs.soton.ac.uk/challenges/VOC/databases.html/ [Accessed: 3rd September 2020].
38. Microsoft Common Object in Context (COCO) (2018) dataset, Microsoft, Available at: https://cocodataset.org/#download [Accessed: 3rd September 2020]
39. Mapillary Vistas Research Dataset, Available at: https://www.mapillary.com/dataset/vistas?pKey=cc5dEAyQECBFF9MN3MbdZA&lat=20&lng=0&z=1.5 [Accessed: 3rd September 2020]
40. Acdemic Environment proposed dataset, Available at: https://mega.nz/folder/YOYCzBrT#Yj-Dq93-TiNLULNiNNeIUA [Accessed: 3rd March 2021]
41. Artacho, B. and Savakis, A. (2019) "Waterfall Atrous Spatial Pooling Architecture for Efficient Semantic Segmentation." *Sensors, MDPI* 19(24): 1–17
42. Wang, X., Han, T.X., and Yan, S. (2009) "An HOG-LBP human detector with partial occlusion handling", IEEE International Conference on Computer Vision (pp. 32–39), Kyoto, Japan, 29th September–2nd October 2009.
43. Yunchao, G., Liwei, W., Guo, R., and Lazebnik, S. (2014) "Multiscale Orderless Pooling of Deep Convolutional Activation Features", 13th European Conference on Computer Vision (ECCV) (pp. 392–407), Zurich, Switzerland, 6th–12th Sept. 2014.

44. Li, J., Liang, X., Shen, S., Xu, T., Feng, J., Yan, S. (April, 2018) "Scale-Aware Fast R-CNN for Pedestrian Detection." *IEEE Transaction on Multimedia* 20(4): 985–996
45. Wang, K. and Zhou, W. (April, 2019). "Pedestrian and Cyclist Detection Based on Deep Neural Network Fast R-CNN." *International Journal of Advanced Robotic Systems", SAGE* 16(2): 1–10.
46. Pobar, M. and Ivasic-Kosm, M. (2018) "Mask R-CNN and Optical flow-based method for detection and marking of handball actions", 11th International Congress on Image and Signal Processing, BioMedical Engineering and Informatics (pp. 1–6), Beijing, China, 13th–15th Oct. 2018.
47. Minkesh, A., Worranitta, K. and Taizo, M.i. (2019) "Human extraction and scene transition utilizing Mask R-CNN", International Conference on Computer Vision and Pattern Recognition (CVPR) (pp. 1–6), California, United States, 16th–20th June 2019.
48. Gawande, U., and Hajari, K., and Golhar, Y. (April, 2020) "Pedestrian Detection and Tracking in Video Surveillance System: Issues, Comprehensive Review, and Challenges." *Recent Trends in Computational Intelligence*(pp. 1–24). IntechOpen.
49. Gawande, U., Zaveri, M., and Kapur, A. (Feb, 2013) "Bimodal Biometric System: Feature Level Fusion of Iris and Fingerprint." *ScienceDirect, Elsevier* 2013(3): 7–8.
50. Silberman, N., Sontag, D., and Fergus, R. (2014) "Instance Segmentation of Indoor Scenes Using a Coverage Loss", European Conference on Computer Vision (ECCV), Zurich, Switzerland, 6th–12th September 2014.
51. Paisitkriangkrai S., Shen C., and Hengel A. (2014) "Strengthening the effectiveness of pedestrian detection with spatially pooled features", 13th European Conference on Computer Vision (ECCV), Springer, Zurich, Switzerland (pp. 546–561), 6th–12th September 2014.
52. Dollar, P., Tu, Z., Perona, P., and Belongie, S. (2009) "Integral channel features", British Machine Vision Conference (BMVC), London, UK (pp. 1–11), 7th–10th September 2009.
53. Gawande, U. and Golhar, Y. (April, 2018). "Biometric Security System: A Rigorous Review of Unimodal and Multimodal Biometrics Techniques." *International Journal of Biometrics (IJBM), InderScience* 10(2): 142–175.
54. Wang, J. and Li, G., (May 2019) "Accelerate Proposal Generation in R-CNN Methods for Fast Pedestrian Extraction." *The Electronic Library, Emerald* 37(3): 1–19.
55. Zhang, Z., Fidler, S., and Urtasun, R. (June, 2020) "Instance-Level Segmentation for Autonomous Driving with Deep Densely Connected MRFs." *Computer Vision and Pattern Recognition (CVPR)* 19(3): 98–118.
56. Wang, X., Wang, M., and Li, W. (March, 2014) "Scene-Specific Pedestrian Detection for Static Video Surveillance." *IEEE Transactions on Pattern Analysis and Machine Intelligence (TPAMI)* 36(2): 361–374.
57. Tian, Y., Luo, P., Wang, X., and Tang, X. (2015) "Pedestrian detection aided by deep learning semantic tasks", Computer Vision and Pattern Recognition (CVPR) (pp. 1–10), Boston, Massachusetts, 8th–10th, June 2015.
58. Zhang, S., Benenson, R., and Schiele, B. (2015) "Filtered channel features for pedestrian detection", Computer Vision and Pattern Recognition (CVPR) (pp. 1751–1760), Boston, Massachusetts, 8th–10th June 2015.
59. Saberian, M., Cai, Z., and Vasconcelos, N. (2015) "Learning complexity-aware cascades for deep pedestrian detection", International Conference on Computer Vision (ICCV) (pp. 1–10), Santiago, Chile, 7th–13th December 2015.
60. Gawande, U., Hajari, K., and Golhar, Y. (Feb, 2020) "Deep Learning Approach to Key Frame Detection in Human Action Videos." *Recent Trends in Computational Intelligence 1*: 1–17.

18

Crop Yield Forecast Using a Hybrid Framework of Deep CNN with RNN Technique

S. Radha Rammohan[1,4], V. R. Niveditha[2,4], K. Amandeep Singh[2,4], and T. Yuvarani[3,4]

[1]*Professor, Department of Computer Applications*
[2]*Research scholar, Department of Computer Science and Engineering*
[3]*Assistant professor, Department of Computer Science and Engineering*
[4]*Dr.M.G.R Educational and Research Institute, Maduravoyal, Chennai*

18.1 Introduction

Agriculture is the primary resource for an extensive area of about 58% of India's population. Agricultural production mainly aims to generate a high yield for crops [1]. Prediction of the crop on a global scale and regional scale is highly important for the agriculture management sector, crop farmers, trading of food, and carbon cycle research [2,3]. Prediction of crop production is a national priority in order to maintain the high demand and food supply to the people. For example, India has major crop production of rice and wheat. The Ministry of Statistics and Program Implementation (MOSPI) conducted three surveys on the improvement of crop statistics related to crop yield from 2011 to 2018, to improve technical guidance for obtaining reliable and timely crop yield estimation. Also, low-populated countries also consider crop yield to be an important area to attain economic stability through agro-trade.

In general, crop production is predicted by several variables, such as water use, planting, area, soil quality, weather conditions, disease occurrences, and so on [4–6]. Figure 18.1 shows the top factors. Of these, weather monitoring, soil parameters, and crop area are considered in this chapter. Data can be gathered using engineering technologies such as wireless sensor networks, remote sensing, etc. Generally, environmental weather and soil quality data are obtained from wireless sensor networks. Data related to soil quality and weather monitoring are collected from the sensors, and the obtained data act as an input to be trained with deep learning algorithms [7]. Deep learning techniques are used to train large data, and various convolution techniques are used to obtain maximum

DOI: 10.1201/9781003166702-18

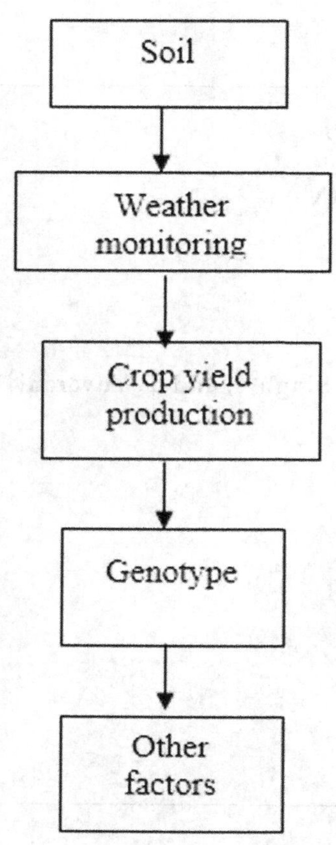

FIGURE 18.1
Various Factors Affecting Yield Prediction of Crops.

accuracy [8]. The main advantage of using deep learning techniques in agriculture applications is that the data get trained with high-level features in a hierarchical and incremental manner, which eliminates the need for domain expertise to generalize the obtained output [9].

This chapter introduces a deep learning system focused on environmental data based on soil and weather practices that will benefit a new model and techniques to predict paddy production. The deep learning algorithm belongs to the description methods based on class, with numerous representation levels, and every non-linear framework converts representation to slightly more abstract levels [9,10]. Deep learning techniques are used to train large data, and various convolution techniques are used to obtain maximum accuracy [11].

This chapter proposes a hybrid deep convolutional neural network (CNN) with a recurrent neural network (RNN) model that can process the data with multiple arrays, namely signals, images, and videos. The hybrid model is composed of CNN with RNN, which usually contains several convolution layers, pooling layers, and fully connected (FC) layers. There are certain proposal parameters for CNNs, namely filter number, size of the filter, stride, and padding type. A filter is a weight matrix with which we convert the weights of data inputs. To preserve the dimension of the input space, padding is the method of adding zeros to the input. The stride can be used to move the filter. RNNs are used to capture time dependencies for tasks containing sequential data [9,12]. In their hidden units, RNNs retain the history of all the past elements of a sequence, called a *state vector*. By using this information, they process unique components of the input sequence at a time. RNNs are very effective at sequence modeling, but owing to disappearing and exploding, gradient training the data has proved to be very challenging [13]. RNNs are augmented by long short-term memory (LSTM) cells to solve this problem, which are carefully engineered recurrent neurons that offer better output in a wide variety of uses through sequence modeling [14,12,11]. Deep learning methods have been applied in many studies of crop yield prediction. This proposed work discusses the current vulnerabilities that encourage the development of smart agriculture and contribute to an increase in food production. In order to predict paddy crop yield across 987 locations in the Tanjore districts between 1995 and 2018, Khaki and Wang developed a deep neural network model [15].

The remainder of the chapter is structured as follows. A literature analysis of current works is discussed in section 18.2. Section 18.3 describes the agriculture dataset. Section 18.4 briefs about the hybrid deep CNN and RNN framework model for forecasting crop yield based on weather prediction and yield prediction. Section 18.5 presents the experimental findings based on output estimation, measurement parameters, properties of data distribution, and precision measurements. Section 18.6 provides a conclusion.

18.2 Literature Review

This section is divided into two sub-sections, which discuss the development of deep learning techniques as well as crop yield prediction by monitoring various factors along with deep learning algorithms to process the volumetric data.

In crop monitoring, traditional machine learning techniques can be used for prediction. However, the ability to learn the optimal features in the data is limited. In 1990, Hepner compared machine learning algorithms with neural network classifications to find useful neural nets for a minimalistic training dataset [16]. With the latest advancements in technology, the training and deployment of image classification using deep learning techniques became more feasible. Convolution Networks (ConvNets), with deep learning related to non-linear operations composed of multiple levels, such as many hidden layers of neural networks with CNN and optimal features, were observed during training with convolutional layers of the network [17]. The artificial neural network (ANN) architecture is another algorithm in deep learning trying to deploy networks with good design for better generalization capability. The key benefit of using ANN is the ability to change the environment, its robustness, and the ability to escape local Optimum [18].

Deep learning algorithms have enhanced performance for feature extraction. Using deep learning algorithms in many fields, such as semantic recognition and images, was proposed by Long [19]. Among all deep learning methods, CNN has the highest impact on the performance of image classification and regression tasks, as observed by Krichevsky and Liu [20,21]. With convolution and pooling techniques, abstract features can be extracted with the series of operations received from the size of input images (width × height × depth). Kuwata proposed two inner-layer neural networks for estimating crop yield [22]. The objective was to extract key features from data using deep learning techniques, and it was expected that the input had less dependency on data.

Even though the data were inadequate, deep learning can be predictable, aimed at good quality appraisal related to the crop field. Other techniques of deep learning, such as the

Bayesian Network (BN) and Gaussian technique, are explained in Bi et al. [23] and You et al. [24]. Prediction of crop disease is very uncertain. Hence, Bi et al. proposed the BN to find crop diseases. The BN often calculates conditional probability inputs and mostly deals with uncertain factors. In this model, the self-learning function desires to produce the output of observed crop disease [23]. A crop yield prediction with a dimensionality reduction approach based on histograms was proposed by J. You; the author also presented the Gaussian process framework to remove spatially correlated errors in remote sensing applications [24].

Yang et al. [25] proposed the DCNN model for estimating crop production, which learns essential spatial features from higher spatial resolution images. Zhong et al. [26] identified crop production in a particular region using an ANN and deep CNN model, which enabled them to identify crops automatically, using no samples. Sharma et al. [27] described how crop disease is predicted with weather monitoring by ANN with forward feed by back-propagation. The weather parameters are minimum and maximum temperature, rainfall, and maximum and minimum humidity. In the complex environment, ANN can learn to provide non-linear solutions. Accuracy of the result using the ANN network is 90.909% for a 58 testing dataset. Lavreniuk et al. (2018) [28] proposed a deep learning approach developed for crop classification on the maps of in-situ data based on a neural network. In this method, the overall result for crop maps was 85.9%. The technique is feasible, and the auto encoders are used to learn the data without labels and fine tuning. Finally, the neural network is applied to the in-situ data. Desai et al. [29] proposed using a time series through the DCNN network for automatic rice production. Koirala et al. [30] described estimating mango fruit production using a two-stage CNN method. Zhou et al. [31] described the spatial distribution of the crops from sentinel-2A extracted with the comparison of CNN and support vector machine (SVM) using remote sensing images. The accuracy was 95.6% for CNN, which was superior to SVM. The observations were non-linear; mostly CNN worked due to the weight-sharing capability and automatic feature extraction. Deep reinforcement learning (DRL) allows an agent to be developed that can be analyzed as meta-learning to an environment [32]. DRL discoveries have applications in numerous fields, such as robotics, agriculture, energy management, healthcare, and game theory, as a general way of resolving optimization issues over the trial-and-error method [33–37]. You et al. [38] discussed how to predict soybean production using a CNN and RNN model based on sensed images before crops are taken the harvest. Russello [39] described using a CNN model that outperforms spatiotemporal features of 3D convolution based on satellite images for drop production identification. Zhaung et al. [40] proposed a CNN model used for planting area from Landsat-8 Multispectral remote sensing images. The obtained accuracy was 96.24%, where CNN used two convolutional layers, two maximum pooling layers, and two FC layers. CNN has a superior capability to extract subtle visual features superior to the human vision. Bu et al. [41] proposed DRL be used for a smart agriculture Internet of Things system for increasing crop growth. DRL was designed with the incremental model to increase performance.

18.3 Data

This study investigates prediction of paddy crop production in the southern part of Tamilnadu nearby the Tanjore district. Tamilnadu is one of the major paddy crop yielders in the southern part of India. The production of paddy crops is measured in terms of the cultivated area in hectares, the output of paddy crops in terms of tones, and the yield acquired in terms of kg/hectare. There were three sets of training data: crop genome, production efficiency, and weather and soil. The experimental hybrids based on each had 18,354 genetic symbols; the dataset of genotypes included genetic information. The yield output database included 137,343 samples for various forms of planting in various decades and locations for the observed yield, control yield, and yield difference. The Indian Meteorological Department submitted climate data from its portal metadata tool. The dataset included weather and soil parameters, along with the quantity of fertilizer application inspired by the study. Information applicable to daily climatic variables, such as temperature, evapotranspiration of reference crops, possible evapotranspiration, precipitation, moisture, and different event parameters, such as occurrence of ground frost, durable temperature range, and air temperature, was used. The parameters of the soil included phosphorus, nitrogen, and potassium based on macronutrients; topsoil density; and soil PH. The analysis took into account distinctive hydrochemical parameters based on micro-nutrient content in the groundwater.

18.4 Methodology

18.4.1 Data Preprocessing

Data from the genotype were implicit in {–1, 0, 1} numbers, demonstrating aa, aA, and AA allele frequencies, respectively. Around 37% of the data on the genotype had missing values. Next, we will describe how we preprocessed the genotype data to fix this problem before using the data in the neural network model. Initially, non-missing values were used to eliminate genetic markers below the 97% call rate. Values that were less heterozygous and less informative were then also removed – genetic markers below 1% of the lowest allele frequency. As an effect, we decreased the number of genetic markers from 19,465 to 627. We tried several imputation approaches, namely mean, median, and most regular, to attribute the absent data in the enduring portion of the genotype data. We found the median method led to better specific forecasting. These complete datasets were used for yield and environment, which did not have missing data.

18.4.2 Hybrid Deep CNN with RNN Framework

As illustrated in Figure 18.2, this chapter proposes a hybrid model for prediction of paddy yield that incorporates a CNN with an FC layer and an RNN. The framework is supposed to reflect linear and non-linear properties based on environmental data. In order to catch the temporal relationship of weather data, the model of CNN used 1-D convolution, while the hybrid CNN model used 1-D convolution for obtaining the soil data, which were measured at various underground depths. In different application domains, related convolution methods were extensively utilized, which were found to be efficient to increase the reliability of prediction. To integrate the high-level structures of environmental conditions derived by the hybrid CNN model, an FC was used, which often minimized the performance dimensions of the CNN model. The proposed model was intended to capture crop production time dependencies over a number of years.

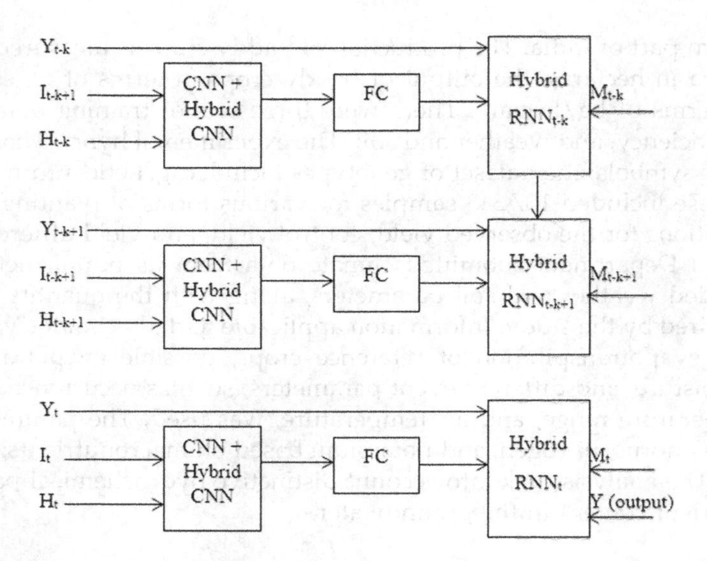

FIGURE 18.2
The Proposed Hybrid Deep CNN-RNN Model.

I_{t-k}, H_{t-k}, Y_t, Y_{t-k}, Y_{t-k+1} represents input variables, including environmental data such as weather and soil, average yield, and yield performance.

Two conflicting findings inspired the use of the hybrid RNN model. For the past four decades, paddy yield has shown a growing trend, as illustrated in Figure 18.3. This has been attributed to improving management and genetics practices.

For this prediction analysis, genotype data were not publicly accessible. The impact of the genotype should, therefore, be implicitly expressed in the model based on data available. This type of analysis is referred to as ANN, in which a directed graph represents the temporal dependencies of nodes. Thus, this research work developed a distinct hybrid RNN model to capture the result of genetic progress as the temporal dynamic behavior of crop production. The LSTM cells were used for enhancing the hybrid RNN. They were carefully engineered recurrent neurons to capture input

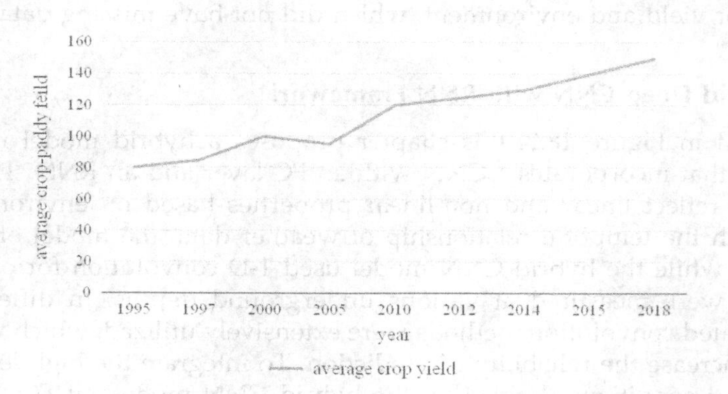

FIGURE 18.3
Average Paddy Crop Yield from 1995 to 2018.

dependencies with time. LSTM networks did not have to define the non-linear functions to be measured and compared to other time series models. They have shown superior performance for sequence modeling in a wide variety of applications.

The model of hybrid RNN is composed of k LSTM cells using data from years t – k to t to predict a county's crop yield for year t. The input data contain average crop yield and performance of the FC layer. Significant factors analyzed by the hybrid CNN model, including weather and soil data, are included in the input to the cell. The hybrid CNN and FC model is the only exception. It was specifically designed to transmit soil data directly to the LSTM cells. While soil data is normally static over time, soil data subscriptions allow the probability of varying soil conditions. The hybrid RNN model can be used to input historical average production to predict crop yield even without soil or weather data using historical crop yield patterns. Let us assume t represents yield prediction of the desired year. Then, it is possible to replace the average yield in year t. In the test process, Y_{t-1} can be substituted instead of Y_t, and I_t represents weather data based on the unseen portion, which can be replaced by forecasting the weather data. During the training phase, all input data were accessible, though, so the substitution was irrelevant.

For training of the proposed hybrid deep CNN-RNN model, we used the following hyper-parameters. In the CNN models, average pooling with steps of 2 was carried out by down-sampling. An FC layer, which has 100 neurons for paddy yield prediction, was followed by performance of CNN. An FC layer with 40 neurons was followed by output of the hybrid CNN model. We considered that a 5-year yield dependence had a time span of 5 years based on the hybrid RNN layer, including 64 hidden units in the RNN layer, which contains LSTM cells. This was designed to have the best performance after trying various network structures. In the Xavier method [42], all weights were initialized. We used a 25 mini-batch scale based on good Stochastic Gradient. The learning rate is 0.03%, which was divided by 2 per each 5,000 iterations in Adamoptimizer [43]. For the extreme 24,000 iterations, the model was trained. For the CNNs and the FC layer, we used the Rectified Linear Unit (ReLU) activation function. There was a linear activation mechanism in the output layer. In Python, using the Tensor flow library [44], we implemented the proposed model, and training time on a CPU required about an hour (i8–3,890, 7.8 GHz).

18.4.2.1 Weather Prediction Using the Hybrid Model

Prediction of weather is unavoidable, but a priori [45] paddy yield prediction is uncertain since soil and weather play an essential part in the forecast of yield. Our proposed method described weather prediction using weather data from 1995 to 2018 to predict the 2019 weather variables. This research work was trained by neural network for the 72 weather variables, which were used throughout all regions, to predict the 2019 meteorological parameters using historical data from 1995 to 2018. There were two reasons for 2,247 locations to be aggregated: (1) In the southern region, many locations were uniform so we could make the implicit assumption about the prediction models, and it was reasonable across all locations. (2) Integrating historical data makes it possible to train 72 neural networks with adequate data. The purpose of using neural networks is that nonlinearity for weather prediction that occurs in the existence of weather data can be captured. These non-linearities can be learned through data without needing the non-linear model to be defined by prior assessment.

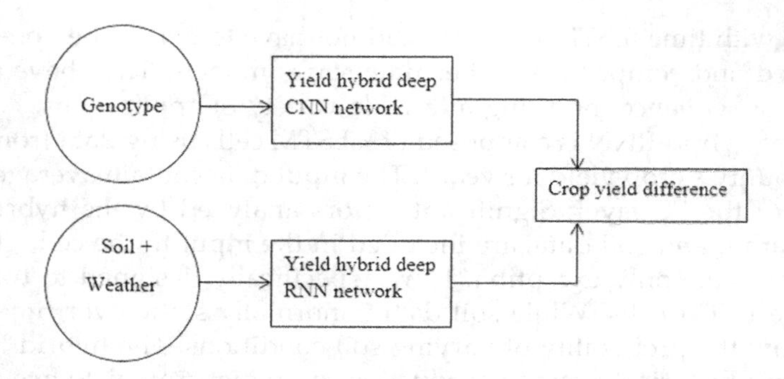

FIGURE 18.4
Predicting Yield Difference.

18.4.2.2 Yield Prediction Using the Hybrid Model

Two deep neural networks were equipped: one for hybrid CNN paddy yield and the other for hybrid RNN paddy yield. Prediction of yield difference was based on the difference in their outputs. Figure 18.4 describes these models. This system was found to be more effective for yield differences than using a single neural network since the genotype and environmental properties were closely applicable to produce and check yields than their differences [45]. In the training phase, the following hyper-parameters were used. Each layer of the neural network had 50 neurons and 21 hidden layers. These proportions were considered to offer the preeminent balance between forecast reliability and restricted overfitting after attempting deeper network structures. We utilized regularization for all hidden layers in order to prevent overfitting. To decrease the impact of redundant functions, this framework also added regularization to the first layer.

This section provided a comparative analysis obtained using the hybrid deep CNN with RNN model to forecast paddy yield. Comparison was made with existing models.

18.5 Results and Discussion

Analyzing requirements of several performance actions or monitoring evaluation metrics can be used to determine the effectiveness of a learning model. The hybrid model is validated for the proposed work in terms of output assessment and comparison of existing algorithms based on accuracy, evaluation metrics, and data distribution.

18.5.1 Performance Estimation

Data from time series modeling and forecasting are complicated and challenging. Dynamically splitting time series data for cross-validation does not perform well. As there is an implied dependence on past observation, which can contribute to a temporal dependency issue, leakage to lag variables from the response variable is bound to occur at the same time. The non-stationary outcome performs in the information space based on frequent changes in mean and variance. The cross-validation is done in such situations in a forward-changing manner. We used 5 fold cross validation on the prediction of data

TABLE 18.1

5 folds Cross Validation for Hybrid CNN and RNN Model

Model	Train data	Test data	Correlation	R^2score
1	1995–2000	2003–2005	0.71	0.851
2	1995–2004	2007–2012	0.85	0.773
3	1995–2007	2013–2015	0.63	0.733
4	1995–2010	2010–2016	0.75	0.873
5	1995–2011	2015–2018	0.85	0.943

based on previous data and forecasts of the forward-looking data. Table 18.1 illustrates the proposed model of 5 fold cross validation.

We used Python Scikit-Learn library for cross-validation. We used min-max scaling to normalize the dataset, which preprocessed the dataset. We introduced the train_-test_split feature from Sklearn to divide the training and test sets using the sub-library range of models. Using the Sklearn library in the cross-val-score function, we achieved hyperparameter modification for cross-validation in selecting the most suitable "K". In this case, by setting the limit based on n splits as 5, we split the data into "K" subsets. The test size parameter was used for the training and validation data set, which was 70% training and 30% testing the data. The process of K-fold cross-validation was trained using this model and determined error metric. The r^2 score was represented as error metric to obtain the best value, which was appended in each iteration, defining the overall accuracy.

18.5.2 Evaluation Metrics

The suggested reasonability of the hybrid deep CNN with RNN model was calculated by testing the model toward several implementation steps or specific performance metrics. The proposed hybrid framework determined various evaluation metrics such as mean squared error (MSE), mean absolute error (MAE), median absolute error (MedAE), determination coefficient (R^2), mean squared logarithmic error (MSLE), explained variance score (Exp. Var.), determination coefficient (R^2), and root mean squared error (RMSE) [46]. Training and validation were designed to forecast yielding to ensure a reasonable analysis of the model error metrics. Optimization of the hyper-parameter for the proposed solution and the other models was demonstrated for the respective models via a manual selection approach. The primary goal of the manual selection based on hyper-parameters was to adjust the performance of the network to fit the difficulty of the target assignment. The hyper-parameters, such as the optimizer, learning rate, number of hidden units, dropout values, and activation function, were measured by the degree to which the test error was minimized by the training process and cost function. Using three input layers and hidden layers, with each layer containing 8 neurons, the FC layer and an output layer presenting the paddy yield value were used to construct the hybrid deep CNN with RNN model. The input layer was made of 30 neurons describing the parameters of the paddy crop dataset. In the hidden layers, the RNN utilized a ReLU activation mechanism for processing. The performance of the deep learning models on both training and validation datasets was compared with the metrics described earlier in the evaluation, as shown in Table 18.2.

TABLE 18.2
Evaluation Metrics Based on Other Deep Learning Models

Evaluation metric	Model training dataset					Model validation dataset				
	Proposed Model	CNN	RNN	ANN	LSTM	Proposed Model	CNN	RNN	ANN	LSTM
MAE	0.18	0.22	0.25	0.25	0.22	0.12	0.24	0.21	0.23	0.21
MSE	0.03	0.06	0.068	0.9	0.04	0.02	0.04	0.06	0.06	0.08
RMSE	0.13	0.16	0.24	0.34	0.33	0.16	0.16	0.22	0.23	0.26
R^2	0.8	0.79	0.69	0.59	0.49	0.88	0.79	0.64	0.65	0.49
MSLE	0.002	0.006	0.004	0.006	0.02	0.002	0.003	0.003	0.007	0.01
Exp.var	0.84	0.64	0.64	0.594	0.54	0.869	0.674	0.62	0.641	0.37
Med AE	0.18	0.17	0.143	0.187	0.18	0.10	0.12	0.134	0.14	0.18

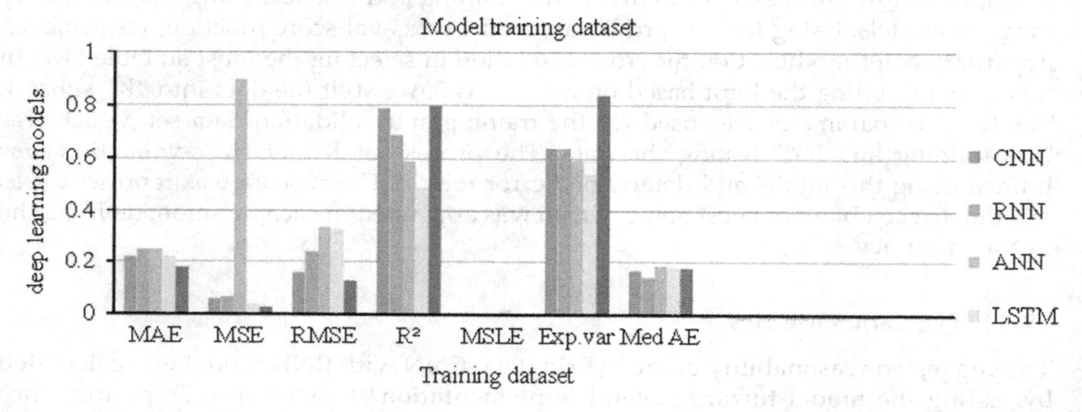

FIGURE 18.5
The Evaluation Metrics Based on Training Dataset Based on the Existing Deep Training Model.

Figures 18.5 and 18.6 illustrate the model error metrics for training and validation that were constructed to forecast yield.

18.5.3 Data Distribution Properties

The probability density function (PDF) of the real paddy yield data and the experimental models was observed in order to decide if the proposed hybrid deep CNN with RNN model retained the actual distributional properties of the data. The PDF was an empirical expression that described a continuous random variable's probability distribution against some distinct random variable. The area under the curve described the interval where the expected variable decreased by graphically defining the PDF. The complete region of the graph interval was equal to the probability of the occurrence based on the continuous random variable. It helped to estimate the possibilities based on the consequence of effects. Figure 18.7 displays the proposed hybrid deep CNN with RNN model for PDF based on real paddy yield and expected paddy yield.

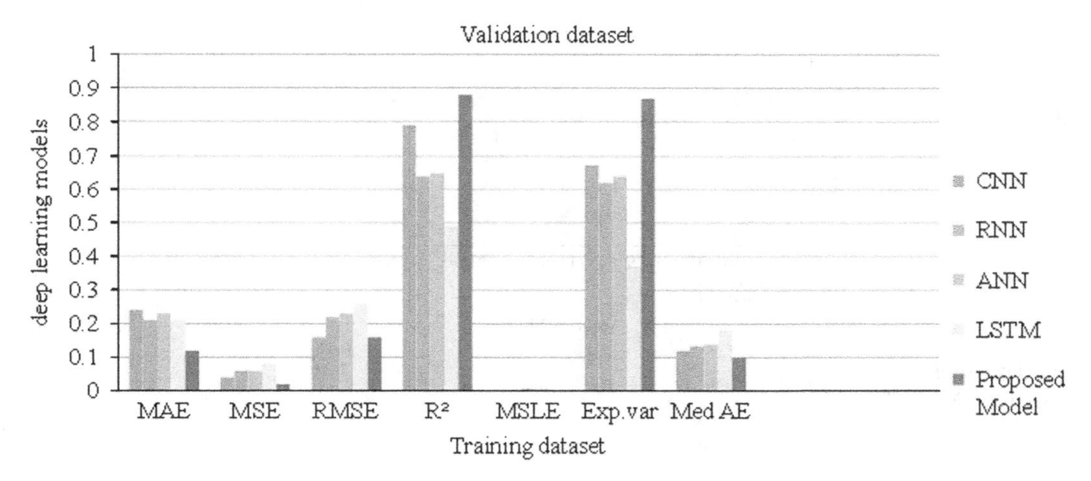

FIGURE 18.6
The Evaluation Metrics Based on Validation Dataset Based on the Existing Deep Training Model.

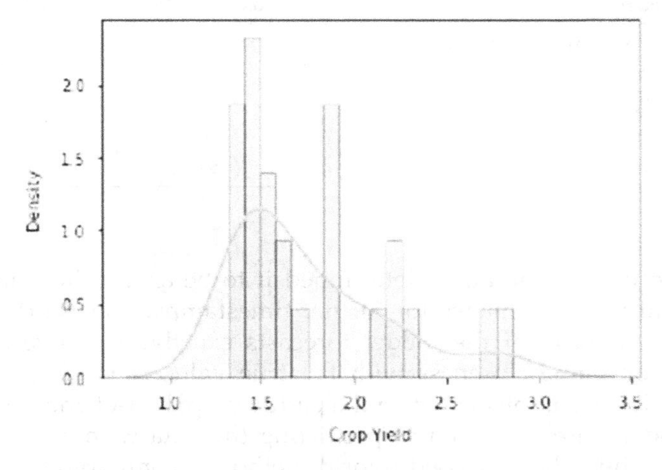

FIGURE 18.7
Probability Density Function of Proposed Hybrid Deep CNN with RNN Model.

18.5.4 Accuracy Measures

The most powerful feature of these models was enforced or incorporated in Python and checked under identical software and hardware conditions to ensure meaningful comparison. The error metric is often used to estimate during a system execution process for level of output. The variables produced estimated the error measure, which was the variation between the real and the expected values. The accuracy was calculated using magnitude of residual spread. The proposed system outperformed with a higher accuracy of 95.7% than the other deep learning models, and the error measures of MAPE value was less in the proposed model compared to other deep learning models. Figure 18.8 and Table 18.3 illustrate the accuracy of the proposed hybrid models with other deep learning models.

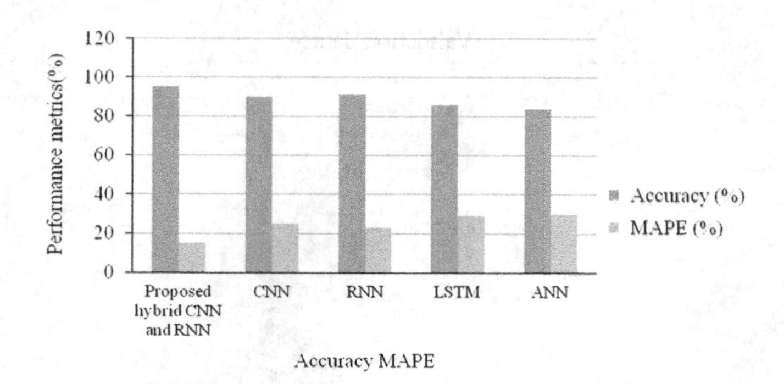

FIGURE 18.8
Accuracy MAPE Measure.

TABLE 18.3

Comparison of Proposed Hybrid Models and Other Deep Learning Models Based on Accuracy and MAPE

Model	Accuracy (%)	MAPE (%)
Proposed hybrid CNN and RNN	95.7	15
CNN	90.12	25
RNN	91.2	23
LSTM	86	29
ANN	84	30

The model accuracy evaluation determined it to be an excellent model for data representation and its performance for further timestamps. The metric accuracy defined that the prediction ratio of the model forecasts exactly. Moreover, the accuracy indicated prediction value closeness with the true value. Therefore, the results of the experimental value accomplished from the paddy crop dataset and the proposed hybrid CNN-RNN model were identified as predicting the data with high accuracy and precision at 95.7%. Thus, the proposed hybrid method is comparatively better than other deep learning methods, such as RNN, CNN, ANN, and LSTM, in computational cost and time complexness.

18.6 Conclusion

This chapter introduced a hybrid deep learning technique to predict crop yield, which accurately predicted paddy yield over the Tanjore district in India. It was completely dependant upon yield performance, crop yield prediction, and the environment, like soil and weather. This research focused on providing significant results to illustrate yield production. However, the proposed hybrid CNN-RNN technique performed better than other deep learning methods, namely RNN, CNN, ANN, and LSTM. Hence, the proposed technique involved CNN in designing to seizure the dependencies of internal temporal weather data, and spatial dependencies of soil data were dignified at various depths of the land. Similarly, the purpose of RNN was to seize the crop yield with incremental trends year by year because of continual improvement over practices of plant breeding and management. Thus, the model performance was relatively sensitive for several variables inclusive of performance like crop yield, soil, and weather, which predicted yield in untested environments and were utilized for predicting future yield of crops. The experimental results accomplished from the paddy crop dataset and the proposed hybrid CNN-RNN model were identified as predicting the data with high accuracy and precision at 95.7%. Moreover, this proposed hybrid method assisted in minimizing the expert dependence and earlier information in advancing paddy yield prediction models.

References

1. Vandendriessche, H. J. and Ittersum, M. K. V. (1995) "Crop Models and Decision Support Systems for Yield Forecasting and Management of the Sugar Beet Crop." *European Journal of Agronomy* 4(3): 269–279.
2. Yun, J. (2003, July) "Predicting Regional Rice Production in South Korea using Spatial Data and Crop-Growth Modelling," *Agricultural Systems* 77(1): 23–38.
3. Tao, F., et al. (2005, May) "Remote Sensing of Crop Production in China by Production Efficiency Models: Models Comparisons, Estimates and Uncertainties," *Ecological Modelling* 183(4): 385–396.
4. Hoogenboom, G. (June, 2000) "Contribution of Agrometeorology to the Simulation of Crop Production and Its Applications." *Agricultural and Forest Meteorology* 103(1): 137–157.
5. Yong, H., et al. (2007) "Application of Artificial Neural Network on Relationship Analysis between Wheat Yield and Soil Nutrients,". Presented at the IEEE 27th annual conference on Engineering in medicine and biology, Shanghai, China.
6. Green, T. R., et al. (April, 2007) "Relating Crop Yield to Topographic Attributes using Spatial Analysis Neural Networks and Regression." *Geoderma* 139(1–2): 23–37.
7. Wolfert S., et al. (May, 2017) "Big Data in Smart Farming – A Review," *Agricultural Systems* 153(1): 69–80.
8. Kamilaris, A. and Prenafeta-Boldú, F. X. (July, 2018) "Deep Learning in Agriculture: A Survey." *Computers and Electronics in Agriculture* 147(1): 70–90.
9. LeCun, Y., Bengio, Y., and Hinton, G. (2015) "Deep Learning," *Nature* 521(7553): 436.
10. Goodfellow, I., Bengio, Y., and Courville, A. (2016) "Deep Learning", 1. Springer, The MIT Press. ISBN :0262035618.
11. Pham, V., Bluche, T., Kermorvant, C., and Louradour, J. (2014) "Dropout Improves Recurrent Neural Networks for Handwriting Recognition," in 14th International Conference on Frontiers in Handwriting Recognition (pp. 285– 290). Crete, Greece: IEEE.
12. Sherstinsky, (2018) "A,Fundamentals of recurrent neural network (rnn) and long short-term memory (lstm) network", arXiv preprint arXiv:1808.03314,2018.
13. Bengio, Y., Simard, P., and Frasconi, P. (1994) "Learning Long-Term Dependencies with Gradient Descent Is Difficult." *IEEE Transactions on Neural Networks* 5: 157–166. doi: 10.1109/72.279181
14. Hochreiter, S. and Schmidhuber, J. (1997) "Long Short-Term Memory." *Neural Computation* 9: 1735–1780. doi: 10.1162/neco.1997.9.8.1735.
15. Khaki, S. and Wang, L. (2019) "Crop Yield Prediction using Deep Neural Networks." *Frontiers in Plant Science* 10: 621. doi: 10.3389/fpls.2019.00621

16. Hepner G. F., et al. (April, 1990) "Artificial Neural Network Classification using a Minimal Training Set: Comparison to Conventional Supervised Classification." *Photogrammetric Engineering and Remote Sensing* 56(4): 469–473.
17. Yamashita R., et al. (June, 2018) "Convolutional Neural Networks: An Overview and Application in Radiology," *Insights into Imaging* 9(4): 611–629.
18. Palnitkar R. M., and Cannady, J. (2004) "A Review of Adaptive Neural Networks," in Proc. at IEEE SoutheastCon. IEEE.
19. Long J., et al. (2015) "Fully convolutional networks for semantic segmentation," Presented at the IEEE Conf. on Computer Vision Pattern Recognition (pp. 3431–3440). Boston, MA.
20. Krizhevsky A., Sulskever I., and Hinton G. E. (2012) "ImageNet classification with deep convolutional neural networks," Proc. of the 25th International Conference on Neural Information Processing Systems, vol. 1 (pp. 1097– 1105).
21. Liu X., et al. (2015) "AgeNet: deeply learned regressor and classifier for robust apparent age estimation," Presented at the Proc. of IEEE Int. Conf. on Computer Vision (pp. 258–266), Santiago, Chile.
22. Kuwata K. and Shibasaki, R. (2015) "Estimating crop yields with deep learning and remotely sensed data," Presented at the IEEE Int. Geoscience and Remote Sensing Symp. (IGARSS) (pp. 858–861), Milan, Italy.
23. Bi, C. and Chen, G. (2010) "Bayesian networks modelling for Crop Diseases," *Computer and Computing Technologies in Agriculture IV* (pp. 312–320). Springer Berlin Heidelberg
24. You J., et al. (2017) "Deep Gaussian process for crop yield prediction based on remote sensing data.", Proc. of 31st Conf. on Association for the Advancement of Artificial Intelligence (pp. 4559–4565), San Francisco, CA.
25. Yang, Q., Shi, L., Han, J., Zha, Y., and Zhu, P. (Apr, 2019) "Deep Convolutional Neural Networks for Rice Grain Yield Estimation at the Ripening Stage using UAV Based Remotely Sensed Images." *Field Crops Research* 235: 142–153.
26. Zhong, L., Hu, L., Zhou, H., and Tao, X. (Nov, 2019) "Deep Learning Based Winter Wheat Mapping using Statistical Data as Ground References in Kansas and Northern Texas, US," *Remote Sensing of Environment* 233: Art. no. 111411.
27. Sharma, P., et al. (2018) "Prediction of Potato Late Blight Disease based upon weather parameters using Artificial Neural Network Approach", Presented in the 9th International Conference on Computing, Communication and Networking Technologies (ICCCNT) (pp. 1–13). Bangalore, India.
28. Lavreniuk, M. and Kussul, N., et al. (2018) "Deep learning crop classification approach based on sparse coding of time series of satellite data," Presented at the IEEE Int. Geoscience and Remote Sensing Symposium (pp. 4812–4815), Valencia, Spain.
29. Desai, S. V., Balasubramanian, V. N., Fukatsu, T., Ninomiya, S., and Guo, W. (Dec, 2019) "Automatic Estimation of Heading Date of Paddy Rice using Deep Learning." *Plant Methods* 15(1): 76.
30. Koirala, A., Walsh, K. B., and McCarthy, W. Z. (2019) "Deep Learning for Realtime Fruit Detection and Orchard Fruit Load Estimation: Benchmarking of `MangoYOLO." *Precision Agriculture* 20(6): 1107–1135.
31. Zhou, Z., et al. (2018) "Crops Classification from Sentinel-2A Multi-spectral Remote Sensing Images Based on Convolutional Neural Networks," Presented at the IEEE Int. Geoscience and Remote Sensing Symposium (pp. 5300–5303), Valencia, Spain.
32. Diallo, E. A. O., Sugiyama, A., and Sugawara, T. "Coordinated behavior of cooperative agents using deep reinforcement learning." *Neurocomputing, to be published.*
33. Bu, F. and Wang, X. (Oct, 2019) "A Smart Agriculture IoT System Based on Deep Reinforcement Learning." *Future Generation Computer System* 99: 500–507.
34. Liu, Z., Yao, C., Yu, H., and Wu, T. (Aug, 2019) "Deep reinforcement learning with its application for lung cancer detection in medical Internet of Things." *Future Generation Computer System* 97: 1–9.

35. Liu, Y., Guan, X., Li, J., Sun, D., Ohtsuki, T., Hassan, M. M., and Alelaiwi, A. "Evaluating smart grid renewable energy accommodation capability with uncertain generation using deep reinforcement learning." *Future Generation Computer System*, to be published.
36. Plasencia, A., Shichkina, Y., Suárez I., and Ruiz, Z. (2019) "Open Source Robotic Simulators Platforms for Teaching Deep Reinforcement Learning Algorithms." *Procedia Computer Science* 150: 162–170.
37. Silver, D., Huang, A., Maddison, C. J., Guez, A., Sifre, L., van den Driessche, G., Schrittwieser, J., Antonoglou, I., Panneershelvam V., Lanctot, M., Dieleman, S., Grewe, D., Nham, J., Kalchbrenner, N., Sutskever, I., Lillicrap, T., Leach, M., Kavukcuoglu, K., Graepel, T., and Hassabis, D. (2016) "Mastering the Game of Go with Deep Neural Networks and Tree Search." *Nature* 529(7587): 484–489.
38. You, J., Li, X., Low, M., Lobell, D., and Ermon, S. (2017) "Deep Gaussian Process for Crop Yield Prediction Based on Remote Sensing Data," in Thirty-First AAAI Conference on Artificial Intelligence (San Francisco, CA) (pp. 4559–4566).
39. Russello, H. (2018) *Convolutional Neural Networks for Crop Yield Prediction using Satellite Images*. IBM Center for Advanced Studies.
40. Zhou, Z., et al. (2017) "Peanut planting area change monitoring from remote sensing images based on deep learning," Presented at the IEEE Int. Conference on Systems and Informatics (ICSAI) (pp. 4812–4815), Hangzhou, China.
41. Bu, F. and Wang, X. (2019) "A Smart Agriculture IoT System Based on Deep Reinforcement Learning," *Future Generation Computer Systems*99: 500–507.
42. Glorot, X., and Bengio, Y. (2010) "Understanding the difficulty of training deep feed forward neural networks," in Proceedings of the Thirteenth International Conference on Artificial Intelligence and Statistics (pp. 249–256), Sardinia, Italy.
43. Kingma, D. P., and Ba, J. A. (2014) "A method for stochastic optimization."International Conference Proceedings on Learning Representations.
44. Abadi, M., Barham, P., Chen, J., Chen, Z., Davis, A., Dean, J., et al. (2016) "TensorFlow: A system for large scale machine learning," in 12th USENIX Symposium on Operating Systems Design and Implementation, vol. 16 (pp.265– 283), Savannah,GA.
45. Khaki, S. and Wang, L., (2019) "Crop Yield Prediction using Deep Neural Networks." *Frontiers in Plant Science* 10.
46. Elavarasan, D. and Vincent, P. M. D. (2020) "Crop Yield Prediction Using Deep Reinforcement Learning Model for Sustainable Agrarian Applications." IEEE.

Index